G. M. Henkin · J. Leiterer

Andreotti-Grauert Theory by Integral Formulas

Mathematical Research Mathematische Forschung

Wissenschaftliche Beiträge
herausgegeben von der
Akademie der Wissenschaften der DDR
Karl-Weierstraß-Institut für Mathematik

Band 43

Andreotti-Grauert Theory by Integral Formulas

by

G. M. Henkin
and
J. Leiterer

Andreotti-Grauert Theory by Integral Formulas

by

Gennadi Markovič Henkin
and
Jürgen Leiterer

Akademie-Verlag Berlin 1988

Autoren:
Prof. Dr. Gennadi Markovič Henkin
Zentrales Ökonomisch-Mathematisches Institut
der Akademie der Wissenschaften der UdSSR, Moskau
Prof. Dr. Jürgen Leiterer
Karl-Weierstraß-Institut für Mathematik
der Akademie der Wissenschaften der DDR, Berlin

Die Titel dieser Reihe werden vom Originalmanuskript der Autoren reproduziert.

ISBN 3-05-5000510-4

ISSN 0138-3019

Erschienen im Akademie-Verlag Berlin, DDR-1086 Berlin,Leipziger Str.3-4

Lizenznummer: 202 · 100/433/88
Printed in the German Democratic Republic
Gesamtherstellung: VEB Kongreß- und Werbedruck, 9273 Oberlungwitz
Lektor: Dr. Reinhard Höppner
LSV 1035
Bestellnummer: 763 840 7 (2182/43)

03600

PREFACE

The contemporary analysis on complex manifolds has been developed by means of the theory of coherent analytic sheaves (see, for instance, [Gunning/Rossi 1965 and Grauert/Remmert 1971,1977,1984]) and/or harmonic analysis (see, for instance, [Chern 1956, Kohn 1964, Hörmander 1966, Morrey 1966, Wells 1973]). However, the first fundamental contributions to the function theory of several complex variables, obtained in the period 1936-51 by K.Oka, were based on the classical constructive method of integral representations (see [Oka 1984]).

In the seventies this constructive approach has had a come-back in order to obtain in a strengthened form (with uniform estimates) the main results of the theory of functions on complex manifolds. A systematic development of the function theory in C^n which uses integral formulas as the principal tool was given in the books [Henkin/Leiterer 1984 and Range 1986].

The constructive method of integral representations is working with success also in several other fields of complex analysis: Cauchy-Riemann cohomology of complex manifolds, holomorphic vector bundles on complex manifolds, analysis on Cauchy-Riemann manifolds, Radon-Penrose transform, inverse scattering problem.

The authors intend to write a book where these fields will be presented from the viewpoint of integral formulas. The present monograph is a tentative version of the first part of that book. Here we develop in detail the basic facts on the Cauchy-Riemann cohomology of complex manifolds, where the emphasis is on finiteness, vanishing, and separation theorems for a class of complex manifolds which lies between the Stein, and the compact manifolds. Theorems A and B of Oka-Cartan for Stein manifolds as well as the finiteness theorems of Kodaira for compact, and Grauert for pseudoconvex manifolds appear as special cases of more general theorems.

The theory developed in the present monograph was mainly obtained in the articles [Andreotti/Grauert 1962, Andreotti/Vesentini 1965, Andreotti/Norguet 1966, Kohn/Rossi 1965, and Hörmander 1965] (it is astonishing that these remarkable results did not as yet enter into books). The novelty added here consists in new proofs based on integral formulas. As in the case of the theory of functions in $\mathbb{C}^n$, this makes it possible to prove all basic facts in a strengthened form: uniform estimates for solutions of the Cauchy-Riemann equation for differential forms on strictly q-convex and strictly q-concave domains, uniform approximation and uniform interpolation for the $\bar{\partial}$-cohomology classes on strictly q-convex domains, solution of the E.Levi problem for the $\bar{\partial}$-cohomology with uniform estimates, the Andreotti-Vesentini separation theorem with uniform estimates etc. A part of these results with uniform estimates was obtained already in the seventies [Fischer/Lieb 1974, Ovrelid 1976, Henkin 1977, Lieb 1979]. Some of these reusults are new.

These results with uniform estimates admit important applications in the theory of holomorphic vector bundles and the theory of the tangent Cauchy-Riemann equation. Such applications will be the subject of the following parts of the pending book mentioned above - elements of this are contained in the articles [Ajrapetjan/Henkin 1984 and Henkin/Leiterer 1986]. Moreover, in the following parts of that book, we intend to present some developments of the Andreotti-Grauert theory in connection with the Radon-Penrose transform - elements of this can be found in [Henkin 1983, Henkin/Poljakov 1986 and Leiterer 1986].

Note. Further on the book [Henkin/Leiterer 1984] will be refered to as [H/L]. The present monograph may be considered as a continuation of [H/L]. However, without proof, we use only results from the elementary Chapter 1 of [H/L]. Moreover, all basic results of Chapter 2 of [H/L], which is devoted to the theory of functions on completely pseudoconvex manifolds (= Stein manifolds, after solution of the E.Levi problem), are obtained anew, as the special case q=n-1 of Chapter 3 of the present work.

Acknowledgments. We thank G.Schmalz for reading portions of the manuscript, catching many errors. We wish to thank also the Akademie--Verlag Berlin, and particulary Dr.R.Höppner, for support and cooperation.

CONTENTS

CHAPTER I. INTEGRAL FORMULAS AND FIRST APPLICATIONS

Summary. This chapter is devoted to the general theory of integral representation formulas for solutions of the Cauchy-Riemann equation in $\mathbb{C}^n$. In Sect. 0 we introduce notations and standard definitions. In Sect. 1, first we recall basic facts about the Martinelli-Bochner-Koppelman formula, where, for some of the proofs we refer to Chapter 1 in [H/L]. Then, by means of this formula, we prove the regularity of the $\bar\partial$-operator as well as the Kodaira finiteness theorem on compact complex manifolds. Sect. 2 is devoted to the Cauchy-Fantappie formula - a generalization of the Martinelli-Bochner-Koppelman formula. (For a direct proof of this formula, again we refer to Chapter 1 in [H/L]. Notice also that in Sect. 3 of the present work the full proof of a more general formula will be given.) Then, as an application of the Cauchy-Fantappie formula, we prove the Poincaré $\bar\partial$-lemma. At the end of Sect. 2, we recall the arguments which lead from the Poincaré $\bar\partial$-lemma and the regularity of the $\bar\partial$-operator to the Dolbeault isomorphism and the theorem on smoothing of the $\bar\partial$-cohomology. In Sect. 3 we prove a generalization of the Cauchy-Fantappie formula, which will be called the piecewise Cauchy-Fantappie formula. This formula is especially useful for domains with a boundary which consists of several pieces each of which has its own "advantages" as, for instance, q-convexity, q-concavity, Levi flatness. In Chapters III and IV, appropriate special cases of this formula will be used to solve the Cauchy-Riemann equation with uniform estimates.

0. Generalities about differential forms and currents

0.1. Some notations. $\mathbb{C}$ is the complex plane, and $\mathbb{C}^n$ is the n-dimensional complex Euclidean space. If $x \in \mathbb{C}^n$, then by $x_1, \ldots, x_n$ we denote the canonical complex coordinates of x. We write

$$\langle x,y\rangle = x_1y_1+\ldots+x_ny_n \quad \text{and} \quad |x| = (|x_1|^2+\ldots+|x_n|^2)^{1/2}$$

for $x, y \in \mathbb{C}^n$. $\mathbb{R}$ is the real line, and $\mathbb{R}^n$ is the n-dimensional real Euclidean space.

The word "domain" means an arbitrary (not necessarily connected) open set. We write $A \subset\subset B$ to say that A is a relatively compact subset of B. The word "neighborhood" always means an open neighborhood.

The notion of a differential form (or a form) will be used for differential forms with measurable complex-valued coefficients. The degree of a differential form f will be denoted by deg f, and its support by supp f. Continuous forms will be called also C^0 forms. C^k forms ($k=1,2,\ldots,\infty$) are forms with k times continuously differentiable coefficients.

A form f defined on a complex manifold is called an (r,s)-form (or a form of bidegree (r,s)) if, with respect to local holomorphic coordinates $z_1, \ldots, z_n$,

$$f = \sum_{|J|=r, |K|=s}' f_{JK} dz^J \wedge d\bar{z}^K. \tag{0.1}$$

Here the summation is over all strictly increasing r-tuples $J=(j_1, \ldots, j_r)$ and all strictly increasing s-tuples $K=(k_1, \ldots, k_s)$ in $\{1, \ldots, n\}$, $dz^J = dz_{j_1} \wedge \ldots \wedge dz_{j_r}$, $d\bar{z}^K = d\bar{z}_{k_1} \wedge \ldots \wedge d\bar{z}_{k_s}$, and the coefficients f_{JK} are complex-valued functions.

As usual, by d we denote the exterior differential operator. On complex manifolds, by $\bar\partial$ we denote the Cauchy-Riemann operator, and we set $\partial := d - \bar\partial$, i.e. if f is as in (0.1), then

$$\bar\partial f := \sum_{|J|=r, |K|=s}' \sum_{l=1}^{n} \frac{\partial f_{JK}}{\partial \bar{z}_l} d\bar{z}_l \wedge dz^J \wedge d\bar{z}^K$$

and

$$\partial f := \sum_{|J|=r, |K|=s}' \sum_{l=1}^{n} \frac{\partial f_{JK}}{\partial z_l} dz_l \wedge dz^J \wedge d\bar{z}^K,$$

where

$$\frac{\partial}{\partial \bar{z}_l} := \frac{1}{2}\left(\frac{\partial}{\partial t_l} + i\frac{\partial}{\partial t_{l+n}} \right) \quad \text{and} \quad \frac{\partial}{\partial z_l} := \frac{1}{2}\left(\frac{\partial}{\partial t_l} - i\frac{\partial}{\partial t_{l+n}} \right)$$

with $t_l :=$ real part of z_l, and $t_{l+n} :=$ imaginary part of z_l.

Remark. The main subject considered in this monograph are differential forms with values in holomorphic vector bundles (for the notion of forms with values in vector bundles, see, for instance [Wells 1973]). Since any (s,r)-form with values in a holomorphic vector bundle E may be identified with a (0,r)-form with values in the holomorphic vector bundle $\Lambda^{s,0}\otimes E$, where $\Lambda^{s,0}$ is the sth exterior power of the holomorphic cotangent bundle, in the most cases, we shall restrict ourselves to (0,r)-forms (but with coefficients in arbitrary holomorphic vector bundles).

0.2. Integration with respect to a part of the variables. Suppose X, Y are real C^1 manifolds, and f is a differential form on $X\times Y$. Let $m = \dim_{\mathbb{R}} X$, $n = \dim_{\mathbb{R}} Y$, and let $y_1,\dots,y_n$ be local C^1 coordinates in some open $U \subseteq Y$. Consider the unique representation

$$f(x,y) = \sum_I{}' f_I(x,y)\wedge dy^I, \qquad x\in X,\ y\in U,$$

where the summation is over all strictly increasing r-tuples $I=(i_1,\dots,i_r)$ in $\{1,\dots,n\}$ with $r \le \deg f$, $dy^I := dy_{i_1}\wedge\dots\wedge dy_{i_r}$, and $f_I(x,y)$ is a differential form of degree $\deg f - r$ on X which depends on $y\in U$. If X is oriented and the integrals $\int_X f_I(x,y)$ exist for all fixed $y\in U$ and any strictly increasing r-tuple I in $\{1,\dots,n\}$ with $r = \deg f - m$, then we define

$$\int_X f(x,y) = \sum_I{}' \left(\int_{x\in X} f_I(x,y)\right) dy^I, \qquad y\in U,$$

where the summation is over all strictly increasing r-tuples in $\{1,\dots,n\}$ with $r = \deg f - m$. The result of this integration is a differential form of degree $\deg f - m$ on U.

This form is independent of the choice of the local coordinates $y_1,\dots,y_n$. Therefore $\int_X f(x,y)$ is well-defined for all $y\in Y$. Notice that, by this definition, $\int_X f(x,y) = 0$ if f does not contain monomials which are of degree m in x.

0.3. The absolute value $|f|$ of a differential form f of maximal degree. Let X be an oriented real C^1 manifold of real dimension m, and let f be a differential form of degree m on X. Then by $|f|$ we denote the differential form on X which is defined as follows: If $x_1,\dots,x_m$ are positively oriented C^1 coordinates in some open $U \subseteq X$ and F is the complex-valued function on U with $f = F dx_1\wedge\dots\wedge dx_m$, then

$$|f| := |F| dx_1\wedge\dots\wedge dx_m \qquad \text{on } U.$$

We remark that if f is integrable, then

$$\left| \int_X f \right| \leq \int_X |f|.$$

If g is a second differential form of degree m on X, then we write

$$|g| \leq |f| \quad \text{on } X$$

if the following condition is fulfilled: If $x_1,\dots,x_m$ are positively oriented C^1 coordinates in some open $U \subseteq X$, and G, F are the functions on U with $g = G dx_1 \wedge \dots \wedge dx_m$ and $f = F dx_1 \wedge \dots \wedge dx_m$, then $|G| \leq |F|$ on U.

0.4. The Riemannian norm of a differential form at a point. Let f be a differential form of degree m defined on a domain $D \subseteq \mathbb{R}^n$. If m=0, i.e. f is a complex valued function, then we write $\|f(z)\| = |f(z)|$, $z \in D$. If m>0, $x_1,\dots,x_n$ are the canonical coordinates in $\mathbb{R}^n$, and

$$f = \sum_{1 \leq i_1 < \dots < i_m \leq n} f_{i_1,\dots,i_m} dx_{i_1} \wedge \dots \wedge dx_{i_m},$$

then we define

$$\|f(z)\| = \left(\sum_{1 \leq i_1 < \dots < i_m \leq n} \left| f_{i_1,\dots,i_m}(z) \right|^2 \right)^{1/2}, \qquad z \in D.$$

Now let f be an arbitrary real C^1 manifold. Then we choose a locally finite open covering $\{U_j\}$ of X together with C^1 coordinates $x_1^j,\dots,x_n^j$ on U_j. Further, we choose a continuous partition of unity $\{\chi_j\}$ subordinated to $\{U_j\}$. If f is a differential form in a neighborhood of some point $z \in X$ and

$$f(z) = \sum_{1 \leq i_1 < \dots < i_m \leq n} f^j_{i_1,\dots,i_m}(z) dx^j_{i_1}(z) \wedge \dots \wedge dx^j_{i_m}(z).$$

for all j with $z \in U_j$, then we define

$$\|f(z)\| = \sum_j \chi_j(z) \left[\sum_{1 \leq i_1 < \dots < i_m \leq n} |f^j_{i_1,\dots,i_m}(z)|^2 \right]^{1/2}.$$

Of course, this definition depends on the choice of the local coordinates as well as on the choice of the partition of unity. Any norm obtained in this way will be called a Riemannian norm on X (Riemannian metric). If $\|\cdot\|_1$ and $\|\cdot\|_2$ are two Riemannian norms on a C^1 manifold X, then, clearly, for any compact set $K \subset\subset X$, one can find constants $c>0$ and $C<\infty$ such that

$$c\|f(x)\|_1 \leq \|f(x)\|_2 \leq C\|f(x)\|_1 \tag{0.2}$$

for all $x \in K$ and all differential forms f on X. In the present monograph, in all cases where we meet Riemannian norms, estimate (0.2) ensures independence of the special choice of this norm. Therefore we shall use the following

<u>Convention.</u> The cotangent bundle of any real C^1 manifold is assumed to be endowed with an (arbitrary but fixed) Riemannian norm. □

Finally, we want to generalize the notion of a Riemannian norm to forms with values in vector bundles. Let E be a vector bundle of rank N over a real C^1 manifold X. Then a <u>Riemannian norm in E</u> is given by a locally finite open covering $\{U_j\}$ of X together with a family of vector bundle trivializations

$$E|U_j = U_j \times \mathbb{C}^N \tag{0.3}$$

as well as a C^1 partition of unity $\{\chi_j\}$ subordinated to $\{U_j\}$. If f is an E-valued differential form (i.e. a measurable section of $\Lambda^m \otimes E$, where m is the degree of f and Λ^m the mth exterior power of the cotangent bundle of X) in a neighborhood U_z of some point $z \in X$ and $(f_1^j, \dots, f_N^j)$ is the vector of differential forms on $U_z \cap U_j$ which represents $f|U_z \cap U_j$ via (0.3), then we define the Riemannian norm $\|f(z)\|$ of $f(z)$ by

$$\|f(z)\| = \sum_j \chi_j(z) \left(\sum_{k=1}^{N} \|f_k^j(z)\|^2\right)^{1/2}, \quad z \in X. \tag{0.4}$$

(The norms $\|f_k^j(z)\|$ are well-defined, since, by our Convention, X is already endowed with a Riemannian norm.) The same arguments as above now lead to the following convention (which will be used throughout this monograph):

<u>Convention.</u> If E is a vector bundle over a real C^1 manifold X, then we assume that E is endowed with an (arbitrary but fixed) Riemannian norm.

<u>0.5. Determinants of matrices of differential forms.</u> Let $A = (a_{ij})_{i,j=1}^n$ be a quadratic matrix whose elements a_{ij} are differential forms. Then we define a differential form $\det A$ by setting

$$\det A = \sum_{\sigma} \operatorname{sgn}(\sigma) a_{\sigma(1),1} \wedge \dots \wedge a_{\sigma(n),n},$$

where the summation is over all permutations σ of $\{1,\dots,n\}$ and $\operatorname{sgn}(\sigma)$

is the signature of σ. Under this definition the usual relations between row operations (the first index is the row index) and determinants hold true. Therefore, as in the case of usual determinants one obtains that if A is an nxn matrix of differential forms and Z is an nxn matrix of complex-valued functions, then det ZA = det Z det A. Notice that, in general, det AZ $\neq$ det A det Z. Further, it is possible that det A = 0 although some of the columns of A are equal. For references, in the following proposition we collect some useful facts about such determinants (the proofs are obvious):

<u>0.6. Proposition.</u> (i) The determinant of a quadratic matrix of differential forms is a multi-linear map of the columns with respect to linear combinations whose coefficients are complex-valued functions.

(ii) If A is an nxn-matrix of differential forms and if, for some $1 \le k < l \le n$, $a_{ik}=z_i b_k$ and $a_{il}=z_i b_l$ for all i, where b_k, b_l are arbitrary differential forms and z_i are complex-valued functions, then det A = 0.

(iii) Let a_i and a_j be two columns in a quadratic matrix A of differential forms such that all forms in a_i have the same degree d_i and all forms in a_j are of the same degree d_j. Further, let $\tilde{A}$ be the matrix obtained from A by interchanging these two columns. Then

$$\det \tilde{A} = \begin{cases} \det A & \text{if } d_i - d_j \text{ is even} \\ -\det A & \text{if } d_i - d_j \text{ is odd.} \end{cases}$$

<u>0.7. Definition.</u> Let $1 \le m \le n$ be integers, let $a_1, \dots, a_m$ be column vectors of differential forms of the length n, and let $s_1, \dots, s_m$ be integers with $s_1 + \dots + s_m = n$. If all s_i are non-negative, then we set

$$\det(\overbrace{a_1}^{s_1}, \dots, \overbrace{a_m}^{s_m}) = \det(\underbrace{a_1, \dots, a_1}_{s_1 \cdot \text{times}}, \dots, \underbrace{a_m, \dots, a_m}_{s_m \text{ times}}),$$

and if at least one of the integers s_i is negative, then we write

$$\det(\overbrace{a_1}^{s_1}, \dots, \overbrace{a_m}^{s_m}) = 0.$$

<u>0.8. The forms $\omega(u)$ and $\omega'(v)$.</u> Let X be a real C^1 manifold, and let $u=(u_1, \dots, u_n)$, $v=(v_1, \dots, v_n)$ be two C^n-valued C^1 maps defined on X. Then we set

$$\omega(u) = du_1\wedge\ldots\wedge du_n = \frac{1}{n!}\det(\overbrace{du}^{n})$$

and

$$\omega'(v) = \sum_{j=1}^{n}(-1)^{j+1}v_j dv_1\wedge\ldots\underset{j}{\wedge}\ldots\wedge dv_n = \frac{1}{(n-1)!}\det(v,\overbrace{dv}^{n-1})$$

on X, where $\underset{j}{\wedge}$ means that dv_j must be omitted.

If X, Y are two real C^1 manifolds and the points in X and Y are denoted by x and y, respectively, then the exterior differential operator with respect to x on $X\times Y$ will be denoted by d_x, the exterior differential operator with respect to y on $X\times Y$ will be denoted by d_y, and the "full" exterior differential operator d on $X\times Y$ then will be denoted also by $d_{x,y}$. Analogously, we use the notations $\omega_x(u)$, $\omega'_x(v)$, $\omega_{x,y}(u)$, $\omega'_{x,y}(v)$, and, if X, Y are complex manifolds, the notations ∂_x, $\bar\partial_x$, $\partial_{x,y}$, $\bar\partial_{x,y}$.

0.9. Proposition. If X is a real C^1 manifold, $v:X\longrightarrow\mathbb{C}^n$ is a C^1 map, and $F:X\longrightarrow\mathbb{C}$ is a C^1 function, then $\omega'(Fv) = F^n\omega'(v)$.

Proof. By Proposition 0.6 (i) and (ii) we have

$$\det(Fv,\overbrace{d(Fv)}^{n-1}) = \det(Fv,\overbrace{(dF)v+Fdv}^{n-1}) = \det(Fv,\overbrace{Fdv}^{n-1}) = F^n(v,\overbrace{dv}^{n-1}). \square$$

0.10. Proposition. If X is a real C^1 manifold and u, v are two $\mathbb{C}^n$-valued C^1 maps on X, then the form

$$\frac{\omega'(v)\wedge\omega(u)}{\langle v,u\rangle^n} = \omega'\left(\frac{v}{\langle v,u\rangle}\right)\wedge\omega(u)$$

(this equation holds true by Proposition 0.9) is closed for all $x\in X$ with $\langle v(x),u(x)\rangle \neq 0$. In particular, if moreover $\langle v,u\rangle=1$ on X, then $\omega'(v)\wedge\omega(u)$ is closed on X.

Proof. We can assume that $\langle v,u\rangle = 1$ on X (otherwise we have to replace v by $v/\langle v,u\rangle$). Then $0 = d\langle v,u\rangle = \sum v_i du_i + \sum u_i dv_i$. Hence then the forms $dv_1,\ldots,dv_n,du_1,\ldots,du_n$ are linearly dependent, and therefore $d\omega'(v)\wedge\omega(u) = \omega(v)\wedge\omega(u) = 0$. $\square$

0.11. Spaces of differential forms. Here we collect the definitions of some spaces of differential forms which will be often used in this monograph. Let X be a complex manifold, and $M\subseteq X$ a subset which is contained in the closure of the set of inner points of M.

For any differential form f on M, we set (cf. Sect. 0.4 for the definition of the Riemannian norm $\|f(z)\|$)

$$\|f\|_{0,M} = \sup_{z \in M} \|f(z)\| \tag{0.5}$$

and

$$\|f\|_{\alpha,M} = \|f\|_{0,M} + \sup_{z,x \in M} \frac{\|f(z)-f(x)\|}{|z-x|^{\alpha}} \quad \text{if } 0<\alpha<1. \tag{0.6}$$

If $0<\alpha<1$, then a form f on M is called <u>α-Hölder continuous</u> on M if

$$\|f\|_{\alpha,K} < \infty \qquad \text{for all compact sets } K \subseteq M. \tag{0.7}$$

The notion of a <u>C^{α} form</u> (resp. of a <u>form of class C^{α}</u>) on M will be used for any $0\le\alpha\le\infty$, where:

C^0 stands for "continuous";

for $k=1,2,\ldots,\infty$, we say f is a <u>C^k form on M</u> if f is a C^k form in the interior of M such that all derivatives of order $<k+1$ of f admit a continuous extension onto M;

if $0<\alpha<1$ and k is a non-negative integer, then we say f is a <u>$C^{k+\alpha}$ form</u> on M if f is a C^k form on M such that all derivatives of order $\le k$ of f are α-Hölder continuous on M.

We use the following notations:

$L_*^{\infty}(M)$ is the space of all <u>bounded</u> differential forms on M. Notice that if (and only if) M is relatively compact in X, then $L_*^{\infty}(M)$ does not depend on the choice of the Riemannian norm on X;

$C_*^{\alpha}(M)$ is the space of all C^{α} forms on M ($0\le\alpha\le\infty$);

$Z_*^{\alpha}(M)$ is the subspace of all $f\in C_*^{\alpha}(M)$ with $\bar\partial f=0$ in the interior of M ($0\le\alpha\le\infty$);

$E_*^{\beta\to\alpha}(M)$ is the subspace of all $f\in Z_*^{\alpha}(M)$ such that $f=\bar\partial u$ for some $u\in C^{\beta}(M)$ ($0\le\alpha,\beta\le\infty$);

$E_*^{\alpha}(M) := E_*^{\alpha\to\alpha}(M)$ ($0\le\alpha\le\infty$).

If A is one of the symbols L^{∞}, C^{α}, Z^{α}, $E^{\beta\to\alpha}$ or E^{α}, then $A_m(M)$ is the subspace of all forms of degree m in $A_*(M)$ ($1\le m\le 2n$), and $A_{s,r}(M)$ is the subspace of all (s,r)-forms in $A_*(M)$ ($1\le s,r\le 2n$).

By $C^{\alpha}\mathcal{O}(M)$ ($0\le\alpha\le\infty$) we denote the space of all complex-valued C^{α} functions on M which are holomorphic in the interior of M. If M is open, then $\mathcal{O}(M)$ is the space of all holomorphic functions on M.

If B is an arbitrary space of differential forms on M, then by $[B]_0$

we denote the subspace of all forms in B which have compact support.

Some of the spaces introduced above will be considered as topological vector spaces. Throughout this monograph, we use the following conventions with regard to these topologies:

The space $L^{\infty}_{*}(M)$ as well as its subspaces $L^{\infty}_{m}(M)$ and $L^{\infty}_{s,r}(M)$ will be considered as Banach spaces endowed with the norm $\|\cdot\|_{0,M}$.

For $0\le\alpha<1$, the spaces $C^{\alpha}_{*}(M)$, $C^{\alpha}_{m}(M)$ and $C^{\alpha}_{s,r}(M)$ will be viewed as Fréchet spaces with the system of half-norms $\|\cdot\|_{\alpha,K}$, where K runs over the compact subsets of M. Notice that if M is compact, then these spaces are Banach spaces with the norm $\|\cdot\|_{\alpha,M}$.

If M is open, then the spaces $C^{\infty}_{*}(M)$, $C^{\infty}_{m}(M)$ and $C^{\infty}_{s,r}(M)$ will be considered as Fréchet spaces endowed with the topology of uniform convergence on the compact sets together with all derivatives, whereas the spaces $[C^{\infty}_{*}(M)]_0$, $[C^{\infty}_{m}(M)]_0$ and $[C^{\infty}_{s,r}(M)]_0$ will be regarded as topological vector spaces endowed with the following topology:

A sequence f_j converges to f if and only if there is a compact set $K\subset\subset M$ such that supp $f_j \subseteq K$ for all j, and f_j converges to f together with all derivatives uniformly on M. (0.8)

Now let E be a holomorphic vector bundle over X. Then we apply all definitions and conventions from above also to E-valued sections resp. E-valued differential forms. The spaces obtained in this way will be denoted by $\mathcal{O}(M,E)$, $C^{\alpha}\mathcal{O}(M,E)$, $C^{\alpha}_{*}(M,E)$, $[C^{\infty}_{*}(M,E)]_0$ etc.

0.12. $C^{\alpha}\mathcal{O}$ vector bundles. Let X be a complex manifold, and $M\subseteq X$ a subset which is contained in the closure of the interior of M. Further, let E be a complex vector bundle of rank N over M.

We say E is a $C^{\alpha}\mathcal{O}$ vector bundle over M ($0\le\alpha\le\infty$) if there are given a relatively open (with repect to M) covering $\{U_j\}$ of M as well as a family of vector bundle trivializations

$$h_j: E|U_j \longrightarrow U_j\times\mathbb{C}^N$$

such that the corresponding transition functions are matrices with elements from $C^{\alpha}\mathcal{O}(U_i\cap U_j)$.

Correspondingly, we shall use such notions as "$C^{\alpha}\mathcal{O}$ homomorph", "$C^{\alpha}\mathcal{O}$ trivial" etc.

Example: The restriction to M of an arbitrary holomorphic vector bundle over a neighborhood of M is a $C^{\alpha}\mathcal{O}$ vector bundle for all $0\le\alpha\le\infty$.

Analogously as in Sect. 0.4, also in $C^{\alpha}\mathcal{O}$ vector bundles, we introduce the notion of a Riemannian norm, and, throughout this monograph, we use the following

Convention. If E is a C^{α} vector bundle over M ($0\le\alpha\le\infty$), then we assume that E is endowed with an (arbitrary but fixed) Riemannian norm. □

Some of the definitions given in Sect. 0.11 will be applied also to differential forms with values in $C^{\alpha}\mathcal{O}$ vector bundles. For instance, we need the following definition:

If M is compact, and E is a $C^{\alpha}\mathcal{O}$ vector bundle over M with $0\le\alpha\le\infty$, then by $C^0_*(M,E)$ we denote the Banach space of all continuous E-valued forms over M endowed with the norm $\|\cdot\|_{0,M}$ defined by (0.5), and, for any $0<\beta<1$ with $\beta\le\alpha$, by $C^{\beta}_*(M,E)$ we denote the Banach space of all β-Hölder continuous E-valued forms over M endowed with the norm $\|\cdot\|_{\beta,M}$ defined by (0.6).

The corresponding subspaces of forms of degree m resp. bidegree (s,r) will be denoted by $C^{\beta}_m(M,E)$ resp. $C^{\beta}_{s,r}(M,E)$ ($0\le\beta\le1$).

0.13. Currents. Let X be an n-dimensional complex manifold.

The space of all continuous linear maps

$$T:[C^{\infty}_*(X)]_0 \longrightarrow \mathbb{C}$$

will be denoted by $C^{cur}_*(X)$. The elements of this space are called currents on X. Recall some definitions from the theory of currents (for a presentation of this theory, see, for instance, [Griffiths/Harris 1978] and [Čirka 1979]):

A current T on X is called of degree m ($0\le m\le 2n$) if

$$Th = 0 \qquad \text{for all } h\in[C^{\infty}_l(X)]_0 \text{ with } l \neq 2n-m. \tag{0.9}$$

The degree of T will be denoted by deg T. The space of all currents of degree m on X will be denoted by $C^{cur}_m(X)$.

A current T on X is called of bidegree (r,s) (or an (r,s)-current), where $0\le r,s\le n$, if

$$Th = 0 \qquad \text{for all } h\in[C^{\infty}_{l,m}(X)]_0 \text{ with } (l,m) \neq (n-r,n-s). \tag{0.10}$$

If f is a differential form with locally integrable coefficients on X, then by <f> we denote the current on X which is defined by

$$\langle f\rangle h = \int_X f\wedge h, \qquad h\in[C_*^{\infty}(X)]_0. \tag{0.11}$$

Notice that if f is of degree m resp. bidegree (r,s), then $\langle f\rangle$ is also of degree m resp. bidegree (r,s), because, by definition,

$$\int_X g = 0$$

if g is a form of degree $<2n$.

If T is a current of degree m on X, then the currents dT, ∂T and $\bar\partial T$ are defined by

$$\begin{aligned}(dT)h &= (-1)^{m+1}T(dh),\\ (\partial T)h &= (-1)^{m+1}T(\partial h),\\ (\bar\partial T)h &= (-1)^{m+1}T(\bar\partial h), \qquad h\in[C_*^{\infty}(X)]_0.\end{aligned} \tag{0.12}$$

By $Z_*^{cur}(X)$ we denote the space of all $\bar\partial$-closed currents on X, i.e. the space of all $T\in C_*^{cur}(X)$ with $\bar\partial T=0$. We set $E_*^{cur}(X) = \bar\partial C_*^{cur}(X)$,

$$Z_{s,r}^{cur}(X) = Z_*^{cur}(X)\cap C_{s,r}^{cur}(X) \quad\text{and}\quad E_{s,r}^{cur}(X) = E_*^{cur}(X)\cap C_{s,r}^{cur}(X) \quad (0\le s,r\le n).$$

Now let E be a holomorphic vector bundle over X, and E^* the dual bundle of E.

Then the space of all continuous linear maps

$$T:[C_*^{\infty}(X,E^*)]_0 \longrightarrow \mathbb{C}$$

will be denoted by $C_*^{cur}(X,E)$. The elements in $C_*^{cur}(X,E)$ will be called <u>E-valued currents</u> on X.

Notice that the E-valued currents on X may be viewed as "current sections" in E over X, i.e. if $\{U_j\}$ is an open covering of X together with a collection of holomorphic trivializations

$$E|U = X\times\mathbb{C}^N \qquad (N = \text{rank } E)$$

of E, and $\{g_{ij}\}$ is the corresponding cocycle of transition functions, then any E-valued current T over X can be identified with a collection of column vectors $T_j=(T_j^1,\dots,T_j^N)$ of "usual" currents T_j^k over U_j such that

$$T_i = g_{ij}T_j \qquad \text{on } U_i\cap U_j.$$

If f is an E-valued differential form with locally integrable coefficients on X, then by $\langle f\rangle$ we denote the E-valued current on X defined by

$$\langle f\rangle h = \int f\wedge h, \qquad h\in[C_*^\infty(X,E^*)]_0. \tag{0.13}$$

If T is an E-valued current on X, then we say <u>T is of class C^α</u> (or <u>T is a C^α-form</u>) on X if $T=\langle f\rangle$ for some $f\in C_*^\alpha(X,E)$ $(0\le\alpha\le\infty)$.

The notions of the <u>degree</u> and <u>bidegree</u> will be used (in the canonical sense) also for arbitrary E-valued currents. By $C_m^{cur}(X,E)$ we denote the space of all E-valued currents of degree m on X, and by $C_{s,r}^{cur}(X,E)$ we denote the space of all E-valued (s,r)-currents on X.

Since E is holomorphic, the $\bar\partial$-operator has an invariant meaning also for E^*-valued forms. Therefore, this operator can be defined also for arbitrary E-valued currents: if T is an E-valued current of degree m on X, then we set

$$(\bar\partial T)h = (-1)^{m+1}(\bar\partial h), \qquad h\in[C_*^\infty(X,E^*)]_0. \tag{0.14}$$

Notice that if T is of bidegree (s,r), then $\bar\partial T$ is of bidegree (s,r+1). Further notice that if f is an E-valued differential form on X such that $\bar\partial f$ has locally integrable coefficients, then, by Stokes' theorem,

$$\bar\partial\langle f\rangle = \langle\bar\partial f\rangle. \tag{0.15}$$

By $Z_*^{cur}(X,E)$ we denote the space of all $\bar\partial$-closed E-valued currents on X, i.e. the space of all $T\in C_*^{cur}(X,E)$ with $\bar\partial T=0$. We set $Z_{s,r}^{cur}(X,E) = Z_*^{cur}(X/E)\cap C_{s,r}^{cur}(X,E)$, $E_*^{cur}(X,E) = \bar\partial C_*^{cur}(X,E)$, and $E_{s,r}^{cur}(X,E) = E_*^{cur}(X,E)\cap C_{s,r}^{cur}(X,E)$ for all $0\le s,r\le n$.

<u>0.14.</u> If T is an E-valued current of degree s on X, then the <u>support</u> of T is, by definition, the smallest closed set $M\subseteq X$ such that $Th=0$ for all $h\in[C_*^\infty(X,E^*)]_0$ with $\operatorname{supp} h\subseteq X\setminus M$. The support of a current T will be denoted by supp T.

For references we notice the following

<u>Proposition.</u> Let T be an E-valued current on X. If supp T is compact, then T is uniquely defined and continuous on the Fréchet space $C_*^\infty(X,E^*)$, that is, then there exists a uniquely determined continuous linear map

$$\tilde T : C_*^\infty(X,E^*) \longrightarrow \mathbb{C}$$

such that $\tilde{T}h=Th$ for all $h\in[C_*^{\infty}(X,E^*)]_0$.

<u>Proof.</u> Uniqueness: $[C_*^{\infty}(X,E^*)]_0$ is dense in $C_*^{\infty}(X,E^*)$ with respect to the topology of $C_*^{\infty}(X,E^*)$. Existence: Take an arbitrary C^{∞} function χ on X such that supp $\chi \subset\subset X$ and $\chi = 1$ in a neighborhood of supp T. Set $\tilde{T}h = T(\chi h)$ for $h\in[C_*^{\infty}(X,E^*)]_0$. □

<u>Remark.</u> Conversely, any continuous linear functional on $C_*^{\infty}(X,E^*)$ defines an E-valued current with compact support on X. Therefore, the space of E-valued currents with compact support on X can be identified with the dual space of the Frechet space $C_*^{\infty}(X,E^*)$.

1. The Martinelli-Bochner-Koppelman formula and the Kodaira finiteness theorem

<u>1.1. Almost C^1 boundaries.</u> Let $G \subseteq \mathbb{C}^n$ be a domain, X a subset of G, and let $\hat{X}$ be the set of all $x\in X$ such that, for some neighborhood U of x, $X\cap U$ is a closed C^1 submanifold of real codimension 1 in U.

We say X is a <u>closed almost C^1 hypersurface</u> in G if X is closed in G and the following two conditions are fulfilled:

(i) For any $\varepsilon>0$, every compact set $K \subseteq X\setminus\hat{X}$, and each neighborhood $U \subseteq \mathbb{C}^n$ of K, there exists a neighborhood $V \subseteq U$ of K with smooth boundary such that the (2n-1)-dimensional Euclidean volume of ∂V is $\leq\varepsilon$.

(ii) For each open $U \subset\subset G$, the (2n-1)-dimensional Euclidean volume of $U \cap \hat{X}$ is finite.

Let X be a closed almost C^1 hypersurface in G. Then the points in $\hat{X}$ resp. $X\setminus\hat{X}$ will be called the <u>smooth</u> resp. the <u>non-smooth</u> points of X. A function on X is said to be of <u>class C^1</u> if it admits an extension into a neighborhood of X in $\mathbb{C}^n$ which is of class C^1. A <u>differential form</u> (or a <u>form</u>) on X is, by definition, a differential form on $\hat{X}$. Such a form f is called <u>bounded</u> resp. <u>continuous</u> on X if there exists a bounded resp. continuous differential form F in some neighborhood of X such that $f=F|\hat{X}$. We say X is oriented if $\hat{X}$ is oriented. If, in this case, f is a form on X, then we write

$$\int_X f := \int_{\hat{X}} f,$$

provided the integral on the right hand side exists.

If $D \subset\subset \mathbb{C}^n$ is a domain, then we say the boundary of D is <u>almost C^1</u> if ∂D is a closed almost C^1 hypersurface in $\mathbb{C}^n$.

<u>1.2. Orientation.</u> We use the following orientation of $\mathbb{C}^n$: If $z_1,\dots,z_n$ are the canonical complex coordinates in $\mathbb{C}^n$ and $t_1,\dots,t_{2n}$ are the real coordinates in $\mathbb{C}^n$ with $z_j = t_j + it_{j+n}$ $(j=1,\dots,n)$, then the form $dx_1\wedge\dots\wedge dx_{2n}$ is positive. If $D \subset\subset \mathbb{C}^n$ is a domain with almost C^1 boundary, then we orient ∂D (i.e., more precisely, the smooth part of ∂D) by the orientation which is induced from D (so that Stokes' formula holds true).

<u>1.3. The Martinelli-Bochner integrals $B_{\partial D}$ and B_D.</u> The form (cf. Sect. 0 for the notations)

$$\hat{B} = \hat{B}(z,x) := \frac{1}{(2\pi i)^n} \frac{\det(\bar{x}-\bar{z}, \overbrace{d\bar{x}-d\bar{z}}^{n-1}) \wedge \omega(x)}{|x-z|^{2n}}$$

defined for all $(z,x) \in \mathbb{C}^n \times \mathbb{C}^n$ with $z \neq x$ is called the <u>Martinelli-Bochner kernel</u>. By $\hat{B}_r = \hat{B}_r(z,x)$ $(r=0,\dots,n-1)$ we denote the sum of all monomials in $\hat{B}$ which are of bidegree $(0,r)$ in z and hence of bidegree $(n,n-r-1)$ in x. It is easy to see that $\hat{B} = \sum_0^{n-1} \hat{B}_r$, and

$$\hat{B}_r(z,x) = \frac{(-1)^r}{(2\pi i)^n}\binom{n-1}{r}\frac{\det(\bar{x}-\bar{z}, \overbrace{d\bar{z}}^{r}, \overbrace{d\bar{x}}^{n-r-1}) \wedge \omega(x)}{|x-z|^{2n}} \tag{1.1}$$

for $r=0,\dots,n-1$.

Now let $D \subset\subset \mathbb{C}^n$ be a domain with almost C^1 boundary. Then, for any bounded differential form f on ∂D, we define by

$$(B_{\partial D} f)(z) = \int_{x \in \partial D} f(x) \wedge \hat{B}(z,x), \qquad z \in D, \tag{1.2}$$

a C^∞ form $B_{\partial D} f$ in D. Further, since the singularity of $\hat{B}(z,x)$ at $x=z$ is of order $2n-1$ and hence integrable, for any bounded differential form f on D, we can define

$$(B_D f)(z) = \int_{x \in D} f(x) \wedge \hat{B}(z,x), \qquad z \in D. \tag{1.3}$$

It is easy to see that so we obtain a continuous differential form $B_D f$ in D. Moreover, the following proposition holds true:

1.4. Proposition. (i) For any $0<\alpha<1$, the Martinelli-Bochner integral B_D defines a bounded linear operator

$$B_D: L^{\infty}_{*}(D) \longrightarrow C^{\alpha}_{*}(\bar{D}).$$

(ii) If $f \in L^{\infty}_{*}(D)$ is of class C^k for some $k \in \{0,1,\dots,\infty\}$, then

$$B_D f \in \bigcap_{0<\alpha<1} C^{k+\alpha}(D).$$

(iii) B_D defines a continuous linear operator

$$B_D: [C^{\infty}_{*}(D)]_0 \longrightarrow C^{\infty}_{*}(D). \quad \square$$

Proof. See, for instance, Lemma 1.8.5 and its proof in [H/L]. □

1.5. Proposition. Let $D \subset\subset \mathbb{C}^n$ be a domain with almost C^1 boundary. Then:

(i) For any bounded $(0,r)$-form f on ∂D, $B_{\partial D} f$ is of bidegree $(0,r)$, and

$$B_{\partial D} f = \int_{x \in \partial D} f(x) \wedge \hat{B}_r(\cdot, x) \qquad \text{if } 0 \leq r \leq n-1, \tag{1.4}$$

and $B_{\partial D} f = 0$ if $r=n$.

(ii) For any bounded $(0,r)$-form f on D, $B_D f$ is of bidegree $(0,r-1)$, and

$$B_D f = \int_{x \in D} f(x) \wedge \hat{B}_{r-1}(\cdot, x) \qquad \text{if } 1 \leq r \leq n, \tag{1.5}$$

and $B_D f = 0$ if $r=0$.

Proof. By degree reasons, $f \wedge \hat{B}_s = 0$ if $s \leq r-2$, and, by definition of the integration with respect to a part of the variables (cf. Sect. 0.1),

$$\int_{x \in \partial D} f(x) \wedge \hat{B}_s(\cdot, x) = 0 \qquad \text{if } s \geq r.$$

Since $\hat{B} = \sum_0^{n-1} \hat{B}_s$, this proves part (i). The proof of part (ii) is similar. □

<u>1.6. Theorem</u> (Martinelli-Bochner formula). Let $D \subset\subset \mathbb{C}^n$ be a domain with almost C^1 boundary. Then, for any continuous complex-valued function f on $\bar{D}$ such that $\bar{\partial}f$ is also continuous on $\bar{D}$, we have the representation

$$f = B_{\partial D}f - B_D\bar{\partial}f \qquad \text{in } D. \tag{1.6}$$

<u>Proof.</u> See, for instance, Sect. 1.9 in [H/L].

<u>1.7. Theorem</u> (Koppelman formula). Let $D \subset\subset \mathbb{C}^n$ be a domain with almost C^1 boundary, and let f be a continuous (0,r)-form on $\bar{D}$ such that $\bar{\partial}f$ is also continuous on $\bar{D}$ ($1 \le r \le n$). Then

$$(-1)^r f = B_{\partial D}f + \bar{\partial}B_D f - B_D\bar{\partial}f \qquad \text{in } D. \tag{1.7}$$

In particular, then $\bar{\partial}B_D f$ is continuous in D.

<u>Proof.</u> See, for instance, Sect. 1.11 in [H/L].

<u>Remark.</u> For r=0, by Proposition 1.5 (ii), $B_D f=0$ and hence (1.7) coincides with (1.6).

<u>1.8. The Martinelli-Bochner operator B' for forms and currents with compact support.</u> For any continuous differential form f with compact support in $\mathbb{C}^n$, the Martinelli-Bochner integral $B_D f$ can be defined also with $D=\mathbb{C}^n$:

$$(B_{\mathbb{C}^n}f)(z) := \int_{x\in\mathbb{C}^n} f(x)\wedge\hat{B}(z,x), \qquad z\in\mathbb{C}^n.$$

If f is a continuous (0,r)-form with compact support in $\mathbb{C}^n$ such that $\bar{\partial}f$ is also continuous in $\mathbb{C}^n$ ($0 \le r \le n$), then the Koppelman formula (1.7) takes the form

$$(-1)^r f = \bar{\partial}B_{\mathbb{C}^n}f - B_{\mathbb{C}^n}\bar{\partial}f. \tag{1.8}$$

We want to generalize this to currents with compact support in $\mathbb{C}^n$. By $\hat{\beta}(z,x)$ we denote the form with

$$\hat{B}(z,x) = (-1)^{n+1}\,\hat{\beta}(z,x)\wedge\omega(x), \tag{1.9}$$

i.e.

$$\hat{\beta}(z,x) := \frac{(-1)^{n+1}}{(2\pi i)^n} \frac{\det(\bar{x}-\bar{z}, \overbrace{d\bar{x}-d\bar{z})}^{n-1}}{|x-z|^{2n}}, \qquad z,x\in\mathbb{C}^n. \tag{1.10}$$

Now, for any continuous differential form f with compact support in $\mathbb{C}^n$, we define

$$(\beta f)(z) = \int_{x\in\mathbb{C}^n} f(x)\wedge\hat{\beta}(z,x)\wedge\omega(z), \qquad z\in\mathbb{C}^n. \tag{1.11}$$

It follows from Proposition 1.4 (iii) that, in this way, a continuous linear map

$$\beta: [C^{\infty}_{*}(\mathbb{C}^n)]_0 \longrightarrow C^{\infty}_{*}(\mathbb{C}^n)$$

is defined. Hence, for any current T with compact support in $\mathbb{C}^n$, the composition $T\circ\beta$ is well-defined and continuous on $[C^{\infty}_{*}(\mathbb{C}^n)]_0$ (cf. Sect. 0.14). Thus, for any current T with compact support in $\mathbb{C}^n$, by setting

$$(B'T)h = T(\beta h) = T\Big(\int_{x\in\mathbb{C}^n} h(x)\wedge\hat{\beta}(\cdot,x)\wedge\omega(\cdot)\Big)$$

for $h\in[C^{\infty}_{*}(\mathbb{C}^n)]_0$, we obtain a current B'T on $\mathbb{C}^n$.

1.9. Proposition. Let T be a current with compact support in $\mathbb{C}^n$. Then:

(i) If T is of bidegree (0,r), $r\geq 1$, then B'T is of bidegree (0,r-1).

(ii) If deg T = 0, i.e. T is a distribution, then B'T=0.

(iii) If T is of bidegree (s,r) with $s\neq 0$, then B'T=0.

Proof. If $h\in[C^{\infty}_{l,m}(\mathbb{C}^n)]_0$, then from (1.10) and (1.11) one obtains the following two assertions:

a) βh is of bidegree (n,m-1) if l=n and $1\leq m\leq n$;

b) $\beta h=0$ if $l\neq n$ or m=0.

In particular, the bidegree of any monomial in βh is of the form (n,k) with k<n. Hence (ii) and (iii) hold true.

To prove part (i), we assume that T is of bidegree (0,r) with $1\leq r\leq n$. We have to show that $T(\beta h) = 0$ if $h\in[C^{\infty}_{l,m}(\mathbb{C}^n)]$ with $(l,m) \neq (n,n-r+1)$. For $l\neq n$ this follows from assertion b). If l=n but $m\neq n-r+1$, then, by assertion a), the bidegree of βh is $\neq(n,n-r)$. Since T is of bidegree $(0,r)$, this implies that $T(\beta h)=0$. □

1.10. Proposition. For any continuous $(0,r)$-form f with compact support in $\mathbb{C}^n$ $(0\leq r\leq n)$, we have

$$B'\langle f\rangle = (-1)^r \langle B_{\mathbb{C}^n} f\rangle.$$

Proof. Let $h\in[C^{\infty}_{n,n-r+1}(\mathbb{C}^n)]_0$. Then

$$(B'\langle f\rangle)h = \langle f\rangle(\beta h) = \int_{z\in\mathbb{C}^n} f(z)\wedge\int_{x\in\mathbb{C}^n} h(x)\wedge\hat{\beta}(z,x)\wedge\omega(z)$$

$$= (-1)^{r+1}\int_{x\in\mathbb{C}^n}\int_{z\in\mathbb{C}^n} f(z)\wedge\hat{\beta}(z,x)\wedge\omega(z)\wedge h(x).$$

Since

$$\hat{\beta}(z,x)\wedge\omega(z) = (-1)^n\hat{\beta}(x,z)\wedge\omega(z) = -\hat{B}(x,z),$$

this implies

$$(B'\langle f\rangle)h = (-1)^r\langle B_{\mathbb{C}^n} f\rangle h. \quad \square$$

1.11. Proposition. For any current T with compact support in $\mathbb{C}^n$, the current B'T is real-analytic in $\mathbb{C}^n \setminus \operatorname{supp} T$, i.e. there is a real-analytic form g on $\mathbb{C}^n \setminus \operatorname{supp} T$ such that

$$T\big|_{(\mathbb{C}^n \setminus \operatorname{supp}T)} = \langle g\rangle.$$

Proof. It is sufficient to show that such a form g exists over each open $U \subset\subset \mathbb{C}^n \setminus \operatorname{supp} T$. Let such an open set U be fixed. Then we choose a C^{∞} function χ on $\mathbb{C}^n$ such that $\chi = 1$ in a neighborhood of supp T and $\chi = 0$ in some neighborhood of $\bar{U}$.

By Proposition 1.9, we may assume that T is of bidegree $(0,r)$ with $1\leq r\leq n$. Denote by $\hat{\beta}_{n-r}(z,x)$ the sum of all monomials in $\hat{\beta}(z,x)$ which are of bidegree $(0,n-r)$ in z. Then, for any $g\in[C^{\infty}_{n,n-r+1}(U)]_0$,

$$\beta g = \int_{x\in\mathbb{C}^n} g(x)\wedge\hat{\beta}_{n-r}(z,x)\wedge\omega(z). \tag{1.12}$$

Further, let $b_{i_1\ldots i_{r-1}}(z,x)$ be the $(0,n-r)$-forms in z depending on x such that

$$\hat{\beta}_{n-r}(z,x) = \sum_{1\le i_1<\ldots<i_{r-1}\le n} b_{i_1\ldots i_{r-1}}(z,x)\wedge d\bar{x}_{i_1}\wedge\ldots\wedge d\bar{x}_{i_{r-1}}. \tag{1.13}$$

Set

$$\tilde{b}_{i_1\ldots i_{r-1}}(z,x) = \chi(z)(1-\chi(x))b_{i_1\ldots i_{r-1}}(z,x)\wedge\omega(z).$$

Since $\chi = 0$ in a neigborhood of $\bar{U}$, then the assignment

$$U\ni x \longrightarrow \tilde{b}_{i_1\ldots i_{r-1}}(\cdot,x) \tag{1.14}$$

is a real-analytic map from U into $C^{\infty}_{n,n-r}(\mathbb{C}^n)$. Hence, the $(0,r-1)$-form

$$g := \sum_{1\le i_1<\ldots<i_{r-1}\le n} g_{i_1\ldots i_{r-1}} d\bar{x}_{i_1}\wedge\ldots\wedge d\bar{x}_{i_{r-1}}$$

with

$$g_{i_1\ldots i_{r-1}}(x) := T\left(\tilde{b}_{i_1\ldots i_{r-1}}(\cdot,x)\right)$$

is real-analytic in U.

Now let $h\in[C^{\infty}_{n,n-r+1}(U)]_0$. Since $\chi = 0$ in a neighborhood of supp h and $\chi = 1$ in a neighborhood of supp T, then we have the relation $(B'T)h = T(\beta h) = T[\chi\beta((1-\chi h)h)]$. By (1.12) and (1.13), this implies that

$$(B'T)h = T\int_{x\in U} h(x)\wedge\chi(\cdot)(1-\chi(x))\hat{\beta}_{n-r}(\cdot,x)\wedge\omega(\cdot)$$

$$= \sum_{1\le i_1<\ldots<i_{r-1}\le n} T\int_{x\in U} h(x)\wedge\tilde{b}_{i_1\ldots i_{r-1}}(\cdot,x)\wedge d\bar{x}_{i_1}\wedge\ldots\wedge d\bar{x}_{i_{r-1}}.$$

Since the maps (1.14) are continuous as maps with values in $C^{\infty}_{n,n-r}(\mathbb{C}^n)$, the integrals on the right hand side may be interchanged with T. So we obtain

$$(B'T)h = \pm \sum \int_{x \in U} h(x) \wedge g_{i_1 \ldots i_{r-1}}(x) d\bar{x}_{i_1} \wedge \ldots \wedge d\bar{x}_{i_{r-1}} = \pm \langle g \rangle h,$$

i.e. $B'T = \pm\langle g \rangle$. □

1.12. Theorem (Koppelman formula for currents with compact support). For every current T with compact support in $\mathbb{C}^n$, we have the representation

$$T = \bar{\partial}(B'T) + B'(\bar{\partial}T). \tag{1.15}$$

Proof. By Proposition 1.9 we may assume that T is of bidegree (0,r) with $1 \le r \le n$. Let $h \in [C^{\infty}_{n,n-r}(\mathbb{C}^n)]_0$. Denote by h/ω the form with $h(x) = (h/\omega)(x) \wedge \omega(x)$. Then, by (1.8),

$$(-1)^{n-r} h/\omega = \bar{\partial} B_{\mathbb{C}^n}(h/\omega) - B_{\mathbb{C}^n} \bar{\partial}(h/\omega).$$

Further, since $\hat{\beta}(z,x) \wedge \omega(x) = (-1)^{n+1} \hat{B}(z,x)$, we have

$$(\beta h)(z) = \int_{x \in \mathbb{C}^n} (h/\omega) \wedge \hat{\beta}(z,x) \wedge \omega(x) \wedge \omega(z)$$

$$= (-1)^{n+1} B_{\mathbb{C}^n}(h/\omega)(z) \wedge \omega(z).$$

Together this implies that

$$(-1)^{r+1} h = \bar{\partial}(\beta h) - \beta(\bar{\partial} h).$$

Applying the current T to this equation and taking into account that B'T is of bidegree (0,r-1) (Proposition 1.9), we obtain

$$\begin{aligned} Th &= (-1)^{r+1}\big(T[\bar{\partial}(\beta h)] - T[\beta(\bar{\partial} h)]\big) \\ &= (\bar{\partial}T)(\beta h) + (-1)^r (B'T)(\bar{\partial} h) \\ &= [B'(\bar{\partial}T)]h + [\bar{\partial}(B'T)]h. \quad \square \end{aligned}$$

<u>1.13. Theorem</u> (Regularity of the $\bar\partial$-operator). Let E be a holomorphic vector bundle over a complex manifold X, and let T be an E-valued distribution on X, i.e. $T\in C_0^{cur}(X,E)$. If

$$\bar\partial T = \langle f\rangle$$

for some $f\in C^k_{0,1}(X,E)$ with $k\in\{0,1,\dots,\infty\}$, then

$$T = \langle g\rangle \qquad \text{for some} \qquad g\in \bigcap_{0<\alpha<1} C^{k+\alpha}_{0,1}(X,E).$$

In particular, if $\bar\partial T=0$, then $T=\langle h\rangle$ for some holomorphic section h of E over X, i.e.

$$Z^{cur}_{0,0}(X,E) = \langle \mathcal{O}(X,E)\rangle.$$

<u>Proof.</u> Since the assertion is local, we may assume that X is a domain in $\mathbb{C}^n$ and E is the trivial line bundle, i.e. $T\in C_0^{cur}(X)$. It is sufficient to prove that, for every open $U\subset\subset X$, there exists a form

$$g\in \bigcap_{0<\alpha<1} C^{k+\alpha}_{0,1}(U)$$

with $T=\langle g\rangle$ on U. To do this, we choose a C^∞ function χ such that $\chi = 1$ in a neighborhood of $\bar U$ and supp $\chi\subset\subset X$. Then, by the Koppelman formula (1.15) and by Proposition 1.9 (ii), we have

$$T = \chi T = B'(\bar\partial(\chi T)) = B'(T\bar\partial\chi) + B'\langle\chi f\rangle \qquad \text{on } U.$$

Since supp $T\bar\partial\chi\subset\subset X\setminus U$, from Proposition 1.11 we get a form $g_\infty\in C^\infty_{0,1}(U)$ with

$$B'(T\bar\partial\chi) = \langle g_\infty\rangle \qquad \text{on } U.$$

Further, observe that, by Proposition 1.4 (ii),

$$B_{\mathbb{C}^n}(\chi f) \quad \bigcap_{0<\alpha<1} C^{k+\alpha}_{0,1}(\mathbb{C}^n)$$

and, by Proposition 1.10,

$$B'\langle\chi f\rangle = -\langle B_{\mathbb{C}^n}(\chi f)\rangle.$$

Hence, the form

$$g := g_{\infty} - B_{\mathbb{C}^n}(\chi f)\big|_U$$

has the required property. □

1.14. Dolbeault cohomology. Let E be a holomorphic vector bundle over an n-dimensional complex manifold X, and let $0\le s,r\le n$ be two integers.

We set (for the notations, see Sects. 0.11 and 0.13)

$$H^{s,r}_{\alpha}(X) = Z^{\alpha}_{s,r}(X)/E^{\alpha}_{s,r}(X) \quad \text{and} \quad H^{s,r}_{\alpha}(X,E) = Z^{\alpha}_{s,r}(X,E)/E^{\alpha}_{s,r}(X,E)$$

$(0\le\alpha\le\infty)$, and

$$H^{s,r}_{cur}(X) = Z^{cur}_{s,r}(X)/E^{cur}_{s,r}(X) \quad \text{and} \quad H^{s,r}_{cur}(X,E) = Z^{cur}_{s,r}(X,E)/E^{cur}_{s,r}(X,E).$$

The space $H^{s,r}_{\infty}(X,E)$ is called the Dolbeault cohomology group (or the $\bar\partial$-cohomology group) of bidegree (s,r) of X with values in E. Further on, in the most cases, we restrict ourselves to the case s=0 (cf. the remark at the end of Sect. 0.1). The space $H^{0,r}_{\infty}(X,E)$ will be called the Dolbeault cohomology group of order r of X with values in E. Usually, the subscript ∞ is omitted, i.e.

$$H^{0,r}(X,E) := H^{0,r}_{\infty}(X,E).$$

Below (Theorem 2.14) we shall see that, for all $0\le\alpha\le\infty$, the natural homomorphism

$$H^{0,r}_{\alpha}(X,E) \longrightarrow H^{0,r}_{cur}(X,E)$$

which is defined by the assignment $C^{\alpha}_{0,r}(X,E) \ni f \longrightarrow \langle f\rangle \in C^{cur}_{0,r}(X,E)$ (cf. (0.13)) is an isomorphism. Therefore, $H^{0,r}(X,E)$ can be represented also by each of the groups $H^{0,r}_{\alpha}(X,E)$ $(0\le\alpha\le\infty)$ and $H^{0,r}_{cur}(X,E)$. From the viewpoint of the method of integral formulas (the principal tool used in the present monograph) it is especially advantageous to work with $H^{0,r}_{0}(X,E)$. Let us demonstrate this in the case of the Kodaira finiteness theorem:

1.15. Theorem (Kodaira finiteness theorem). If E is a holomorphic vector bundle over a compact n-dimensional complex manifold X, then

$$\dim H_0^{0,r}(X,E) < \infty \qquad \text{for all } 0 \le r \le n.$$

To prove this theorem, we use the Koppelman formula (1.8) for forms with compact support, the Hölder estimates from Proposition 1.4 (i), and the fact that compact perturbations of the identity operator in a Banach space are Fredholm. The first step is the following

1.16. Lemma. Let $U \subset\subset D$ be a relatively compact open subset of a domain $D \subset\subset \mathbb{C}^n$. Then there exist linear operators

$$T_r : C^0_{0,r}(\bar{D}) \longrightarrow \bigcap_{0<\alpha<1} C^{\alpha}_{0,r-1}(\bar{U}) \qquad (1 \le r \le n)$$

and

$$K_r : C^0_{0,r}(\bar{D}) \longrightarrow \bigcap_{0<\alpha<1} C^{\alpha}_{0,r}(\bar{U}) \qquad (1 \le r \le n-1)$$

such that, for any $1 \le r \le n$, the following assertions hold true:

(i) For all $0<\alpha<1$, T_r and K_r are bounded as operators acting from the Banach space $C^0_{0,r}(\bar{D})$ into the Banach space $C^{\alpha}_{0,r-1}(\bar{U})$ resp. $C^{\alpha}_{0,r}(\bar{U})$.

(ii) For each $f \in C^0_{0,r}(\bar{D})$ such that $\bar{\partial} f$ is also continuous on $\bar{D}$, over U, we have the representation

$$\bar{\partial} T_r f = \begin{cases} f + K_r f - T_{r+1} \bar{\partial} f & \text{if } 1 \le r \le n-1 \\ f & \text{if } r=n. \end{cases} \qquad (1.16)$$

Proof. Choose a C^{∞} function χ on $\mathbb{C}^n$ with supp $\chi \subset\subset D$ and $\chi = 1$ in U. Set

$$T_r f = (-1)^r B_{\mathbb{C}^n}(\chi f)\big|_{\bar{U}} \qquad \text{and} \qquad K_r f = (-1)^r B_{\mathbb{C}^n}(\bar{\partial}\chi \wedge f)\big|_{\bar{U}}.$$

Then the assertions follow from Proposition 1.4 (i) and the Koppelman formula (1.8). □

By means of the local operators given by Lemma 1.16, now we prove

1.17. Theorem. Let E be a holomorphic vector bundle over a compact n-dimensional complex manifold X. Then there exist linear operators

$$T_r : C^0_{0,r}(X,E) \longrightarrow \bigcap_{0<\alpha<1} C^{\alpha}_{0,r-1}(X,E) \qquad (1\le r\le n)$$

and

$$K_r : C^0_{0,r}(X,E) \longrightarrow \bigcap_{0<\alpha<1} C^{\alpha}_{0,r}(X,E) \qquad (1\le r\le n)$$

such that the following conditions are fulfilled:

(i) For any $0\le\alpha\le 1$, T_r and K_r are bounded as operators acting from the Banach space $C^0_{0,r}(X,E)$ into the Banach space $C^{\alpha}_{0,r-1}(X,E)$ resp. $C^{\alpha}_{0,r}(X,E)$.

(ii) For every $f\in C^0_{0,r}(X,E)$ such that $\bar\partial f$ is also continuous on X, we have the representation

$$\bar\partial T_r f = \begin{cases} f + K_r f - T_{r+1}\bar\partial f & \text{if } 1\le r\le n-1 \\ f + K_n f & \text{if } r=n \end{cases} \qquad (1.17)$$

Proof. Since X is compact, we can find a finite open covering $\{U_j\}$ of X such that each $\bar U_j$ has a neighborhood $\tilde U_j$ which is biholomorphically equivalent to some domain in $\mathbb{C}^n$ and such that E is holomorphically trivial over $\tilde U_j$. Then from Lemma 1.16 we get linear operators

$$T^j_r : C^0_{0,r}(X,E) \longrightarrow \bigcap_{0<\alpha<1} C^{\alpha}_{0,r-1}(\bar U_j,E) \qquad (1\le r\le n)$$

and

$$K^j_r : C^0_{0,r}(X,E) \longrightarrow \bigcap_{0<\alpha<1} C^{\alpha}_{0,r}(\bar U_j,E) \qquad (1\le r\le n)$$

which are bounded as operators acting from $C^0_{0,r}(X,E)$ into $C^{\alpha}_{0,r-1}(\bar U_j,E)$ resp. $C^{\alpha}_{0,r}(\bar U_j,E)$ (for all $0<\alpha<1$) and such that, for every $f\in C^0_{0,r}(X,E)$ with $\bar\partial f\in C^0_{0,r+1}(X,E)$, we have the representation

$$\bar\partial T^j_r f = \begin{cases} f + K^j_r f - T^j_{r+1}\bar\partial f & \text{if } 1\le r\le n-1 \\ f & \text{if } r=n. \end{cases}$$

Choose a C^{∞} partition $\{\chi_j\}$ of unity subordinated to $\{U_j\}$. Setting

$$T_r = \sum_j \chi_j T_r^j,$$

and

$$K_r = \begin{cases} \sum_j \bar\partial\chi_j \wedge T_r^j + \sum_j \chi_j K_r^j & \text{if } 1 \le r \le n-1 \\ \sum_j \bar\partial\chi_j \wedge T_n^j & \text{if } r=n, \end{cases}$$

we complete the proof. □

<u>1.18. Proof of Theorem 1.15.</u> First consider the case r=0. Since X is compact, $\mathcal{O}(X,E)$ is a Banach space with the norm $\|\cdot\|_{0,X}$. Since, by Montel's theorem, any bounded set in $\mathcal{O}(X,E)$ is relatively compact, this space is finitely dimensional. It remains to observe that, by Theorem 1.13 (regularity of $\bar\partial$),

$$H_0^{0,0}(X,E) = Z_{0,0}^0(X,E) = \mathcal{O}(X,E).$$

Now let $1 \le r \le n$. Then we fix some $0<\alpha<1$ and observe that, by Theorem 1.17, we can find bounded linear operators

$$T_r : Z_{0,r}^0(X,E) \longrightarrow C_{0,r-1}^{\alpha}(X,E)$$

and

$$K_r : Z_{0,r}^0(X,E) \longrightarrow Z_{0,r}^{\alpha}(X,E)$$

such that

$$\bar\partial \circ T_r = \text{id} + K_r \quad \text{on } Z_{0,r}^0(X,E), \tag{1.18}$$

where id is the identical operator. Since, by Ascoli's theorem, the embedding

$$Z_{0,r}^{\alpha}(X,E) \longrightarrow Z_{0,r}^0(X,E)$$

is compact, K_r is compact as an operator from $Z^0_{0,r}(X,E)$ into itself. Hence, $\bar{\partial}\circ T_r$ is a Fredholm operator in $Z^0_{0,r}(X,E)$. Since the image of this operator is contained in $E^0_{0,r}(X,E)$, it follows that $E^0_{0,r}(X,E)$ is finitely codimensional in $Z^0_{0,r}(X,E)$. □

2. Cauchy-Fantappie formulas, the Poincaré $\bar{\partial}$-lemma the Dolbeault isomorphism and smoothing of the $\bar{\partial}$-cohomology

2.1. Leray data and Leray maps. Let $D \subset\subset \mathbb{C}^n$ be a domain with almost C^1 boundary. A $\mathbb{C}^n$-valued C^1 map $v=(v_1,\dots,v_n)$ defined on $D\times\partial D$ will be called a Leray datum for D if

$$\langle v(z,x), x-z\rangle \neq 0 \qquad \text{for all } z\in D \text{ and } x\in\partial D.$$

If, moreover,

$$\langle v(z,x), x-z\rangle = 1 \qquad \text{for all } z\in D \text{ and } x\in\partial D,$$

then v will be called a Leray map for D.

Remark. In [H/L] the notion of a Leray map is used for arbitrary Leray data . Notice that, of course, any Leray datum $w(z,x)$ leads to the Leray map $v(z,x):= w(z,x)/(x-z)$. In Sect. 3.4 we shall consider generalizations of both the notion of a Leray data and the notion of a Leray map. Then the difference between them becomes more substantial.

2.2. The Cauchy-Fantappie integral L^v. Let $D \subset\subset \mathbb{C}^n$ be a domain with almost C^1 boundary, and let v be a Leray map for D.

Then we set

$$\hat{L}^v = \hat{L}^v(z,x) = \frac{1}{(2\pi i)^n}\det\Big(v(z,x),\overbrace{\bar{\partial}_{z,x}v(z,x)}^{n-1}\Big)\wedge\omega(x) \tag{2.1}$$

for $z\in D$ and $x\in\partial D$, and for any bounded differential form f on ∂D, by

$$(L^v f)(z) = \int_{x\in\partial D} f(x)\wedge\hat{L}^v(z,x), \qquad z\in D, \tag{2.2}$$

we define a continuous differential form $L^v f$ in D.

By $\hat{L}^v_r = \hat{L}^v_r(z,x)$ ($0 \le r \le n-1$) we denote the sum of all monomials in $\hat{L}^v$ which are of bidegree $(0,r)$ in z and hence of bidegree $(n,n-r-1)$ in x. It is easy to see that

$$\hat{L}^v_r(z,x) = \frac{\binom{n-1}{r}}{(2\pi i)^n} \det(v(z,x), \overbrace{\bar{\partial}_z v(z,x)}^{r}, \overbrace{\bar{\partial}_x v(z,x)}^{n-r-1}) \wedge \omega(x) \tag{2.3}$$

for $r=0,\dots,n-1$.

<u>2.3. Proposition.</u> Let $D \subset\subset \mathbb{C}^n$ be a domain with almost C^1 boundary, and let v be a Leray map for D. Then, for any bounded $(0,r)$-form f on ∂D, $L^v f$ is of bidegree $(0,r)$, and we have

$$L^v f = \int_{x \in \partial D} f(x) \hat{L}^v_r(\cdot, x) \qquad \text{if } 0 \le r \le n-1, \tag{2.4}$$

and $L^v f = 0$ if $r=n$.

<u>Proof.</u> Repetition of the proof of Proposition 1.5. □

<u>2.4. The Cauchy-Fantappie integrals R^v.</u> Let $D \subset\subset \mathbb{C}^n$ be a domain with almost C^1 boundary, and let v be a Leray map for D. Then, for all $z \in D$, $x \in \partial D$ and $0 \le t \le 1$, we set

$$v^0 = v^0(z,x,t) = t\frac{\bar{x}-\bar{z}}{|x-z|^2} + (1-t)v(z,x) \tag{2.5}$$

and

$$\hat{R}^v = \hat{R}^v(z,x,t) = \frac{1}{(2\pi i)^n} \det\Big(v^0(z,x,t), \overbrace{(\bar{\partial}_{z,x} + d_t) v^0(z,x,t)}^{n-1}\Big) \wedge \omega(x). \tag{2.6}$$

Further, then, for any bounded differential form f on ∂D, by

$$(R^v f)(z) = \int_{(x,t) \in \partial D \times [0,1]} f(x) \wedge \hat{R}^v(z,x,t), \qquad z \in D, \tag{2.7}$$

we define a continuous differential form $R^v f$ in D.

By $\hat{R}^v_r = \hat{R}^v_r(z,x,t)$ ($r=0,\dots,n-1$) we denote the sum of all monomials in $\hat{R}^v$ which are of bidegree $(0,r)$ in z. Then

$$\hat{R}_r^v = \frac{\binom{n-2}{r}}{(2\pi i)^n} \det(v^0, \overbrace{\bar{\partial}_z v^0}^{r}, \overbrace{(\bar{\partial}_x + d_t) v^0}^{n-r-1}) \wedge \omega(x) \tag{2.8}$$

for $r=0,\ldots,n-1$.

2.5. Proposition. Let $D \subset\subset \mathbb{C}^n$ be a domain with almost C^1 boundary, and let v be a Leray map for D. Then, for any bounded (0,r)-form f on ∂D, $R^v f$ is of bidegree (0,r-1), and one has

$$R^v f = \int_{(x,t)\in\partial D\times[0,1]} f(x)\wedge\hat{R}_{r-1}^v(\cdot,x) \qquad \text{if } 1\le r\le n-1 \tag{2.9}$$

and $R^v f = 0$ if $r=0$ or $r=n$.

Proof. Repetition of the proof of Proposition 1.5. □

2.6. The operators T^v. If $D \subset\subset \mathbb{C}^n$ is a domain with almost C^1 boundary and v is a Leray map for D, then, for any continuous differential form f on $\bar{D}$, we set

$$T^v f = B_D f + R^v f.$$

2.7. Theorem (Cauchy-Fantappie formula for functions). Let $D \subset\subset \mathbb{C}^n$ be a domain with almost C^1 boundary, and let v be a Leray map for D. Then, for any continuous complex-valued function f on $\bar{D}$ such that $\bar{\partial} f$ is also continuous on $\bar{D}$, we have the representation

$$f = L^v f - T^v \bar{\partial} f \qquad \text{on } D. \tag{2.10}$$

Proof. In Sect. 3 we prove the more general Theorem 3.11. A direct proof of Theorem 2.7 is given, for instance, in Sect. 1.9 of [H/L]. □

Supplement to Theorem 2.7. The proof of Theorem 2.7 given in Sect. 1.9 of [H/L] does not use any information about the kind of dependence of v(z,x) on z. So, actually, there the following more general statement is proved:

Let $D \subset\subset \mathbb{C}^n$ be a domain with almost C^1 boundary, and let, for some point $z_0\in D$, a $\mathbb{C}^n$-valued C^1 map v on ∂D be given such that $\langle v(x),x-z_0\rangle=1$ for all $x\in\partial D$. Then, for any continuous function f on $\bar{D}$ such that $\bar{\partial} f$ is also continuous on $\bar{D}$, we have the relation

$$f(z_0) = -(B_D f)(z_0) + \frac{1}{(2\pi i)^n} \int_{x\in\partial D} f(x)\det(v(x),\overbrace{\bar\partial v(x)}^{n-1})\wedge\omega(x)$$

$$- \frac{1}{(2\pi i)^n} \int_{\substack{x\in\partial D\\ 0\le t\le 1}} f(x)\det(v^0(x,t),\overbrace{(\bar\partial_x+d_t)v^0(x,t)}^{n-1}),$$

where

$$v^0(x,t) := t\frac{\bar x-\bar z}{|x-z|^2} + (1-t)v(x).$$

<u>Remark.</u> In [H/L], (2.10) is called the Leray formula.

<u>Remark.</u> If $v(z,x)=(\bar x-\bar z)/|x-z|^2$, then $L^v=B_{\partial D}$ and $R^v=0$. Hence then the Cauchy-Fantappie formula (2.10) coincides with the Martinelli-Bochner formula (1.6).

<u>2.8. Theorem</u> (Cauchy-Fantappie formula for forms). Let $D\subset\subset\mathbb{C}^n$ be a domain with almost C^1 boundary, v a Leray map for D, and f a continuous (0,r)--form on $\bar D$ such that $\bar\partial f$ is also continuous on $\bar D$ ($0\le r\le n$). Then

$$f = L^v f + \bar\partial T^v f - T^v\bar\partial f. \tag{2.11}$$

In particular, then $\bar\partial T^v f$ is continuous in D.

<u>Remark.</u> In [H/L], (2.11) is called the Koppelman-Leray formula.

<u>Remark.</u> If $v(z,x)=(\bar x-\bar z)/|x-z|^2$, then $L^v=B_{\partial D}$ and $R^v=0$. Hence then (2.11) coincides with the Koppelman-formula (1.7).

<u>2.9. Corollary.</u> Let $D \subset\subset \mathbb{C}^n$ be a domain with almost C^1 boundary, and let v be a Leray map for D such that, for some $0\le r\le n$, at least one of the following three conditions is fulfilled:

(i) $r=n$;

(ii) $v(z,x)$ depends holomorphically on $z_1,\dots,z_{n-r+1}$;

(iii) $v(z,x)$ depends holomorphically on $x_1,\dots,x_{r+2}$.

Then, for each continuous (0,r)-form f on $\bar D$ such that $\bar\partial f$ is also continuous on $\bar D$, $L^v f=0$ and hence the Cauchy-Fantappie formula (2.11) takes the form

$$(-1)^r f = \bar\partial T^v f - T^v\bar\partial f \quad \text{on } D. \tag{2.12}$$

Proof. For r=n the assertion follows immediately from Proposition 2.3. If $0\le r\le n-1$, then, by this proposition,

$$(L^v f)(z) = \frac{\binom{n-1}{r}}{(2\pi i)^n} \int_{x\in\partial D} f(x)\wedge \det(v(z,x),\overbrace{\bar\partial_z v(z,x)}^{r},\overbrace{\bar\partial_x v(z,x)}^{n-r-1})\wedge\omega(x).$$

It remains to observe that, obviously, each of the conditions (ii) and (iii) implies the equation

$$\det(v(z,x),\overbrace{\bar\partial_z v(z,x)}^{r},\overbrace{\bar\partial_x v(z,x)}^{n-r-1}) = 0.$$

2.10. **Remark.** Let $D \subset\subset \mathbb{C}^n$ be a domain with C^2 boundary, U a neighborhood of $\bar D$, and $\varrho:U\longrightarrow \mathbb{R}$ a C^2 function with $D = \{z\in U:\ \varrho(z)<0\}$.
Denote by $\nabla^{\mathbb{C}}\varrho:U\longrightarrow\mathbb{C}^n$ and $\nabla^{\mathbb{R}}\varrho:U\longrightarrow \mathbb{R}^{2n}$ the holomorphic resp. real gradient of ϱ, i.e.

$$\nabla^{\mathbb{C}}\varrho(z):= \left(\frac{\partial\varrho(z)}{\partial z_1},\dots,\frac{\partial\varrho(z)}{\partial z_n}\right), \tag{2.13}$$

and

$$\nabla^{\mathbb{R}}\varrho(z):= \left(\frac{\partial\varrho(z)}{\partial t_1},\dots,\frac{\partial\varrho(z)}{\partial t_{2n}}\right), \tag{2.13'}$$

$z\in U$, where $t=(t_1,\dots,t_{2n}):\mathbb{C}^n\longrightarrow \mathbb{R}^{2n}$ are the real coordinates in $\mathbb{C}^n$ with $z_j=t_j(z)+it_{j+n}(z)$, $z\in\mathbb{C}^n$. Then, for all $z\in U$ and $x\in\mathbb{C}^n$,

$$2\,\mathrm{Re}\langle\nabla^{\mathbb{C}}\varrho(z),x\rangle = \langle\nabla^{\mathbb{R}}\varrho(z),t(x)\rangle, \tag{2.14}$$

where Re:= "real part of".

Proof.

$$2\mathrm{Re}\langle\nabla^{\mathbb{C}}\varrho(z),x\rangle = \mathrm{Re}\sum_{j=1}^{n}\left(\frac{\partial\varrho(z)}{\partial t_1} - i\frac{\partial\varrho(z)}{\partial t_{j+n}}\right)(t_j(x) + it_{j+n}(x))$$

$$= \sum_{j=1}^{n}\frac{\partial\varrho(z)}{\partial t_j}t_j(x) + \sum_{j=1}^{n}\frac{\partial\varrho(z)}{\partial t_{j+n}}t_{j+n}(x) = \langle\nabla^{\mathbb{R}}\varrho(z),t(x)\rangle.$$

2.11. Observation. Let $D \subset\subset \mathbb{C}^n$ be a convex domain with C^2 boundary, U a neighborhood of $\bar{D}$, and $\varrho: U \longrightarrow \mathbb{R}$ a C^2 function with

$$D = \{z \in U:\ \partial(z) < 0\}.$$

Further, let $t = (t_1, \ldots, t_{2n}): \mathbb{C}^n \longrightarrow \mathbb{R}^{2n}$ be the real coordinates in $\mathbb{C}^n$ with $z_j = t_j(z) + i t_{j+n}(z)$. Then by Remark 2.10, for any $z \in \partial D$, the complex plane

$$\{z \in \mathbb{C}^n:\ \langle \nabla^{\mathbb{C}} \varrho(x), x-z \rangle = 0\}$$

is contained in the (real) tangent plane

$$\{z \in \mathbb{C}^n:\ \langle \nabla^{\mathbb{R}} \varrho(x), t(x-z) \rangle = 0\}$$

of ∂D at x. Since D is convex and hence all tangent planes of ∂D are contained in $\mathbb{C}^n \setminus D$, it follows that

$$\langle \nabla^{\mathbb{C}} \varrho(x), x-z \rangle \neq 0 \qquad \text{for all } x \in \partial D \text{ and } z \in D,$$

i.e. $\nabla^{\mathbb{C}} \varrho$ is a Leray datum for D and the corresponding Leray map

$$v(\varrho) = v(\varrho)(x,z) := \frac{\nabla^{\mathbb{C}} \varrho(x)}{\langle \nabla^{\mathbb{C}} \varrho(x), x-z \rangle} \tag{2.15}$$

depends holomorphically on z. Therefore, by Corollary 2.9 (ii), for any continuous $(0,r)$-form f on $\bar{D}$ $(1 \le r \le n)$ such that $\bar{\partial} f$ is also continuous on $\bar{D}$, we have the representation

$$(-1)^r f = \bar{\partial} T^{v(\varrho)} f - T^{v(\varrho)} \bar{\partial} f \qquad \text{on } D. \tag{2.16}$$

In particular, if f is a continuous $\bar{\partial}$-closed $(0,r)$-form on $\bar{D}$ $(1 \le r \le n)$, then the equation $\bar{\partial} u = f$ can be solved in D with

$$u := (-1)^r T^{v(\varrho)} f. \tag{2.17}$$

2.12. Theorem (Poincaré $\bar{\partial}$-lemma for forms). If $D \subset\subset \mathbb{C}^n$ is a convex domain with C^2 boundary, then, for any continuous $\bar{\partial}$-closed $(0,r)$-form f on $\bar{D}$, formula (2.17) gives a solution of the equation $\bar{\partial} u = f$ in D.

This solution is of class C^{α} in D for all $0 < \alpha < 1$, and if, moreover, f is of class C^k for some $k \in \{1, 2, \ldots, \infty\}$, then this solution is even of class $C^{k+\alpha}$ for all $0 < \alpha < 1$.

Proof. This follows immediately from Observation 2.11 and Proposition 1.4 (ii). □

2.13. Theorem (Poincaré $\bar\partial$-lemma for currents). Let $D \subset\subset \mathbb{C}^n$ be a convex domain, and let T be a $\bar\partial$-closed current on D. Then, for any open $U \subset\subset D$, there exists a current S on U such that

$$\bar\partial S = T \quad \text{on } U.$$

Proof. Choose a C^∞ function χ such that $\chi = 1$ in neighborhood of $\bar{U}$ and supp $\chi \subset\subset D$. Then, by Theorem 1.12,

$$\chi T = \bar\partial(B'\chi T) + B'(\bar\partial\chi \wedge T).$$

Thus it remains to solve the equation $\bar\partial u = B'(\bar\partial\chi\wedge T)$ on U. This can be done in view of Theorem 2.12, since, by Proposition 1.11, $B'(\bar\partial\chi\wedge T)$ is of class C^∞ in a neighborhood of $\bar{U}$, and, without loss of generality, U is convex and with C^∞ boundary. □

In the following theorem and its proof we assume some familiarity with the general theory of sheaves (long exact cohomology sequences, the equation $H^q(\mathcal{F})=0$ $(q\geq 1)$ for sheaves $\mathcal{F}$ which admit a partition of unity). For this material many references are available (see, for instance: [Hirzebruch 1966, Wells 1973, Grauert/Remmert 1977, Griffiths/Harris 1978]). If E is a holomorphic vector bundle over a complex manifold X, then by $\mathcal{O}^E$ we denote the sheaf of germs of local holomorphic sections of E, i.e. $\mathcal{O}^E$ is the map which attaches to any open $U \subseteq X$ the space $\mathcal{O}(U,E)$: $\mathcal{O}^E(U) := \mathcal{O}(U,E)$ for all open $U \subseteq X$.

2.14. Theorem (Dolbeault isomorphism and smoothing of the Dolbeault cohomology). Let E be a holomorphic vector bundle over an n-dimensional complex manifold X. Then

$$H^r(X,\mathcal{O}^E) = 0 \qquad \text{if} \quad r \geq n+1, \tag{2.18}$$

and, for each $r \in \{0,\dots,n\}$, there exist isomorphisms

$$\delta^r_\alpha : H^{0,r}_\alpha(X,E) \longrightarrow H^r(X,\mathcal{O}^E) \qquad (0\leq\alpha\leq\infty) \tag{2.19}$$

and

$$\delta^r_{cur} : H^{0,r}_{cur}(X,E) \longrightarrow H^r(X,\mathcal{O}^E) \tag{2.20}$$

such that, for all $0 \leq \alpha \leq \infty$, the diagram

$$\begin{array}{ccc} H^{0,r}_{\alpha}(X,E) & \xrightarrow{\quad \delta^r_{\alpha} \quad} & \\ \Big\downarrow {\scriptstyle \langle\rangle} & & H^r(X, \mathcal{O}^E) \\ H^{0,r}_{cur}(X,E) & \xrightarrow{\quad \delta^r_{cur} \quad} & \end{array} \tag{2.21}$$

is commutative. In particular, for all $0 \leq \alpha \leq \infty$ and $r=0,\dots,n$, the natural homomorphism

$$H^{0,r}_{\alpha}(X,E) \xrightarrow{\langle\rangle} H^{0,r}_{cur}(X,E) \tag{2.22}$$

induced by the assignment $Z^{\alpha}_{0,r}(X,E) \ni f \xrightarrow{\langle\rangle} \langle f\rangle \in Z^{cur}_{0,r}(X,E)$ (cf. Sect. 0.13) is an isomorphism.

<u>Remark.</u> The isomorphism (2.21) is called the Dolbeault isomorphism, and the inverse map of the isomorphism (2.22) is refered to as "smoothing of the Dolbeault cohomology".

<u>Proof of Theorem 2.14.</u> We consider the following sheaves: $(Z^{\alpha}_{0,r})^E$, $(C^{cur}_{0,r})^E$, and $(Z^{cur}_{0,r})^E$ $(0 \leq \alpha \leq \infty;\ r=0,\dots,n)$ are the sheaves defined by

$$(Z^{\alpha}_{0,r})^E(U) = Z^{\alpha}_{0,r}(U,E),$$

$$(C^{cur}_{0,r})^E(U) = C^{cur}_{0,r}(U,E),$$

$$(Z^{cur}_{0,r})^E(U) = Z^{cur}_{0,r}(U,E)$$

for all open $U \subseteq X$. Further, if $U \subseteq X$ is an open set, then by $\tilde{C}^{\alpha}_{0,r}(U,E)$ $(0 \leq \alpha \leq \infty;\ r=0,\dots,n)$ we denote the subspace of all $f \in C^{\alpha}_{0,r}(U,E)$ such that $\bar{\partial} f$ is also of class C^{α} in U. The corresponding sheaves will be denoted by $(\tilde{C}^{\alpha}_{0,r})^E$, i.e. $(\tilde{C}^{\alpha}_{0,r})^E(U) = \tilde{C}^{\alpha}_{0,r}(U,E)$ for all open $U \subseteq X$;

Notice that if f is a section of $(\tilde{C}^{\alpha}_{0,r})^E$ and χ is an arbitrary C^{∞} function, then χf is again a section of $(\tilde{C}^{\alpha}_{0,r})^E$. Therefore, the sheaves $(\tilde{C}^{\alpha}_{0,r})^E$ admit a partition of unity and so, for all $0 \leq \alpha \leq \infty$ and $0 \leq r \leq n$,

$$H^q(X,(\tilde{C}^{\alpha}_{0,r})^E) = 0 \qquad \text{if} \quad q\geq 1. \tag{2.23}$$

The same is true for the sheaves $(C^{cur}_{0,r})^E$:

$$H^q(X,(C^{cur}_{0,r})^E) = 0 \qquad \text{if} \quad q\geq 1. \tag{2.23'}$$

Now we observe that, by Theorem 1.13 (regularity of $\bar\partial$), the map

$$\mathcal{O}(X,E) \xrightarrow{\langle\rangle} Z^{cur}_{0,0}(X,E)$$

is an isomorphism, and, for all $0\leq\alpha\leq\infty$, we have the equation $\mathcal{O}(X,E) = Z^{\alpha}_{00}(X,E)$. Hence the assertion of the theorem holds true for r=0.

Let $r\geq 1$. Fix $0\leq\alpha\leq\infty$. In view of Theorem 2.12 (Poincaré $\bar\partial$-lemma for forms), for all s=1,...,n, the sequence of sheaves

$$0 \longrightarrow (Z^{\alpha}_{0,s-1})^E \longrightarrow (\tilde{C}^{\alpha}_{0,s-1})^E \xrightarrow{\bar\partial} (Z^{\alpha}_{0,s})^E \longrightarrow 0$$

is exact. Let

$$0\longrightarrow Z^{\alpha}_{0,s-1}(X,E)\longrightarrow \tilde{C}^{\alpha}_{0,s-1}(X,E)\xrightarrow{\bar\partial} Z^{\alpha}_{0,s}(X,E)\xrightarrow{\delta^0_{\alpha,s}}$$

$$\xrightarrow{\delta^0_{\alpha,s}} H^1(X,(Z^{\alpha}_{0,s-1})^E)\longrightarrow H^1(X,(\tilde{C}^{\alpha}_{0,s-1})^E)\xrightarrow{\bar\partial} H^1(X,(Z^{\alpha}_{0,s})^E)\xrightarrow{\delta^1_{\alpha,s}}$$

$$\cdots\cdots\cdots\cdots\cdots\cdots$$

$$\xrightarrow{\delta^{q-1}_{\alpha,s}} H^q(X,(Z^{\alpha}_{0,s-1})^E)\longrightarrow H^q(X,(\tilde{C}^{\alpha}_{0,s-1})^E)\xrightarrow{\bar\partial} H^q(X,(Z^{\alpha}_{0,s})^E)\xrightarrow{\delta^q_{\alpha,s}}$$

$$\cdots\cdots\cdots\cdots\cdots\cdots$$

be the corresponding long exact cohomology sequence. Then, by (2.23), it follows that, for s=1,...,n, the coboundary maps

$$\delta^q_{\alpha,s}: H^q(X,(Z^{\alpha}_{0,s})^E) \longrightarrow H^{q+1}(X,(Z^{\alpha}_{0,s-1})^E) \qquad (q\geq 1) \tag{2.24}$$

are isomorphisms, and the coboundary map $\delta^0_{\alpha,s}$ induces an isomorphism

$$\hat{\delta}^0_{\alpha,s}:H^{0,s}_{\alpha}(X,E) \longrightarrow H^1(X,(Z^{\alpha}_{0,s-1})^E) \qquad (1\le s\le n). \qquad (2.25)$$

If $r\ge n+1$, then, in view of the isomorphisms (2.24), we see that the spaces $H^r(X,(Z^{\alpha}_{0,0})^E)$ and $H^{r-n}(X,(Z^{\alpha}_{0,n})^E)$ are isomorphic. Since $(Z^{\alpha}_{0,n})^E = (\tilde{C}^{\alpha}_{0,n})^E$ and, by Theorem 1.13 (regularity of $\bar{\partial}$),

$$(Z^{\alpha}_{0,0})^E = \mathcal{O}^E, \qquad (2.26)$$

this implies that, for $r\ge n+1$, $H^r(X,\mathcal{O}^E)$ is isomorphic to $H^{r-n}(X,(\tilde{C}^{\alpha}_{0,n})^E)$. By (2.23), this yields (2.18).

For the remainder of this proof we may assume that $1\le r\le n$. Set

$$\delta^r_{\alpha} := \delta^{r-1}_{\alpha,1}\circ\ldots\circ\delta^1_{\alpha,r-1}\circ\hat{\delta}^0_{\alpha,r}.$$

Since the maps (2.24) and (2.25) are isomorphisms and (2.26) holds true, in this way we obtain an isomorphism

$$\delta^r_{\alpha}:H^{0,r}_{\alpha}(X,E) \longrightarrow H^r(X,\mathcal{O}^E).$$

Now, analogously, we construct the isomorphism δ^r_{cur}:

First observe that, by Theorem 2.13 (Poincaré $\bar{\partial}$-lemma for currents), for $s=1,\ldots,n$, the sequence

$$0 \longrightarrow (Z^{cur}_{0,s})^E \longrightarrow (C^{cur}_{0,s-1})^E \xrightarrow{\bar{\partial}} (Z^{cur}_{0,s})^E \longrightarrow 0$$

is exact. Let

$$\delta^q_{cur,s}:H^q(X,(Z^{cur}_{0,s})^E) \longrightarrow H^{q+1}(X,(Z^{cur}_{0,s-1})^E) \qquad (q\ge 0) \qquad (2.27)$$

be the coboundary homomorphisms from the corresponding long exact cohomology sequence. Then it follows from (2.23') that (2.27) is an isomorphism if $q\ge 1$, and the maps $\delta^0_{cur,s}$ induce isomorphisms

$$\hat{\delta}^0_{cur}:H^{0,s}_{cur}(X,E) \longrightarrow H^1(X,(Z^{cur}_{0,s-1})^E) \qquad (1\le s\le n). \qquad (2.28)$$

Hence, the composition

$$\delta^{r-1}_{cur,1} \circ \ldots \circ \delta^{1}_{cur,r-1} \circ \hat{\delta}^{0}_{cur,r}$$

is an isomorphism from $H^{0,r}_{cur}(X,E)$ onto $H^{r}(X,(Z^{cur}_{0,0})^{E})$. Since, by Theorem 1.13 (regularity of $\bar\partial$), also

$$\mathcal{O}^{E} \xrightarrow{\langle\rangle} (Z^{cur}_{0,0})^{E}$$

is an isomorphism, it follows that

$$\delta^{r}_{cur} := \langle\rangle^{-1} \circ \delta^{r-1}_{cur,1} \circ \ldots \circ \delta^{1}_{cur,r-1} \circ \hat{\delta}^{0}_{cur,r}$$

is an isomorphism from $H^{0,r}_{cur}(X,E)$ onto $H^{r}(X,\mathcal{O}^{E})$. To complete the proof, it remains to observe that, by definition of the coboundary operators in the long exact cohomology sequences of short exact sheaf sequences, obviously, each of the diagrams

$$\begin{array}{ccc} H^{q}(X,(Z^{\alpha}_{0,s})^{E}) & \xrightarrow{\delta^{q}_{\alpha,s}} & H^{q+1}(X,(Z^{\alpha}_{0,s-1})^{E}) \\ \downarrow \langle\rangle & & \downarrow \langle\rangle \\ H^{q}(X,(Z^{cur}_{0,s})^{E}) & \xrightarrow{\delta^{q}_{cur,s}} & H^{q+1}(X,(Z^{cur}_{0,s-1})^{E}) \end{array} \qquad (q \geq 0,\ 1 \leq s \leq n)$$

is commutative and hence (2.21) is commutative. □

The preceding theorem contains, in particular, the following two statements: 1) If f is a C^{α} form $(0 \leq \alpha \leq \infty)$ and the equation $\bar\partial T = \langle f \rangle$ can be solved with a current T, then we can solve $\bar\partial u = f$ even with a C^{α} form u (injectivity of (2.22)). 2) For any $\bar\partial$-closed current T, there exists a current S such that $T - \bar\partial S$ is of class C^{∞} (surjectivity of (2.22) for $\alpha = \infty$). Actually, we have the following more precise

<u>2.15. Corollary.</u> Let E be a holomorphic vector bundle over a complex manifold X. Then:

(i) For any E-valued $\bar\partial$-closed current T on X and every neighborhood U of supp T, there exists an E-valued current S on X such that $T - \bar\partial S$ is of class C^{∞} on X and supp $S \subseteq U$.

(ii) If f is an E-valued C^{k} form on X $(k = 0,1,\ldots,\infty)$ and T is an E--valued current on X such that $\bar\partial T = \langle f \rangle$ on X, then, for any neighborhood U of supp T, there exists a differential form

$$u \in \bigcap_{0<\alpha<1} C_*^{k+\alpha}(X,E)$$

such that $\bar\partial u = f$ on X and supp $u \subseteq U$.

<u>Proof of part (i).</u> Choose a neighborhood V of supp T such that $\bar V \subseteq U$. Further, take C^∞ functions χ_0 and χ_1 such that $\chi_0=1$ in a neighborhood of $X\setminus V$, $\chi_0=0$ in a neighborhood of supp T, $\chi_1=1$ in a neighborhood of $X\setminus U$, and $\chi_1=0$ in a neigborhood of $\bar V$.

Since T is $\bar\partial$-closed and (2.22) is surjective, there exists an E--valued current S_0 on X and a $\bar\partial$-closed C^∞ form f_0 on X such that $T-\bar\partial S_0=\langle f_0\rangle$ on X.

Then $\langle f_0\rangle=-\bar\partial S_0$ on $X \setminus$ supp T. Since (2.22) is injective for $\alpha=\infty$, this implies that there exists an E-valued C^∞ form g_0 on $X \setminus$ supp T with $f_0=\bar\partial g_0$ on $X \setminus$ supp T. Over $X\setminus\bar V$, then

$$\bar\partial(S_0+\langle\chi_0 g_0\rangle) = \langle -f_0+\bar\partial g_0\rangle = 0.$$

Since (2.22) is surjective, this implies that there is an E-valued current R over $X\setminus\bar V$ and a $\bar\partial$-closed E-valued C^∞ form g_1 on $X\setminus\bar V$ such that $S_0+\langle\chi_0 g_0\rangle-\bar\partial R=\langle g_1\rangle$ on $X\setminus V$. Setting $S=S_0+\langle\chi_0 g_0-\chi_1 g_1\rangle-\bar\partial(\chi_1 R)$ we conclude the proof of part (i).

<u>Proof of part (ii).</u> Denote by $\mathcal{F}$ the sheaf on X which is defined as follows: If $V\subseteq X$ is open, then $\mathcal{F}(V)$ is the space of all forms

$$g \in \bigcap_{0<\alpha<1} C_*^{k+\alpha}(V,E)$$

such that also

$$\bar\partial g \in \bigcap_{0<\alpha<1} C_*^{k+\alpha}(V,E).$$

We remark that the sections of $\mathcal{F}$ admit multiplication by C^∞ functions. Hence

$$H^1(X,\mathcal{F}) = 0. \tag{2.29}$$

Take open coverings $\{U_j\}$, $\{V_j\}$, $\{W_j\}$ of X such that $U_j \subset\subset V_j \subset\subset W_j$, and each of the sets V_j and W_j is biholomorphically equivalent to the unit ball in $\mathbb{C}^n$. Then, by Theorem 2.12 (Poincaré $\bar\partial$-lemma for forms), we can find a family

$$u_j \in \bigcap_{0<\alpha<1} C_*^{k+\alpha}(V_j, E)$$

such that $\bar\partial u_j = f$ on V_j. On the other hand, since (2.22) is injective (for $\alpha = k$) and $\bar\partial T = \langle f\rangle$, we can find an E-valued C^k-form g on X such that $\bar\partial g = f$ on X. Set $w_j = u_j - g$ on V_j. Then $w_j \in Z^k(V_j, E)$ and, again by Theorem 2.12, we can solve the equations $w_j = \bar\partial v_j$ with

$$v_j \in \bigcap_{0<\alpha<1} C_*^{k+\alpha}(U_j, E).$$

Then $v_i - v_j \in \mathcal{F}(U_i \cap U_j)$ and hence, by (2.29), we can find a family $g_i \in \mathcal{F}(U_i)$ with $v_i - v_j = g_i - g_j$ on $U_i \cap U_j$. Setting $h = u_i - \bar\partial g_i$ on U_i, so we obtain a form

$$h \in \bigcap_{0<\alpha<1} C_*^{k+\alpha}(X, E)$$

with $\bar\partial h = f$ on X. Then $T - \langle h\rangle$ is $\bar\partial$-closed on X, and since (2.22) is surjective for $\alpha = \infty$, we can find an E-valued current S on X as well as an E-valued C^∞ form u_∞ on X such that $T - \langle h\rangle + \bar\partial S = \langle u_\infty\rangle$. Then $\bar\partial S = \langle h + u_\infty\rangle$ on $X \setminus \operatorname{supp} T$. Since $h + u_\infty$ is of class C^k, repeating the arguments from above, we obtain a form

$$H \in \bigcap_{0<\alpha<1} C_*^{k+\alpha}(X \setminus \operatorname{supp} T\ ,\ E)$$

with $\bar\partial H = h + u_\infty$ on $X \setminus \operatorname{supp} T$. Choose a C^∞ function χ on X with $\chi = 0$ in a neighborhood of supp T and $\chi = 1$ in a neighborhood of $X\setminus U$. Then the form $u := h + u_\infty - \bar\partial(\chi H)$ has the required properties. □

3. Piecewise Cauchy-Fantappie formulas

3.0. Further on, for every collection $K = (k_1, \ldots, k_l)$ of integers, we write $|K| := l$ and $K(\hat{s}) := (k_1, \ldots, k_{s-1}, k_{s+1}, \ldots, k_l)$ $(1<s<l)$.

3.1. Domains with piecewise almost C^1 boundary and frames for such domains. Let $D \subset\subset \mathbb{C}^n$ be a domain. We say ∂D is piecewise almost C^1 if there exist a neighborhood U of ∂D and a finite number of closed almost C^1 hypersurfaces $Y_1, \ldots, Y_N$ in U (cf. Sect. 1.1) such that the following conditions are fulfilled:

(i) $$\partial D = \bigcup_{j=1}^{N} Y_j \cap \partial D;$$

(ii) for all $1 \le i < j \le N$, the points in $Y_i \cap Y_j$ are smooth points of both Y_i and Y_j.

(iii) for all $1 \le k_1 < \ldots < k_l \le N$, the intersection $Y_{k_1} \cap \ldots \cap Y_{k_l}$ is transversal (or empty).

If $D \subset\subset \mathbb{C}^n$ is a domain with piecewise almost C^1 boundary, then any collection $Y=(Y_1,\ldots,Y_N)$ of closed almost C^1 hypersurfaces in a neighborhood of ∂D satisfying conditions (i)-(iii) will be called a <u>frame</u> for D. Notice that, for each domain $D \subset\subset \mathbb{C}^n$ with piecwise almost C^1 boundary, there exist different frames, in particular, ∂D is also a frame for D.

<u>3.2. The manifolds S_K.</u> Let $D \subset\subset \mathbb{C}^n$ be a domain with piecewise almost C^1 boundary, and let $Y=(Y_1,\ldots,Y_N)$ be a frame for D. Then we set $S_j = S_j(Y) = \partial D \cap Y_j$ for $j=1,\ldots,N$. Further, for each ordered collection $K=(k_1,\ldots,k_l)$ of integers $1 \le k_1,\ldots,k_l \le N$, we set

$$S_K = S_K(Y) = \begin{cases} S_{k_1} \cap \ldots \cap S_{k_l} & \text{if } k_1,\ldots,k_l \text{ are different in pairs} \\ \emptyset & \text{otherwise.} \end{cases}$$

Then we have (as relations between sets) the equations

$$\partial D = \sum_{j=1}^{N} S_j \tag{3.1}$$

and

$$\partial S_K = \sum_{j=1}^{l} S_{Kj}, \tag{3.2}$$

where $Kj := (k_1,\ldots,k_l,j)$. We orient the manifolds S_K so that (3.1) and (3.2) hold true as relations between oriented manifolds.

<u>3.3. The manifolds $S_K \times \triangle_K$ and $S_K \times \triangle_{OK}$.</u> Let $D \subset\subset \mathbb{C}^n$ be a domanin with piecewise almost C^1 boundary, $Y=(Y_1,\ldots,Y_N)$ a frame for D, and $S_K=S_K(Y)$.

By $\triangle$ we denote the simplex of all points $(t_0,\ldots,t_N) \in R^{N+1}$ with $t_0,\ldots,t_N \ge 0$ and $\sum_0^N t_j = 1$. We orient $\triangle$ by $dt_1 \ldots dt_N$.

For every strictly increasing collection $K=(k_1,\ldots,k_l)$ of integers $0\le k_1<\ldots<k_l\le N$, we set

$$\triangle_K = \{t\in\triangle: \sum_{s=1}^{l} t_{k_s} = 1\}.$$

We orient the manifolds $\triangle_K$ so that

$$\partial\triangle_K = \sum_{s=1}^{l} (-1)^{s+1}\triangle_{K(s^\wedge)}. \tag{3.3}$$

Then, for any strictly increasing collection $K=(k_1,\ldots,k_l)$ of integers $1\le k_1<\ldots<k_l\le N$, the orientation of $\triangle_{0K}$ (here $0K:=(0,k_1,\ldots,k_l)$) is defined by the form $dt_{k_1}\wedge\ldots\wedge dt_{k_l}$. This implies that the one-point sets $\triangle_j$ $(j=0,\ldots,N)$ are oriented by $+1$.

3.4. Leray maps and Leray data . Let $D\subset\subset\mathbb{C}^n$ be a domain with piecewise almost C^1 boundary, $Y=(Y_1,\ldots,Y_N)$ a frame for D, and $S_K=S_K(Y)$.

A Leray map for (D,Y) (or for D) is, by definition, a map v which attaches to each strictly increasing collection $K=(k_1,\ldots,k_l)$ of integers $1\le k_1<\ldots<k_l\le N$ a $\mathbb{C}^n$-valued C^1 map $v_K = (v_K^1,\ldots,v_K^n)$ defined for $z\in D$, $x\in S_K$, and $t\in\triangle_K$ such that, for any collection $K=(k_1,\ldots,k_l)$ of strictly increasing integers $1\le k_1<\ldots<k_l\le N$ and all $z\in D$, $x\in S_K$, $t\in\triangle_K$, the following two conditions are fulfilled:

$$\langle v_K(z,x,t),x-z\rangle = 1 \tag{3.4}$$

and

$$v_{K(s^\wedge)}(z,x,t) = v_K(z,x,t) \qquad \text{if } t\in\triangle_{K(s^\wedge)}\ (1\le s\le l). \tag{3.5}$$

A Leray datum for (D,Y) (or for D) is, by definition, a family $w=\{w_j\}_{j=1}^N$ of $\mathbb{C}^n$-valued C^1 maps

$$w_j = w_j(z,x) = (w_j^1(z,x),\ldots,w_j^n(z,x))$$

defined for all $z\in D$ and $x\in S_j$ such that

$$\langle w_j(z;x),x-z\rangle = 0 \qquad \text{for all } z\in D \text{ and } x\in S_j\ (j=1,\ldots,N). \tag{3.6}$$

If $w=\{w_j\}$ is a Leray datum for (D,Y), then we define a Leray map for (D,Y) by setting

$$v_K(z,x,t) := \sum_{j\in K} t_j \frac{w_j(z,x)}{\langle w_j(z,x), x-z\rangle} \tag{3.7}$$

for any strictly increasing collection $K=(k_1,\dots,k_l)$ of integers $1\le k_1<\dots<k_l\le N$ and all $z\in D$, $x\in S_K$, $t=(t_1,\dots,t_l)\in\triangle_K$. This Leray map will be called the canonical combination of the Leray datum w.

3.5. The Cauchy-Fantappie integrals L_K^v. Let $D \subset\subset \mathbb{C}^n$ be a domain with piecewise almost C^1 boundary, $Y=(Y_1,\dots,Y_N)$ a frame for D, and v a Leray map for (D,Y). Fix some strictly increasing collections $K=(k_1,\dots,k_l)$ of integers $1\le k_1<\dots<k_l\le N$.

We set

$$\hat{L}_K^v = \hat{L}_K^v(z,x,t) = \frac{1}{(2\pi i)^n}\det\Big(v_K(z,x,t), \overbrace{(\bar\partial_{z,x}+d_t)v_K(z,x,t)}^{n-1}\Big)\wedge\omega(x) \tag{3.8}$$

for $z\in D$, $x\in S_K$, and $t\in\triangle_K$. For any bounded differential form f on S_K, by

$$(L_K^v f)(z) = \int_{(x,t)\in S_K\times\triangle_K} f(x)\wedge\hat{L}_K^v(z,x,t), \qquad z\in D,$$

we define a continuous differential form $L_K^v f$ in D.

By $(\hat{L}_K^v)_r = (\hat{L}_K^v)_r(z,x,t)$ $(0\le r\le n-1)$ we denote the sum of all monomials in $\hat{L}_K^v$ which are of bidegree (0,r) in z. Then

$$(\hat{L}_K^v)_r = \frac{\binom{n-1}{r}}{(2\pi i)^n}\det(v_K, \overbrace{\bar\partial_z v_K}^{r}, \overbrace{(\bar\partial_x+d_t)v_K}^{n-r-1})\wedge\omega(x) \tag{3.9}$$

for $r=0,\dots,n-1$.

3.6. Proposition. Let $D\subset\subset\mathbb{C}^n$ be a domain with piecewise almost C^1 boundary, $Y=(Y_1,\dots,Y_N)$ a frame for D, v a Leray map for (D,Y), and $K=(k_1,\dots,k_l)$ a strictly increasing collection of integers $1\le k_1<\dots<k_l\le N$. Then, for any bounded (0,r)-form f on S_K, $L_K^v f$ is also of bidegree (0,r), and we have

$$L_K^v f = \int_{(x,t)\in S_K\times\triangle_K} f(x)\wedge(\hat{L}_K^v)_r(\cdot,x,t) \qquad \text{if } 0\le r\le n-|K|, \tag{3.10}$$

and $L_K^v f = 0$ if $n-|K|+1\le r\le n$.

Proof. Repetition of the proof of Proposition 1.5. □

3.7. The Cauchy-Fantappie integrals R_K^v. Let $D \subset\subset \mathbb{C}^n$ be a domain with piecewise almost C^1 boundary, $Y=(Y_1,\dots,Y_N)$ a frame for D, v a Leray map for (D,Y), and $K=(k_1,\dots,k_l)$ a collection of strictly increasing integers $1\le k_1<\dots<k_l\le N$.

For all $z\in D$, $x\in S_K$ and $t\in\triangle_{OK}$, we set

$$v_{OK}(z,x,t) = t_0\frac{\bar{x}-\bar{z}}{|x-z|^2} + (1-t_0)v_K(z,x,t'), \tag{3.11}$$

where $t' := (0,t_1/(1-t_0),\dots,t_N/(1-t_0))$. We remark that if v_K is of the form (3.7), then

$$v_{OK}(z,x,t) = t_0\frac{\bar{x}-\bar{z}}{|x-z|^2} + \sum_{j\in K} t_j \frac{w_j(z,x)}{\langle w_j(z,x),x-z\rangle}. \tag{3.11'}$$

Further, we define

$$\hat{R}_K^v = \hat{R}_K^v(z,xt) = \frac{(-1)^{|K|}}{(2\pi i)^n}\det\big(v_{OK}(z,x,t),\overbrace{(\bar\partial_{z,x}+d_t)v_{OK}(z,x,t)}^{n-1}\big)\wedge\omega(x) \tag{3.12}$$

for all $z\in D$, $x\in S_K$ and $t\in\triangle_{OK}$.

Now, for any bounded differential form f on S_K, by setting

$$(R_K^v f)(z) = \int_{(x,t)\in S_K\times\triangle_{OK}} f(x)\wedge\hat{R}_K^v(z,x,t), \qquad z\in D, \tag{3.13}$$

we define a continuous differential form $R_K^v f$ in D.

By $(R_K^v)_r=(R_K^v)_r(z,x,t)$ $(0\le r\le n-1)$ we denote the sum of all monomials in R_K^v which are of bidegree (0,r) in z. Then

$$(\hat{R}_K^v)_r = (-1)^{|K|} \frac{\binom{n-1}{r}}{(2\pi i)^n} \det(v_{0K}, \overbrace{\bar\partial_z v_{0K}}^{r}, \overbrace{(\bar\partial_x + d_t) v_{0K}}^{n-r-1}) \wedge \omega(x) \qquad (3.14)$$

for $r=0,\dots,n-1$.

3.8. Proposition. Let $D \subset\subset \mathbb{C}^n$ be a domain with piecewise almost C^1 boundary, $Y=(Y_1,\dots,Y_N)$ a frame for D, v a Leray map for (D,Y), and $K=(k_1,\dots,k_l)$ a strictly increasing collection of integers $1 \le k_1 < \dots < k_l \le N$. Then, for any bounded (0,r)-form f on S_K, $R_K^v f$ is of bidegree (0,r-1), and we have

$$R_K^v f = \int_{(x,t)\in S_K\times\triangle_{0K}} f(x)\wedge(\hat{R}_K^v)_{r-1}(\cdot,x,t) \qquad \text{if } 1\le r\le n-|K|, \qquad (3.15)$$

and $R_K^v f = 0$ if $r=0$ or $n-|K|+1 \le r \le n$.

Proof. Repetition of the proof of Proposition 1.5. □

3.9. Proposition. Let $D \subset\subset \mathbb{C}^n$ be a domain with piecewise almost C^1 boundary, $Y=(Y_1,\dots,Y_N)$ a frame for D, v a Leray map for (D,Y), and $K=(k_1,\dots,k_l)$ a strictly increasing collection of integers $1 \le k_1 < \dots < k_l \le N$. Then:

(i) For all $z\in D$, $x\in S_K$, and $t\in\triangle_K$, we have

$$(\bar\partial_z + d_{x,t})\hat{L}^v = 0$$

and hence

$$d_{x,t}(\hat{L}_K^v)_0(z,x,t) = 0,$$

$$d_{x,t}(\hat{L}_K^v)_r(z,x,t) = -\bar\partial_z(\hat{L}_K^v)_{r-1}(z,x,t) \qquad \text{if } 1\le r\le n-1,$$

$$\bar\partial_z(\hat{L}_K^v)_{n-1}(z,x,t) = 0.$$

(ii) For all $z\in D$, $x\in S_K$, and $t\in\triangle_{0K}$, we have

$$(\bar\partial_z + d_{x,t})\hat{R}^v = 0$$

and hence

$$d_{x,t}(R_K^v)_0(z,x,t) = 0,$$

$$d_{x,t}(R_K^v)_r(z,x,t) = -\bar\partial_z(\hat R_K^v)_{r-1}(z,x,t) \qquad \text{if } 1\le r\le n-1,$$

$$\bar\partial_z(\hat R_K^v)_{n-1}(z,x,t) = 0.$$

<u>Proof.</u> The proofs of parts (i) and (ii) are similar. We restrict ourselves to part (ii). By condition (3.4) in the definition of a Leray map and by (3.11), we have the relation

$$\sum_{j=1}^{n} (x_j-z_j)v_{OK}^j(z,x,t) = 1.$$

Therefore

$$\sum_{j=1}^{n} (x_j-z_j)(\bar\partial_{z,x}+d_t)v_{OK}^j(z,x,t) = 0$$

and hence

$$(\bar\partial_z+d_{x,t})\hat R_K^v = (\bar\partial_{z,x}+d_t)\frac{1}{(2\pi i)^n}\det(v_{OK},\overbrace{(\bar\partial_{z,x}+d_t)v_{OK}}^{n-1})\wedge\omega(x)$$

$$= \frac{1}{(2\pi i)^n}[\bigwedge_{j=1}^{n}(\bar\partial_{z,x}+d_t)v_{OK}^j]\wedge\omega(x) = 0. \qquad \square$$

<u>3.10. The operators L^v and T^v.</u> Let $D \subset\subset \mathbb{C}^n$ be a domain with piecewise almost C^1 boundary, $Y=(Y_1,\dots,Y_N)$ a frame for D, and v a Leray map for (D,Y). Then, for any continuous differential form f on D, we set

$$L^vf = \sum_K{}' L_K^vf \qquad \text{and} \qquad T^vf = B_Df + \sum_K{}' R_K^vf,$$

where the summations are over all strictly increasing collections $K=(k_1,\dots,k_l)$ of integers $1\le k_1<\dots<k_l\le N$.

<u>3.11. Remark.</u> If f is a continuous differential form of bidegree (0,r) on $\bar D$, then by Propositions 1.5, 3.6 and 3.8, L^vf is also of bidegree (0,r), T^vf is of bidegree (0,r-1), and we have the following relations:

$$L^vf = \sum_{1\le|K|\le n-r}{}' L_K^vf \qquad \text{for all } 0\le r\le n,$$

$$L^v f = 0 \qquad \text{if } r=n,$$

$$T^v F = B_D f + \sum_{1\le|K|\le n-r}' R_K^v f \qquad \text{for all } 0\le r\le n,$$

$$T^v f = B_D f \qquad \text{if } r=n,$$

$$T^v f = 0 \qquad \text{if } r=0,$$

where the summations are over all strictly increasing collections $K=(k_1,\dots,k_l)$ of integers $1\le k_1<\dots<k_l\le N$ with $1\le l\le n-r$.

<u>3.12. Theorem</u> (Piecewise Cauchy-Fantappie formula). Let $D \subset\subset \mathbb{C}^n$ be a domain with piecewise almost C^1 boundary, $Y=(Y_1,\dots,Y_N)$ a frame for D, and v a Leray map for (D,Y). Then, for any continuous (0,r)-form f on $\bar{D}$ such that $\bar{\partial} f$ is also continuous on $\bar{D}$ $(0\le r\le n)$, we have the representation

$$(-1)^r f = L^v f + \bar{\partial} T^v f - T^v \bar{\partial} f \qquad \text{in } D. \tag{3.17}$$

In particular, then $\bar{\partial} T^v f$ is continuous in D.

<u>Remark.</u> For r=n, formula (3.17) coincides with the Koppelman formula (1.7), because then, by Proposition 1.5 and Remark 3.11,

$$L^v f = B_{\partial D} f = 0 \qquad \text{and} \qquad T^v f = B_D f.$$

For r=0, i.e. if f is a function, in view of Remark 3.11, formula (3.17) takes the form

$$f = L^v f - T^v \bar{\partial} f \qquad \text{in } D. \tag{3.18}$$

<u>Proof of Theorem 3.12.</u> Let P be the set of all collections $K=(k_1,\dots,k_l)$ of strictly increasing integers $1\le k_1<\dots<k_l\le n$. Since, by (3.2) and (3.3), for any $K\in P$,

$$\partial(S_K\times\triangle_{OK}) = (-1)^{|K|}(S_K\times\triangle_K) + \sum_{s=1}^{|K|}(-1)^{|K|+s+1} S_K\times\triangle_{OK(s^\wedge)} + \sum_{j=1}^{N} S_{Kj}\times\triangle_{OK}$$

and, by Proposition 3.9, for all $K\in P$ and $v\in[C^{\infty}_{n,n-r}(D)]_0$,

$$d_{z,x,t}[f(x)\wedge\hat{R}_K^v(z,x,t)\wedge v(z)] = \bar{\partial} f(x)\wedge\hat{R}_K^v(z,x,t)\wedge v(z)$$

$$+ (-1)^{r+1} f(x)\wedge \hat{R}_K^v(z,x,t)\wedge\bar{\partial} v(z),$$

it follows from Stokes' formula that

$$\int_D (R_K^v\bar{\partial} f)\wedge v + (-1)^{r+1}\int_D (R_K^v f)\wedge\bar{\partial} v = (-1)^{|K|}\int_{D\times S_K\times\Delta_K} f(x)\wedge\hat{R}_K^v(z,x,t)\wedge v(z)$$

$$+ \sum_{s=1}^{|K|}(-1)^{|K|+s+1}\int_{D\times S_K\times\Delta_{0K(s\hat{})}} f(x)\wedge\hat{R}_K^v(z,x,t)\wedge v(z)$$

$$+ \sum_{j=1}^{N}\int_{D\times S_{Kj}\times\Delta_{0K}} f(x)\wedge\hat{R}_K^v(z,x,t)\wedge v(z)$$

for all $v\in[C^{\infty}_{n,n-r}(D)]_0$ and $K\in P$. Since

$$(-1)^{|K|}\hat{R}_K^v(z,x,t) = \hat{L}_K^v(z,x,t) \qquad \text{if } t\in\Delta_K$$

and since

$$\int_D (R_K^v f)\wedge\bar{\partial} v = (-1)^r\int_D(\bar{\partial} R_K^v)\wedge v,$$

this implies that

$$\left(R_K^v\bar{\partial} f - \bar{\partial} R_K^v f - L_K^v f\right)(z) = \sum_{s=1}^{|K|}(-1)^{|K|+s+1}\int_{S_K\times\Delta_{0K(s\hat{})}} f(x)\wedge\hat{R}_K^v(z,x,t)$$

$$+ \sum_{j=1}^{N}\int_{S_{Kj}\times\Delta_{0K}} f(x)\wedge\hat{R}_K^v(z,x,t), \tag{3.19}$$

for all $K\in P$ and $z\in D$. Now we observe that, for all $1\leq k\leq N$, the restriction of $\hat{R}_k^v(z,x,t)$ to $S_k\times\Delta_0$ is equal to (cf. the proof of Proposition 0.9)

$$- \frac{1}{(2\pi i)^n} \det\Big(\frac{\bar{x}-\bar{z}}{|x-z|^2}, \overbrace{\bar\partial_{z,x}\frac{\bar{x}-\bar{z}}{|x-z|^2}}^{n-1}\Big) \wedge \omega(x) = - \frac{1}{(2\pi i)^n} \frac{\det(\bar{x}-\bar{z}, \overbrace{d\bar{x}-d\bar{z}}^{n-1})}{|x-z|^{2n}} \wedge \omega(x)$$

$$= - \hat{B}(z,x).$$

Therefore

$$\sum_{k=1}^{n} \int_{S_k \times \Delta_0} f(x) \wedge \hat{R}^v_k(z,x,t) = -(B_{\partial D} f)(z),$$

and hence

$$\sum_{K \in P} \sum_{s=1}^{|K|} (-1)^{|K|+s+1} \int_{S_K \times \Delta_{OK(s^\wedge)}} f(x) \wedge \hat{R}^v_K(z,x,t)$$

$$= -(B_{\partial D} f)(z) + \sum_{\substack{K \in P \\ |K| \geq 2}} \sum_{s=1} (-1)^{|K|+s+1} \int_{S_K \times \Delta_{OK(s^\wedge)}} f(x) \wedge \hat{R}^v_K(z,x,t)$$

for all $z \in D$. Thus, after summation over all $K \in P$ with $1 \leq |K| \leq n-r$, from (3.19) we obtain the relation

$$\Big(\sum_{\substack{K \in P \\ 1 \leq |K| \leq n-r}} R^v_K \bar\partial f - \bar\partial \sum_{\substack{K \in P \\ 1 \leq |K| \leq n-r}} R^v_K f - \sum_{\substack{K \in P \\ 1 \leq |K| \leq n-r}} L^v_K f + B_{\partial D} f \Big)(z)$$

$$= \sum_{\substack{K \in P \\ 2 \leq |K| \leq n-r}} \sum_{s=1}^{|K|} (-1)^{|K|+s+1} \int_{S_K \times \Delta_{OK(s^\wedge)}} f(x) \wedge \hat{R}^v_K(z,x,t)$$

$$+ \sum_{\substack{K \in P \\ 1 \leq |K| \leq n-r}} \sum_{j=1}^{N} \int_{S_{Kj} \times \Delta_{OK}} f(x) \wedge \hat{R}^v_K(z,x,t) \qquad (3.20)$$

for all $z \in D$. In view of Remark 3.11 and the Koppelman formula (1.7), the left hand side of (3.20) is equal to

$$[B_{\partial D}f + \bar\partial B_D f - B_D\bar\partial f - L^v f - \bar\partial T^v f + T^v\bar\partial f](z)$$

$$= [(-1)^r f - L^v f - \bar\partial T^v f + T^v\bar\partial f](z).$$

Therefore, we have to prove that the right hand side of (3.20) vanishes. First observe that

$$\sum_{\substack{K\in P\\ |K|=n-r}} \sum_{j=1}^{N} \int_{S_{Kj}\times\triangle_{OK}} f(x)\wedge\hat{R}^v_K(z,x,t) = 0, \qquad z\in D, \tag{3.21}$$

because, for $|K|=n-r$, $\dim_{\mathbb{R}} S_{Kj} = n+r-1$, whereas all monomials in the form $f(x)\wedge\hat{R}^v_K(z,x,t)$ are of degree $\geq n+r$ in x. Further, we notice that, by condition (3.5) in the definition of a Leray map, for all $K\in P$ with $|K|\geq 2$ and any $1\leq s\leq |K|$, one has

$$\hat{R}^v_K(z,x,t) = \hat{R}^v_{K(s\hat{})}(z,x,t) \qquad \text{if } t\in\triangle_{OK(s\hat{})}\,. \tag{3.22}$$

In view of (3.21) and (3.22), the right hand side of (3.20) is equal to

$$\sum_{\substack{K\in P\\ 2\leq |K|\leq n-r}} \sum_{s=1}^{|K|} (-1)^{|K|+s+1} \int_{S_K\times\triangle_{OK(s\hat{})}} f(x)\wedge\hat{R}^v_{K(s\hat{})}(z,x,t)$$

$$+ \sum_{\substack{K\ P\\ 1\leq |K|\leq n-r-1}} \sum_{j=1}^{N} \int_{S_{Kj}\times\triangle_{OK}} f(x)\wedge\hat{R}^v_K(z,x,t). \tag{3.23}$$

In order to show that (3.23) vanishes, for any $K=(k_1,\ldots,k_l)\in P$, we set

$$c(K) = \{1,\ldots,N\}\setminus\{k_1,\ldots,k_l\} \qquad \text{and} \qquad j(K,s) = k_s \ (1\leq s\leq |K|).$$

Then $S_K = (-1)^{|K|+s} S_{K(s\hat{})j(K,s)}$ for all $1\leq s\leq l$, and, by definition (cf. Sect. 3.2), we have $S_{Kj} = \emptyset$ if $j\notin c(K)$. Therefore, (3.23) is equal to

$$-\sum_{l=2}^{n-r} \sum_{\substack{K\in P\\ |K|=l\\ 1\leq s\leq l}} \int_{S_{K(s\hat{})j(K,s)}\times\triangle_{OK(s\hat{})}} f(x)\wedge\hat{R}^v_{K(s\hat{})}(z,x,t)$$

$$+\sum_{l=2}^{n-r} \sum_{\substack{K\in P\\ |K|=l-1\\ j\in c(K)}} \int_{S_{Kj}\times\triangle_{OK}} f(x)\wedge\hat{R}^v_K(z,x,t).$$

To complete the proof it remains to observe that, for any $2\le l\le n-r$, the sets of pairs

$$\{(K,j):\ K\in P,\ |K|=l-1,\ j\in c(K)\}$$

and

$$\{\big(K(s\hat{\ }),j(K,s)\big):\ K\in P,\ |K|=l,\ 1\le s\le l\}$$

are equal. □

<u>3.13. Corollary.</u> Let $D \subset\subset \mathbb{C}^n$ be a domain with piecewise almost C^1 boundary, $Y=(Y_1,\dots,Y_N)$ a frame for D, v a Leray map for (D,Y), and $0\le r\le n$ an integer. Suppose, for some collection $K=(k_1,\dots,k_l)$ of strictly increasing integers $1\le k_1<\dots<k_l\le N$, at least one of the following conditions is fulfilled (recall that $|K|:=l$):

(i) $|K|\ge n-r+1$;

(ii) $r\ge 1$ and $v_K(z,x,t)$ depends holomorphically on $z_1,\dots,z_{n-r+1}$;

(iii) $v_K(z,x,t)$ depends holomorphically on $x_1,\dots,x_{r+|K|+1}$.

Then, for each continuous (0,r)-form f on $\bar{D}$,

$$L^v_K f = 0.$$

In particular, if for <u>each</u> collection $K=(k_1,\dots,k_l)$ of strictly increasing integers $1\le k_1\le\dots\le k_l\le N$, at least one of the conditions (i)-(iii) is fulfilled, then, for any continuous (0,r)-form f on $\bar{D}$ such that $\bar{\partial}f$ is also continuous on $\bar{D}$, the piecewise Cauchy-Fantappie formula (3.17) takes the form

$$(-1)^r f = \bar{\partial}T^v f - T^v\bar{\partial}f \qquad \text{in } D. \tag{3.24}$$

<u>Proof.</u> By Proposition 3.6, $L^v_K f=0$ if $|K|\ge n-r+1$. Now let $|K|\le n-r$ and assume that one of the conditions (ii) or (iii) is fulfilled. Then, by Proposition 3.6,

$$L_K^v f(z) = \int_{(x,t)\in S_K\times\Delta_K} f(x)\wedge(\hat{L}_K^v)_r(z,x,t), \qquad z\in D.$$

In view of (3.9) and since $\dim_{\mathbb{R}} S_K = 2n - |K|$ and $\dim_{\mathbb{R}} \Delta_K = |K| - 1$, this implies that

$$L_K^v f(z) = \frac{\binom{n-1}{r}\binom{n-r-1}{|K|-1}}{(2\pi i)^n} \int_{(x,t)\in S_K\times\Delta_K} f(x)\wedge M(z,x,t)\wedge\omega(x)$$

for all $z\in D$, where

$$M(z,x,t) = \det[v_K(z,x,t),\overbrace{\bar{\partial}_z v_K(z,x,t)}^{r},\overbrace{\bar{\partial}_x v_K(z,x,t)}^{n-|K|-r},\overbrace{d_t v_K(z,x,t)}^{|K|-1}].$$

It remains to observe that, obviously, each of the conditions (ii) and (iii) implies that $M(z,x,t)=0$. □

CHAPTER II. q-CONVEX AND q-CONCAVE MANIFOLDS

Summary. In this chapter we introduce the concepts of q-convex and q-concave manifolds and prove some elementary properties of them.

4. q-convex functions

4.1. The Levi form. If M is a real C^1 manifold and $x \in M$, then by $T_{\mathbb{R},x}(M)$ we denote the (real) tangent space of M at x, and by $T_{\mathbb{C},x}(M)$ the complexified tangent space of M at x.

Now let X be an n-dimensional complex manifold.

For $x \in X$, by $T'_x(X)$ we denote the subspace of $T_{\mathbb{C},x}(X)$ which consists of all holomorphic tangent vectors: if $z_1, \ldots, z_n$ are holomorphic coordinates in a neighborhood of x, then $T'_x(X)$ is spanned by $\partial/\partial z_1, \ldots, \partial/\partial z_n$.

If Y is a real C^1 submanifold of X and $x \in Y$, then we set

$$T'_x(Y) = T'_x(X) \cap T'_{\mathbb{C},x}(Y).$$

$T'_x(Y)$ is called the holomorphic tangent space of Y at x.

Let ϱ be a real-valued C^2 function on X and $x \in X$. Then we define an Hermitian form $L_\varrho(x)$ on $T'_x(X)$ as follows: Choose holomorphic coordinates $z_1, \ldots, z_n$ in a neighborhood of x and set

$$L_\varrho(x)t = \sum_{j,k=1}^{n} \frac{\partial^2 \varrho(x)}{\partial \bar{z}_j \partial z_k} \bar{t}_j t_k \qquad \text{if } t = \sum_{j=1}^{n} t_j \frac{\partial}{\partial z_j}(x).$$

Since

$$\frac{\partial^2 \varrho}{\partial \bar{w}_r \partial w_s} = \sum_{j,k=1}^{n} \overline{\left(\frac{\partial z_j}{\partial w_r}\right)} \frac{\partial^2 \varrho}{\partial \bar{z}_j \partial z_k} \frac{\partial z_k}{\partial w_s}$$

if $w_1,\dots,w_n$ are other holomorphic coordinates, this definition is independent of the choice of holomorphic coordinates. The Hermitian form $L_\varrho(x)$ is called the Levi form of ϱ at x.

If Y is a real C^1 submanifold of X and $x\in Y$, then by $L_{\varrho|Y}(x)$ we denote the restriction of $L_\varrho(x)$ to $T'_x(Y)$.

4.2. The Levi numbers $p^\pm$ of a function. Let H be a Hermitian form. If $\tilde{H}$ is the corresponding Hermitian matrix with respect to some linear coordinates, then by Sylvester's theorem on inertia the number $p^+(\tilde{H})$ of positive eigenvalues of $\tilde{H}$ and the number $p^-(\tilde{H})$ of negative eigenvalues of $\tilde{H}$ are independent of the choice of the coordinates. Put $p^\pm(H)=p^\pm(\tilde{H})$.

Let X be a complex manifold and ϱ a real-valued C^2 function on X. Then, for all $x\in X$, we set

$$p^\pm_\varrho(x) = p^\pm(L_\varrho(x)).$$

If Y is a real C^1 submanifold of X and $x\in Y$, then we set

$$p^\pm_{\varrho|Y}(x) = p^\pm(L_{\varrho|Y}(x)).$$

4.3. Definition. Let X be an n-dimensional complex manifold, and $1\le q\le n$ an integer. A real-valued C^2 function ϱ defined on X will be called q-convex on X if

$$p^+_\varrho(x) \ge q \qquad \text{for all } x\in X.$$

n-convex functions usually are called strictly plurisubharmonic (for such functions see, for instance, Section 1.4 in [H/L]).

Remarks to this definition.

I. Clearly, each q-convex function is r-convex for all $1\le r\le q$.

II. In the literature, it is also used to call a function q-convex if it is (n-q+1)-convex in the sense of Definition 4.3. We prefer Definition 4.3, because, by this definition, a real-valued C^2 function ϱ is q-convex in a point x with $d\varrho(x)\neq 0$ if and only if there exist local holomorphic coordinates $z_1,\dots,z_n$ in a neighborhood of x such that ϱ is linearly convex with respect to $z_1,\dots,z_q$ (cf. Theorem 4.6 below). Notice that the other definition mentioned above is more in accordance with the Andreotti-Grauert finiteness theorem (cf. Theorem 12.16 below).

4.4. Proposition. Let X be an n-dimensional complex manifold, and ϱ a real-valued C^2 function on X. Suppose $1\le q\le n$ is an integer and $x\in X$.

Then:

(i) ϱ is q-convex in a neighborhood of x if and only if there exists a q-dimensional complex submanifold Y of some neighborhood of x such that $x \in Y$ and the restriction of ϱ to Y is strictly plurisubharmonic.

(ii) If $d\varrho(x) \neq 0$ and ϱ is q-convex in a neighborhood of x, then the submanifold Y in part (i) can be chosen so that, moreover, $d(\varrho|Y)(x) \neq 0$.

Proof. First suppose that ϱ is q-convex in a neighborhood of x. Then we can choose a q-dimensional subspace T of $T'_x(X)$ such that $L_\varrho(x)|T$ is positive definite and, moreover (after a small variation), $d\varrho(x)|T \neq 0$. It is clear that then each complex submanifold Y with $x \in Y$ and $T'_x(Y)=T$ has the required properties.

On the other hand, if Y is a q-dimensional complex submanifold such that $x \in Y$ and $\varrho|Y$ is strictly plurisubharmonic, then $p^+_\varrho(x) \geq p^+_{\varrho|Y}(x)=q$. □

4.5. Corollary. 1-convex functions do not have local maxima.

Proof. By Proposition 4.4 (i), this follows from the maximum principle for strictly subharmonic functions of one complex variable. □

4.6. Theorem. Let X be an n-dimensional complex manifold and ϱ a real-valued C^2 function on X. Let $1 \leq q \leq n$ and $y \in X$ such that $d\varrho(y) \neq 0$. Then ϱ is q-convex in some neighborhood of y if and only if there exists a biholomorphic map h from some neighborhood U of y onto some neighborhood of the origin in $\mathbb{C}^n$ such that $h(y)=0$ and the restriction of $\varrho \circ h^{-1}$ to

$$h(U) \cap \{z \in \mathbb{C}^n:\ z_{q+1}=\ldots=z_n=0\}$$

is strictly convex (in the usual linear sense, i.e. the Hessian matrix with respect to the real and imaginary parts of $z_1,\ldots,z_q$ is strictly positive definite).

Proof. In view of Proposition 4.4, we may assume that q=n. In this case the assertion follows from Theorem 1.4.14 and Proposition 1.4.9 (v) in [H/L]. □

4.7. The orientation of complex manifolds. If X is a complex manifold, then we assume that X is oriented so that if $z_1,\ldots,z_n$ are local holomorphic coordinates on X and $x_1,\ldots,x_{2n}$ are the real coordinates with $z_j=x_j+ix_{j+n}$, then $dx_1 \wedge \ldots \wedge dx_{2n}$ is positive.

4.8. Defining functions for oriented hypersurfaces. Let X be an n-dimensional complex manifold and M an oriented real C^1 hypersurface in X (i.e. an oriented C^1 submanifold of real dimension 2n-1). Then we say ϱ is a defining function for X if ϱ is a real-valued C^1 function in a neighborhood U of M such that the following two conditions are fulfilled:

(i) $\varrho(x)=0$ and $d\varrho(x)\neq 0$ for all $x\in M$;

(ii) the orientation of M coincides with the orientation which is induced on M as a part of the boundary of the domain $\{x\in U: \varrho(x)<0\}$.

<u>4.9. Lemma.</u> Let X be an n-dimensional complex manifold and M an oriented real C^2 hypersurface in X. Suppose both ϱ and φ are defining functions for M. Then

$$p^{\pm}_{\varrho|M}(x) = p^{\pm}_{\varphi|M}(x) \quad \text{for all } x\in M. \tag{4.1}$$

Moreover, then there exists a positive C^1 function ψ in a neighborhood V of M such that $\varphi = \psi\varrho$ in V and

$$L_{\varphi|M}(x) = \psi(x)L_{\varrho|M}(x) \quad \text{for all } x\in M. \tag{4.2}$$

<u>Proof.</u> It is easy to see that there is a C^1 function $\psi > 0$ in some neighborhood V of M with $\varphi = \psi\varrho$ on V. Let $x\in M$. Since $\varphi(x)=\varrho(x)=0$ and $\varphi = \psi\varrho$, then, for all $t\in T'_x(X)$,

$$L_{\varphi}(x)t = \psi(x)L_{\varrho}(x) + \overline{t(\psi)}t(\varrho) + t(\psi)\overline{t(\varrho)}.$$

This implies (4.2), because $t(\varrho)=0$ if $t\in T'_x(M)$. It is clear that (4.1) follows from (4.2). $\square$

<u>4.10. The Levi numbers $p^{\pm}$ of an oriented hypersurface.</u> If M is an oriented C^2 hypersurface of a complex manifold, then we define

$$p^{\pm}_M(x) = p^{\pm}_{\varrho|M}(x) \quad \text{for all } x\in M,$$

where ϱ is an (arbitrary) defining C^2 function for M. (By Lemma 4.9, this definition is correct.)

<u>Convention.</u> If $D \subseteq X$ is a domain, then by ∂D we denote the <u>oriented</u> boundary of D. More precisely, then we assume that the C^1 part $(\partial D)_1$ of ∂D is oriented so that the following condition is fulfilled: if ϱ is a defining function for $(\partial D)_1$, then $\varrho(x)<0$ as $D\ni x \longrightarrow (\partial D)_1$. Accordingly we use the notations $p^{+}_{\partial D}(x)$ and $p^{-}_{\partial D}(x)$ if x is a C^2 point of ∂D.

<u>4.11. Lemma.</u> Suppose that the following two conditions are fulfilled:

(a) $G \subseteq \mathbb{R}^k$ is a domain and u is a real-valued C^2 function on G which is convex;

(b) X is an n-dimensional complex manifold and

$$\varrho = (\varrho_1,\ldots,\varrho_k):X \longrightarrow G$$

is a C^2 map such that the functions $\varrho_2,\ldots,\varrho_k$ are stictly plurisubharmonic on X and, for some $1\le q\le n$, ϱ_1 is q-convex.

Then the following two assertions hold true:

(i) If, for all $x\in G$,

$$\min_{1\le i\le k} \frac{\partial u(x)}{\partial x_i} \ge 0 \qquad \text{and} \qquad \max_{1\le i\le k} \frac{\partial u(x)}{\partial x_i} > 0,$$

then $u\circ\varrho$ is q-convex on X.

(ii) If, for all $x\in G$,

$$\max_{1\le i\le k} \frac{\partial u(x)}{\partial x_i} \le 0 \qquad \text{and} \qquad \min_{1\le i\le k} \frac{\partial u(x)}{\partial x_i} < 0,$$

then $-u\circ(-\varrho)$ is q-convex on X.

Proof. In view of Proposition 4.4 we may assume that q=n. In this case, assertion (i) of the lemma coincides with Theorem 1.4.12 (ii) in [H/L]. Assertion (ii) follows from (i). □

4.12. Definition. For any $\beta>0$, we fix some positive C^∞ function χ_β on $\mathbb{R}$ such that, for all $x\in\mathbb{R}$,

$$\chi_\beta(x) = \chi_\beta(-x),$$

$$|x| \le \chi_\beta(x) \le |x| + \beta,$$

$$\left|\frac{d\chi_\beta(x)}{dx}\right| \le 1,$$

$$\frac{d^2\chi_\beta(x)}{dx^2} \ge 0,$$

$$\chi_\beta(x) = |x| \qquad \text{if} \quad |x|\ge\beta/2,$$

and we set

$$\max_\beta(t_1,t_2) = \frac{t_1+t_2}{2} + \chi_\beta\left(\frac{t_1-t_2}{2}\right) \qquad \text{for } t_1,t_2\in\mathbb{R}.$$

4.13. Lemma. For all $\beta>0$, $\max_\beta$ is a convex C^∞ function on $\mathbb{R}^2$ and, moreover, for all $(t_1,t_2)\in\mathbb{R}^2$, the following relations hold true:

$$\min_{i=1,2} \frac{\partial}{\partial t_i} \max_\beta(t_1,t_2) \geq 0, \tag{4.3}$$

$$\max_{i=1,2} \frac{\partial}{\partial t_i} \max_\beta(t_1,t_2) > 0, \tag{4.4}$$

$$\max_\beta(t_1,t_2) = \max_\beta(t_2,t_1), \tag{4.5}$$

$$\max(t_1,t_2) \leq \max_\beta(t_1,t_2) \leq \max(t_1,t_2) + \beta, \tag{4.6}$$

$$\max_\beta(t_1,t_2) = \max(t_1,t_2) \quad \text{if } |t_1-t_2| \geq \beta. \tag{4.7}$$

Proof. Since the Hessian matrix of $\max_\beta$ is

$$\frac{1}{4}\chi_\beta''\left(\frac{t_1-t_2}{2}\right)\begin{pmatrix}1 & -1\\ -1 & 1\end{pmatrix}$$

and since $\chi_\beta'' \geq 0$, we see that $\max_\beta$ is convex. Since the first-order derivatives of $\max_\beta$ are

$$\frac{1}{2} \pm \frac{1}{2}\chi_\beta'\left(\frac{t_1-t_2}{2}\right)$$

and since $|\chi_\beta'| \leq 1$, we obtain the estimates (4.3) and (4.4). Relations (4.5)-(4.7) immediately follow from the correspoonding properties of the function χ_β. $\square$

4.14. Corollary. Let X be an n-dimensional complex manifold, ϱ a strictly plurisubharmonic C^k function on X $(2\leq k\leq\infty)$, and φ a q-convex C^k function on X $(1\leq q\leq n)$. Then, for each $\beta>0$, $\max_\beta(\varrho,\varphi)$ is a q-convex C^k function on X.

Proof. This follows from Lemmas 4.11 and 4.13. $\square$

Finally, let us notice the following obvious but important

4.15. Observation. Let X be an n-dimensional complex manifold, ϱ a q--convex function on X $(1\leq q\leq n)$, $D \subset\subset X$ a domain, and ϱ_j $(j=1,2,\dots)$ a sequence of real-valued C^2 functions on D such that the second-order derivatives of $\varrho_j-\varrho$ tend to zero uniformly on D. Then there exists an index j_0 such that, for any $j>j_0$, the function ϱ_j is q-convex on D.

5. q-convex manifolds

5.1. Definition. Let X be an n-dimensional complex manifold.

A C^2 function $\varrho: X \longrightarrow \mathbb{R}$ is called an exhausting function for X if, there is a compact set $K \subseteq X$ such that, for all $x \in X\setminus K$,

$$\varrho(x) < \sup_{z \in X} \varrho(z),$$

and, for all real numbers

$$\alpha < \sup_{z \in X} \varrho(z),$$

the set $\{z \in X: \varrho(z) < \alpha\}$ is relatively compact in X.

An exhausting function for X will be called bounded or unbounded according as $\sup_X \varrho < \infty$ or $\sup_X \varrho = \infty$.

If ϱ is an exhausting function for X, then we say ϱ is q-convex at infinity if there exists a compact set $K \subset\subset X$ such that ϱ is q-convex on $X\setminus K$ $(1 \leq q \leq n)$. For $q=n-1$, in this case, ϱ will be called also strictly plurisubharmonic at infinity.

X will be called q-convex if there exists an exhausting function for X which is (q+1)-convex at infinity $(0 \leq q \leq n-1)$.

We say X is completely q-convex if there exists an exhausting function for X which is (q+1)-convex everywhere on X $(0 \leq q \leq n-1)$.

(n-1)-convex (resp. completely (n-1)-convex) complex manifolds will be called also pseudoconvex (resp. completely pseudoconvex).

Remarks to this definition.

I. Recall that a complex manifold is completely pseudoconvex if and only if it is Stein (cf., for instance, Sect. 2.13 in [H/L]).

II. It is clear that every compact complex manifold is pseudoconvex. On the other hand, no compact complex manifold is completely 0-convex, because 1-convex functions do not have a maximum (Corollary 4.5).

III. If ϱ is a bounded exhausting function for a complex manifold X, and $c := \sup_X \varrho$, then $\tilde{\varrho} := -\ln(c-\varrho)$ is an unbounded exhausting function for X. Therefore, it follows from Lemma 4.11 (i) that if X admits a bounded exhausting function which is q-convex at infinity (resp. everywhere on X), then X admits also an unbounded exhausting function which is q-convex at infinity (resp. everywhere on X).

Notice that the opposite is not true. Example: The complex plane.

IV. In the literature, it is also used to call a complex manifold q--convex if it is (n-q)-convex in the sense of Definition 5.1 (cf. Remark II following Definition 4.3).

5.2. Lemma (Morse's lemma). Every q-convex complex manifold X admits an unbounded exhausting C^{∞} function without degenerate critical points which is (q+1)-convex at infinity. If X is even completely q-convex, then this function can be chosen to be (q+1)-convex everywhere on X.

Proof. This follows from Observation 4.15, Proposition 0.5 in Appendix B, and Remark III following Definition 5.1. □

5.3. Theorem. If $X \subseteq M$ is a q-convex domain in a complex manifold M ($0 \leq q \leq \dim_{\mathbb{C}} M - 1$), then, for any pseudoconvex domain $Y \subseteq M$, the intersection $X \cap Y$ is q-convex; and for any completely pseudoconvex domain $Y \subseteq M$, the intersection $X \cap Y$ is completely q-convex.

In particular, every q-convex domain in a completely pseudoconvex manifold is completely q-convex.

Proof. As observed in Remark III following Definition 5.1, we can find an unbounded exhausting function φ_X for X which is (q+1)-convex at infinity, and we can find an unbounded exhausting function φ_Y for Y which is strictly plurisubharmonic at infinity. Then, by Corollary 4.14, $\max_1(\varphi_X, \varphi_Y)$ is an exhausting function for $X \cap Y$ which is (q+1)-convex at infinity. If Y is completely pseudoconvex, then φ_Y can be chosen to be strictly plurisubharmonic everywhere on Y, and, for any sufficiently large $C < \infty$, $\max_1(\varphi_X, \varphi_Y + C)$ is an exhausting function for $X \cap Y$ which is q-convex everywhere on $X \cap Y$. □

5.4. Definition. Let $D \subset\subset X$ be a non-empty domain in an n-dimensional complex manifold, and $0 \leq q \leq n-1$ an integer.

(i) D is called strictly q-convex if there exists a (q+1)-convex function ϱ in a neighborhood $U_{\partial D}$ of ∂D such that

$$D \cap U_{\partial D} = \{z \in U_{\partial D} : \varrho(z) < 0\}.$$

(ii) We say D is completely strictly q-convex if there exists a (q+1)-convex function ϱ in a neighborhood $U_{\bar{D}}$ of $\bar{D}$ such that

$$D = \{z \in U_{\bar{D}} : \varrho(z) < 0\}.$$

(iii) D will be called non-degenerate strictly q-convex (resp. non--degenerate completely strictly q-convex) if the function ϱ in (i) (resp. (ii)) can be chosen without degenerate critical points (for the notions of a degenerate and a non-degenerate critical point, see Sect. 0 in Appendix B).

(iv) If q=n-1, then in (i)-(iii) we say also "pseudoconvex" instead of "q-convex".

Remarks to this definition.

I. The case $\partial D = \emptyset$ is admitted. Thus, every compact complex manifold is non-degenerate strictly pseudoconvex in the sense of this definition.

II. It is clear that any strictly q-convex domain is q-convex (in the sense of Definition 5.1).

III. Strictly q-convex domains need not be smooth.

5.5. Proposition. Let X be a complex manifold and ϱ a 1-convex function on X. If the boundary of the domain $D := \{z \in X: \varrho(z) < 0\}$ is of class C^2, then $d\varrho(y) \neq 0$ for all $y \in \partial D$.

Proof. Repetition of the proof of Proposition 1.5.16 in [H/L]. □

5.6. Corollary. Let X be an n-dimensional complex manifold, and $D \subset\subset X$ a strictly q-convex domain ($0 \le q \le n-1$). If the boundary of D is of class C^2 and ϱ is as in Definition 5.4 (i), then $d\varrho(z) \neq 0$ for all $z \in \partial D$.

Proof. This follows immediately from Proposition 5.5. □

5.7. Theorem. Let $D \subset\subset X$ be a strictly q-convex domain in an n-dimensional complex manifold X ($0 \le q \le n-1$). Then:

(i) For each compact set $K \subset\subset D$, there exists a strictly q-convex domain $D' \subset\subset X$ with C^∞ boundary such that $K \subset\subset D' \subset\subset D$.

(ii) For every neighborhood U of $\bar{D}$, there exists a strictly q-convex domain $D' \subset\subset X$ with C^∞ boundary such that $\bar{D} \subset\subset D' \subset\subset U$.

Proof. (i) In view of Morse's Lemma 5.2, we can find an exhausting C^∞ function ϱ for D without degenerate critical points and which is (q+1)-convex at infinity. Since non-degenerate critical points are isolated (cf. Proposition 0.1 in Appendix B), there is a sufficiently large number $\alpha < \sup_D \varrho$ such that the set $D' := \{z \in D: \varrho(z) < \alpha\}$ has the required property.

(ii) By part (i), it is enough to find a strictly q-convex domain $D' \subset\subset U$ (the boundary of which need not be smooth) such that $\bar{D} \subset\subset D'$. To do this we take ϱ and $U_{\partial D}$ as in Definition 5.4 (i) and we choose a C^∞ function χ on X such that $\chi = 0$ on $\bar{D}$ and $\chi > 0$ on $X \setminus \bar{D}$. Then, by Observation 4.15, after shrinking $U_{\partial D}$, for any sufficiently small $\varepsilon > 0$, the function $\varrho_\varepsilon := \varrho + \varepsilon\chi$ is (q+1)-convex on $U_{\partial D}$. Further, it is clear that $\varrho_\varepsilon > 0$ on $U_{\partial D} \setminus \bar{D}$. Hence, for each sufficiently small $\delta > 0$, the set

$$D' := D \cup \{z \in U_{\partial D}: \varrho_\varepsilon(z) < \delta\}$$

has the required property. □

5.8. Lemma (cf. Sects. 4.8 and 4.10 for the notations). Let X be an n-dimensional complex manifold, M an oriented real C^2 hypersurface in X, and ϱ a defining C^2 function for M. Then, for each $y \in M$, we can find a neighborhood U_y of y and a positive number $C_y < \infty$ such that, for all $C \geq C_y$, the function $e^{C\varrho}$ is $(p_M^+(y)+1)$-convex on U_y.

Proof. Let $q := p_M^+(y)$. Choose a q-dimensional subspace T of $T_y'(M)$ such that $L_\varrho(y)|T$ is positive definite. Further, choose a (q+1)-dimensional subspace $\tilde{T}$ of $T_y'(X)$ with $\tilde{T} \cap T_y'(M) = T$, and set

$$K = \{t \in \tilde{T}: \|t\|=1 \text{ and } L_\varrho(y)t < 0\},$$

where $\|\cdot\|$ is some norm on $T_y'(X)$. Then K is compact and $t(\varrho) \neq 0$ for all $t \in K$. Hence, we can find a number $C_y < \infty$ such that

$$\max_{t \in K} |L_\varrho(y)t| < C \min_{t \in K} |t(\varrho)|^2 \quad \text{if } C > C_y.$$

Since

$$L_{e^{C\varrho}}(y)t = e^{C\varrho}\left(CL_\varrho(y)t + C^2|t(\varrho)|^2\right),$$

then

$$L_{e^{C\varrho}}(y)t > 0 \qquad \text{if } 0 \neq t \in \tilde{T} \text{ and } C \geq C_y. \quad \square$$

5.9. Theorem. Let X be an n-dimensional complex manifold, $D \subset\subset X$ a domain with C^2 boundary, and $0 \leq q \leq n-1$. Then D is strictly q-convex if and only if

$$p_{\partial D}^+(x) \geq q \qquad \text{for all } x \in \partial D.$$

Moreover, if this condition is fulfilled and ϱ is a defining C^2 function for ∂D, then, for each sufficiently large positive number C,

$$e^{C\varrho} - 1$$

is a defining function for ∂D which is (q+1)-convex in some neighborhood of ∂D.

Proof. First suppose that D is strictly q-convex, i.e. there is a (q+1)-convex function φ in a neighborhood U of ∂D such that $D \cap U = \{x \in U: \varphi(x) < 0\}$. Since $T_x'(\partial D)$ is of codimension 1 in $T_x'(X)$, then

$$p^+\left(L_{\varphi|\partial D}(x)\right) \geq q \qquad \text{for all } x\in\partial D.$$

Since, by Corollary 5.6, $d\varphi(x) \neq 0$ for all $x\in\partial D$ and hence φ is a defining function for ∂D, this means that $p^+_{\partial D}(x)\geq q$ for all $x\in\partial D$.

Now we suppose that $p^+_{\partial D}(x)\geq q$ for all $x\in\partial D$. If ϱ is a defining C^2 function for ∂D, then it is clear that, for every $C>0$, the function $e^{C\varrho}-1$ is also a defining function for ∂D, and it follows from Lemma 5.8 that this function is $(q+1)$-convex in some neighborhood of ∂D if C is sufficiently large. □

5.10. Corollary. Every relatively compact domain with C^2 boundary in an arbitrary complex manifold is strictly 0-convex.

Proof. The assertion follows from Theorem 5.9, because the inequality $p^+_{\partial D}(x)\geq 0$ is trivial. □

5.11. Corollary. Let $D \subset\subset X$ be a strictly q-convex domain in an n-dimensional complex manifold X ($0\leq q\leq n-1$). If the boundary of D is of class C^k ($2\leq k\leq\infty$), then the function ϱ in Definition 5.4 (i) can be chosen to be also of class C^k.

Proof. If φ is an arbitrary defining C^k function for ∂D, then by Theorem 5.9, for each sufficiently large positive number C, $\varrho := e^{C\varphi} - 1$ has the required properties. □

5.12. Corollary. Let X be an n-dimensional complex manifold and $D \subset\subset X$ a strictly q-convex domain ($0\leq q\leq n-1$). Then, for each compact set $K \subseteq \partial D$ and every neighborhood U of K in X, there exist strictly q-convex domains $D', D'' \subset\subset X$ such that

$$D\cup K \subseteq D' \subseteq D\cup U, \qquad D\setminus U \subseteq D'', \qquad \bar{D}'' \subseteq \bar{D}\cup U.$$

Moreover, if the boundary of D is of class C^k ($2\leq k\leq\infty$), then these domains D' and D" can be chosen also with C^k boundaries.

Proof. Let ϱ and $U_{\partial D}$ be as in Definition 5.4 (i), and let $V \subset\subset U_{\partial D}$ be some smaller neighborhood of ∂D. Choose a neighborhood $W \subset\subset U\cap V$ of K and a non-negative real-valued C^∞ function χ on X such that $\chi > 0$ on K and $\chi = 0$ on $X\setminus W$. Since V is relatively compact in $U_{\partial D}$, we can find $\varepsilon>0$ such that $\varrho + \varepsilon\chi$ and $\varrho - \varepsilon\chi$ are $(q+1)$-convex in V (cf. Observation 4.15). Then the domains

$$D' := D \cup \{z\in V:\ \varrho(z) < \varepsilon\chi(z)\}$$

and

$$D'' := (D\setminus V) \cup \{z\in V:\ \varrho(z) < -\varepsilon\chi(z)\}$$

have the required properties.

If ∂D is of class C^k, then, by Corollary 5.11, we can assume that ϱ is C^k and $d\varrho(z)\neq 0$ for all $z\in U_{\partial D}$. Since V is relatively compact in $U_{\partial D}$, then also $d(\varrho \pm \varepsilon\chi) \neq 0$ on V if ε is sufficiently small. □

5.13. Theorem. Let $D \subseteq X$ be a domain with C^k boundary ($2\le k\le\infty$) in an n-dimensional complex manifold X. Then, for each point $y\in\partial D$ and all $0\le q\le n-1$, the following conditions (i)-(vi) are equivalent:

(i) There exist a neighborhood U of y and a (q+1)-convex C^k function ϱ on U such that $D\cap U = \{x\in U: \varrho(x)<0\}$.

(ii) $p^{+}_{\partial D}(y) \geq q$.

(iii) There exist a neighborhood U of y and a (q+1)-dimensional closed complex submanifold Y of U such that $y\in Y$, the intersection $Y\cap\partial D$ is transversal, and $p^{+}_{\partial D\cap Y}(y) = q$. (Here $\partial D\cap Y$ is considered as a real hypersurface of Y endowed with the orientation induced from $D\cap Y$.)

(iv) There exist a neighborhood $U \subseteq X$ of y and a (q+1)-dimensional closed complex submanifold Y of some neighborhood of $\bar{U}$ such that $y\in Y$, the intersection $Y\cap\partial D$ is transversal, and $D\cap U\cap Y$ is a strictly pseudoconvex domain with C^k boundary in Y.

(v) There exist a neighborhood $V \subset\subset X$ of y and a biholomorphic map h from some neighborhood U of $\bar{V}$ onto a neighborhood h(U) of the origin in $\mathbb{C}^n$ such that: $h(y)=0$, the surface $h(\partial D\cap U)$ intersects transversally the subspace

$$\mathbb{C}^n_{q+1} := \{z\in\mathbb{C}^n: z_{q+2}=\ldots=z_n=0\},$$

and $h(D\cap V)\cap\mathbb{C}^n_{q+1}$ is a strictly convex set with C^k boundary in $\mathbb{C}^n_{q+1}$.

(vi) There exist a neighborhood U of y and a q-dimensional closed complex submanifold Z of U such that $Z\cap\bar{D} = \{y\}$ and

$$\operatorname{dist}(x,y) = O([\operatorname{dist}(x,\bar{D})]^{1/2}) \qquad \text{for } Z\ni x \longrightarrow y, \tag{5.1}$$

where dist is the Euclidean distance with respect to some local holomorphic coordinates in a neighborhood of y.

Proof. It is easy to see that (v) ⟹ (iv) ⟹ (iii) ⟹ (ii) and (v) ⟹ (vi), and it follows immediately from Lemma 5.8 that (ii) ⟹ (i). So it remains to prove (i) ⟹ (v) and (vi) ⟹ (ii).

Proof of (i) ⟹ (v). Let U, ϱ be as in condition (i). By Theorem 4.6, after shrinking U, we can find a biholomorphic map h from U onto a neighborhood of the origin in $\mathbb{C}^n$ such that $h(y)=0$, the intersection of $h(\partial D\cap U)$ with $\mathbb{C}^n_{q+1}$ is transversal, and the restriction $\varrho\circ h^{-1}\,|\mathbb{C}^n_{q+1}\cap h(U)$

is strictly convex. Now it is easy to find a neighborhood $V \subset\subset U$ of y such that $h(V\cap D)\cap \mathbb{C}^n_{q+1}$ is strictly convex and with C^k boundary (for the details, see, for instance, the proof of Lemma 1.5.23 (i) in [H/L]).

Proof of (vi) $\Longrightarrow$ (ii). Without loss of generality we can assume that X is a domain in $\mathbb{C}^n$, $y=0$, and $Z = \{z\in\mathbb{C}^n: z_{q+1}=\ldots=z_n=0\}$. Let ϱ be a defining C^2 function for ∂D. Then it follows from condition (5.1) that

$$|z|^2 = O(\varrho(z)) \qquad \text{as } Z\ni z \longrightarrow 0.$$

Since $\varrho(0)=0$, this implies that $\varrho|Z$ is strictly convex at 0. Since strictly convex functions are strictly plurisubharmonic, it follows that $p^+_{\varrho|Z}(0)=q$. Since $T'_0(Z) \subseteq T'_0(\partial D)$, we conclude that $p^+_{\partial D}(0)\geq q$. ☐

5.14. Theorem. Any strictly q-convex domain $D \subset\subset X$ in an n-dimensional completely pseudoconvex complex manifold X is completely strictly q-convex ($0\leq q\leq n-1$). Moreover, if such a domain D is of class C^k, then the function ϱ in Definition 5.4 (ii) can be chosen to be also of class C^k ($2\leq k\leq\infty$).

Proof. Since D is strictly q-convex, we can find a real-valued C^2 function φ in some neighborhood $U_{\bar{D}}$ of $\bar{D}$ with $D=\{z\in U_{\bar{D}}: \varphi(z)<0\}$ such that, for some compact set $K \subset\subset D$, φ is $(q+1)$-convex in $U_{\bar{D}} \setminus K$. If ∂D is of class C^k, then, by Corollary 5.11, this function φ can be chosen to be also of class C^k. Since X is completely pseudoconvex, we can find a function ψ which is of class C^∞ and strictly plurisubharmonic everywhere on X. Since D is compact, we can choose constants $C<\infty$ and $\varepsilon>0$ so that

$$\max_{z\in\bar{D}} \varepsilon(\psi(z) - C) < 0 \qquad \text{and} \qquad \min_{z\in\bar{D}} \varepsilon(\psi(z) - C) > \max_{z\in K} \varphi(z).$$

Now it follows from Corollary 4.14 that, for sufficiently small $\beta>0$, the function $\varrho := \max_\beta\{(\psi -C), \varphi\}$ is as in Definition 5.4 (ii). ☐

5.15. Theorem. Let X be an n-dimensional complex manifold, and $D \subset\subset X$ a domain such that, for some $0\leq q\leq n-1$, the following "local" condition is fulfilled: For each point $x\in\partial D$, there exists a $(q+1)$-convex function ϱ without degenerate critical points in some neighborhood U of x such that $D\cap U=\{z\in U: \varrho(z)<0\}$. Then D is non-degenerate strictly q-convex.

Proof. Let $S \subseteq \partial D$ be the set of all points where ∂D is not of class C^2. By hypothesis, S is empty or finite, and we can find a neighborhood V of S and a $(q+1)$-convex function φ without degenerate critical points on V such that $V\cap D=\{z\in V: \varphi(z)<0\}$ and $S=\{z\in V: d\varphi(z)=0\}$. Choose a C^2 function $\psi : X\setminus S \longrightarrow \mathbb{R}$ which is, after shrinking to an appropriate neighborhood of $\partial D\setminus S$, a defining function for $\partial D\setminus S$. Further, we take a real-valued C^∞ function χ on X such that $0 < \chi < 1$, supp $\chi \subset\subset V$ and $\chi = 1$ in some neighborhood of S. Set

$$\varrho = \chi\varphi + (1 - \chi)\psi .$$

Then $D=\{z\in X: \varrho(z)<0\}$, and ϱ is (q+1)-convex in some neighborhood of S. It follows from Lemma 5.8 and Lemma 4.11 that, for a sufficiently large constant $C>0$, the function $h:=e^{C\varrho}-1$ is (q+1)-convex in some neighborhood U of ∂D. Clearly, $D\cap U=\{z\in U: h(z)<0\}$. □

5.16. Examples.

I. Let $\mathbb{P}^n$ be the n-dimensional complex projective space. For $z\in\mathbb{P}^n$, by $[z^0:\ldots:z^n]$ we denote the homogeneous coordinates of z. Then, for all $0\le q\le n-1$,

$$D:= \{[z^0:\ldots:z^n]\in \mathbb{P}^n: |z^0|^2+\ldots+|z^q|^2 < |z^{q+1}|^2+\ldots+|z^n|^2\}$$

is a completely strictly q-convex domain with C^∞ boundary. In fact, if

$$\mathbb{P}^q:= \{[z^0:\ldots:z^n]\in \mathbb{P}^n: z^{q+1}=\ldots=z^n=0\}, \tag{5.2}$$

then

$$\varrho:= \frac{|z^0|^2+\ldots+|z^q|^2}{|z^{q+1}|^2+\ldots+|z^n|^2} \tag{5.3}$$

is a (q+1)-convex function on $\mathbb{P}^n\setminus \mathbb{P}^q$ and $D=\{z\in\mathbb{P}^n\setminus \mathbb{P}^q: \varrho(z)<1\}$.

II. If $\mathbb{P}^q$ is a q-dimensional projective subspace of the n-dimensional complex projective space $\mathbb{P}^n$ $(0\le q\le n-1)$, then $\mathbb{P}^n\setminus \mathbb{P}^q$ is completely q-convex. In fact, if $[z^0:\ldots:z^n]$ are homogeneous coordinates in $\mathbb{P}^n$ such that (5.2) holds true, then (5.3) is an exhausting function for $\mathbb{P}^n\setminus \mathbb{P}^q$ which is (q+1)-convex everywhere in $\mathbb{P}^n\setminus \mathbb{P}^q$.

Remark. In [Barth 1970] it was proved that, for any q-dimensional closed complex submanifold X of $\mathbb{P}^n$, the domain $\mathbb{P}^n\setminus X$ is q-convex but not necessarily completely q-convex. Further, in [Barth 1970], examples are given which show that $\mathbb{P}^n\setminus X$ need not be q-convex if X is a q-dimensional analytic set with singularities in $\mathbb{P}^n$.

6. q-concave functions and q-concave manifolds

6.1. Definition. Let X be an n-dimensional complex manifold.

A function ϱ will be called q-concave on X ($1 \leq q \leq n$) if $-\varrho$ is q-convex on X (cf. Definition 4.3).

If ϱ is an exhausting function for X (cf. Definition 5.1), then ϱ will be called q-concave at infinity if there exists a compact set $K \subset\subset X$ such that ϱ is q-concave on $X \setminus K$.

X will be called q-concave ($0 \leq q \leq n-1$) if there exists an exhausting function for X which is (q+1)-concave at infinity.

If X is (n-1)-concave, then X will be called also pseudoconcave.

Remarks to this definition.

I. Since 1-concave functions do not admit local minima (Corollary 4.5), there do not exist complex manifolds which admit exhausting functions which are 1-concave everywhere on X, i.e. there do not exist "completely" 1-concave manifolds.

II. Every n-dimensional compact complex manifold is pseudoconcave.

III. Notice the following corollary of Lemma 4.11 (ii): "If $u: \mathbb{R} \longrightarrow \mathbb{R}$ is a concave and strictly increasing C^2 function, then, for each q-concave function ϱ, $u \circ \varrho$ is q-concave." This implies that each q-concave complex manifold X admits a bounded exhausting function which is (q+1)-concave at infinity. In fact, if ϱ is an unbounded exhausting function for X which is q-convex at infinity, then $\tilde{\varrho} := -e^{-\varrho}$ is an unbounded exhausting function for X which is, by the corollary mentioned above, (q+1)-concave at infinity.

Notice that, in general, the opposite is not true. Example: There do not exist unbounded exhausting functions for the unit disc $\{z \in \mathbb{C}: |z|<1\}$ which are 1-concave at infinity. In fact, assume ϱ is such a function. Then, for some $\varepsilon>0$, $-\varrho$ is subharmonic in "$1-\varepsilon<|z|<1$" and tends to $-\infty$ for $|z| \longrightarrow 1$; this is impossible. On the other hand, if ϱ is a negative C^2 function on the unit disc such that $\varrho(z)=1-1/|z|^2$ for $|z|>1-1/2$, then ϱ is a bounded exhausting function for the unit disc, which is 1-concave at infinity.

IV. As observed at the end of Remark III, the unit disc in $\mathbb{C}$ is 0-concave. This implies (by the Riemann mapping theorem) that any finitely connected bounded domain in the complex plane is 0-concave. Also it is clear that the whole complex plane is 0-concave. In fact, each negative C^2 function ϱ on C with $\varrho(z)=-1/|z|^2$ for $|z|>1$ is an exhausting function for $\mathbb{C}$, which is 1-concave at infinity. However (in distinction to the convex case - any domain in $\mathbb{C}$ is 0-convex), not every domain in $\mathbb{C}$ is 0-concave, because no complex manifold which consists of an infinite number of connected components is 0-concave - this follows from the fact that 1-concave functions do not admit local minima (cf. Corollary 4.5).

V. In the literature, it is also used to call a complex manifold q-concave if it is (n-q)-concave in the sense of Definition 6.1 (cf. Remark III following Definition 4.3 and Remark IV following Definition 5.1). □

6.2. Lemma (Morse's lemma). Every q-concave complex manifold admits a bounded C^{∞} exhausting function without degenerate critical points which is (q+1)-concave at infinity. □

Proof. This follows from Observation 4.15, Proposition 0.5 in Appendix B, and Remark III following Definition 6.1.

6.3. Theorem. Let M be an arbitrary n-dimensional complex manifold and let $X, Y \subseteq M$ be two domains such that X is q-concave ($0 \leq q \leq n-1$) and Y is pseudoconcave. Then $X \cup Y$ is q-concave.

Proof. As observed in Remark III following Definition 6.1, we can find exhausting functions ϱ_X and ϱ_Y for X and Y, respectively, such that ϱ_X is (q+1)-concave at infinity, ϱ_Y is n-concave at infinity, and $\sup \varrho_X = \sup \varrho_Y = 0$. Then, by Corollary 4.14, $-\max_1(-\varrho_X, -\varrho_Y)$ is an exhausting function for $X \cup Y$ which is (q+1)-concave at infinity. □

6.4. Definition. Let $D \subset\subset X$ be an non-empty domain in an n-dimensional complex manifold X, and $0 \leq q \leq n-1$ an integer.

(i) D will be called strictly q-concave if there exists a (q+1)--concave function ϱ in a neighborhood U of ∂D such that

$$D \cap U = \{z \in U: \varrho(z) < 0\}.$$

(Then, in particular, D is q-concave in the sense of Definition 6.1.)

(ii) D is called non-degenerate strictly q-concave if the function ϱ in (i) can be chosen without degenerate critical points (for the notions of a degenerate and a non-degenerate critical point, see Sect. 0 in Appendix B).

(iii) Strictly (n-1)-concave domains will be called also strictly pseudoconcave.

6.5. Proposition. Let D be a strictly q-concave domain in an n-dimensional complex manifold X ($0 \leq q \leq n-1$). If the boundary of D is of class C^2 and ϱ is as in Definition 6.4 (i), then $d\varrho(z) \neq 0$ for all $z \in \partial D$.

Proof. This follows immediately from Proposition 5.5. □

6.6. Theorem. Let X be an n-dimensional complex manifold, and $D \subset\subset X$ a strictly q-concave domain ($0 \leq q \leq n-1$). Then:

(i) For each compact set $K \subset\subset D$, there exists a strictly q-convex domain $D' \subset\subset D$ with C^∞ boundary such that $K \subset\subset D'$.

(ii) For every neighborhood $U_{\bar{D}}$ of $\bar{D}$, there exists a strictly q-concave domain $D' \subset\subset U_{\bar{D}}$ with C^∞ boundary such that $\bar{D} \subset\subset D'$.

Proof. Repetition of the proof of Theorem 5.7. □

6.7. Theorem. Let $D \subset\subset X$ be a domain with C^2 boundary in an n-dimensional complex manifold X, and $0 \leq q \leq n-1$. Then D is strictly q-concave if and only if

$$p^{-}_{\partial D}(x) \geq q \qquad \text{for all } x \in \partial D.$$

Moreover, if this condition is fulfilled and ϱ is an arbitrary defining C^2 function for ∂D, then, for any sufficiently large positive number C, the function

$$-e^{-C\varrho}+1$$

is a defining function for ∂D which is (q+1)-concave in some neighborhood of ∂D.

Proof. Repetition of the proof of Theorem 5.9. □

6.8. Corollary. Every relatively compact domain with C^2 boundary in a complex manifold is strictly 0-concave.

Proof. This follows from Theorem 6.7, because the inequality $p^{-}_{\partial D}(x) \geq 0$ is trivial. □

6.9. Corollary. Let $D \subset\subset X$ be a strictly q-concave domain in an n-dimensional complex manifold X ($0 \leq q \leq n-1$). If the boundary of D is of class C^k ($2 \leq k \leq \infty$), then the function ϱ in Definition 6.4 (i) can be chosen to be also of class C^k.

Proof. Repetition of the proof of Corollary 5.11, using Theorem 6.7 instead of Theorem 5.9. □

6.10. Corollary. Let $D \subset\subset X$ be a strictly q-concave domain in an n-dimensional complex manifold X ($0 \leq q \leq n-1$). Then, for each compact set $K \subseteq \partial D$ and any neighborhood U of K in X, there exist strictly q-concave domains $D', D'' \subset\subset X$ such that

$$D \cup K \subseteq D' \subseteq D \cup U, \qquad D \setminus U \subseteq D'', \qquad \bar{D}'' \subseteq \bar{D} \cup U.$$

Moreover, if the boundary of D is of class C^k ($0 \leq k \leq \infty$), then these domains D' and D" can be chosen also with C^k boundaries.

Proof. Repetition of the proof of Corollary 5.12, using Corollary 6.9 instead of Corollary 5.11. □

6.11. Theorem. Let $D \subset\subset X$ be a domain in an n-dimensional complex manifold such that, for some $0 \le q \le n-1$, the following local condition is fulfilled : For each point $x \in \partial D$, there exists a (q+1)-concave function ϱ without degenerate critical points in some neighborhood U of x such that $D \cap U = \{z \in U: \varrho(z) < 0\}$. Then D is non-degenerate strictly q-concave.

Proof. Repetition of the proof of Theorem 5.15. □

6.12. Examples.

I. If $D \subset\subset X$ is a strictly q-convex domain in an n-dimensional compact complex manifold, then $X \setminus D$ is strictly q-concave ($0 \le q \le n-1$). This follows immediately from the definitions of strict q-convexity and strict q-concavity.

Warning: If D is an arbitrary q-convex domain in a compact complex manifold X, then $X \setminus D$ need not be q-concave. Example: Let Y be an open subset of the Riemannian sphere $\mathbb{P}^1$ which consists of infinitely many connected components such that $Y = \mathbb{P}^1 \setminus \overline{(\mathbb{P}^1 \setminus \bar{Y})}$. Then $D := \mathbb{P}^1 \setminus \bar{Y}$ is 0-convex , but $Y = \mathbb{P}^1 \setminus \bar{D}$ is not 0-concave (cf. Remark IV following Definition 6.1).

II. With the notations from Example I in Sect. 5.16,

$$\mathbb{P}^n \setminus \bar{D} = \{[z^0 : \ldots : z^n] \in \mathbb{P}^n : |z^0|^2 + \ldots + |z^q|^2 > |z^{q+1}|^2 + \ldots + |z^n|^2\}$$

is strictly q-concave, because D is strictly q-convex.

III. Each q-dimensional projective subspace $\mathbb{P}^q$ of $\mathbb{P}^n$ ($0 \le q \le n-1$) admits a basis of strictly q-concave neighborhoods, because $\mathbb{P}^n \setminus \mathbb{P}^q$ is q-concave (cf. Example II in Sect. 5.16).

CHAPTER III. THE CAUCHY-RIEMANN EQUATION ON q-CONVEX MANIFOLDS

Summary. In Sect. 7 a local Cauchy-Fantappie formula for non-degenerate strictly q-convex domains in n-dimensional complex manifolds is constructed, which yields local solutions of $\bar\partial u = f_{0,r}$ if $r \geq n-q$ (Theorem 7.8). In Sect. 9 we prove 1/2-Hölder estimates for these solutions. In Sect. 8, by means of the formula from Sect. 7, we prove a local uniform approximation theorem for continuous $\bar\partial$-closed $(0,n-q-1)$-forms given in a neighborhood of the closure of an n-dimensional non-degenerate strictly q-convex domain. Then, in Sect. 10, using the estimates from Sect. 9, we prove this result for forms given only on the closure of such a domain. In Sect. 11 we show that, via Fredholm operator theory in Banach spaces, the local solutions with Hölder estimates for the $\bar\partial$-equation lead to the following version with uniform estimates of the Andreotti-Grauert finiteness theorem (Theorem 11.2): If D is a non-degenerate strictly q-convex domain in an n-dimensional complex manifold X, and E is a holomorphic vetor bundle over X, then $\dim H^{0,r}_{1/2\longrightarrow 0}(\bar D, E) < \infty$ for all $r \geq n-q$, where $H^{0,r}_{1/2\longrightarrow 0}(\bar D, E) := Z^0_{0,r}(\bar D, E)/E^{1/2\longrightarrow 0}_{0,r}(\bar D, E)$.

In Sect. 12 we introduce the concept of a q-convex extension of a complex manifold X, and prove that, with respect to such extensions, the Dolbeault cohomology classes of order r admit uniquely determined continuations if $r \geq n-q$ (Theorem 12.14), where $n = \dim_{\mathbb{C}} X$, and can be uniformly approximated if $r=n-q-1$ (Theorem 12.11 and Corollary 12.12). Then, as a consequence, we obtain the classical Andreotti-Grauert finiteness theorem (Theorem 12.16): If E is a holomorphic vector bundle over an n-dimensional q-convex manifold X, then $\dim H^{0,r}(X,E) < \infty$ for all $r \geq n-q$, where, in the <u>completely</u> q-convex case, even $H^{0,r}(X,E) = 0$ for all $r \geq n-q$. Also in Sect. 12, we prove the following supplement to Theorem 11.2: If D is a non-degenerate <u>completely</u> q-convex domain in an n-dimensional complex manifold X, and E is a holomorphic vector bundle over X, then $H^{0,r}_{1/2\longrightarrow 0}(\bar D, E) = 0$ for all $r \geq n-q$ (Theorem 12.7).

7. Local solution of $\bar\partial u=f_{0,r}$ on strictly q-convex domains with $r\geq n-q$

Local Cauchy-Fantappie formulas with uniform estimates for non-degenerate strictly q-convex (resp. strictly q-concave) domains form an important tool in this monograph. In the construction of these formulas we use the fact that (q+1)-convex functions without degenerate critical points, with respect to appropriate local holomorphic coordinates, take an especially nice form - functions of this form will be called normalized. Before giving the definition let us recall the notion of the

7.1. Levi polynomial. Let $U \subseteq \mathbb{C}^n$ be an open set and $\varrho: U \longrightarrow \mathbb{R}$ a C^2 function. Then the function

$$F_{\varrho}(z,x) := 2\sum_{j=1}^{n} \frac{\partial\varrho(x)}{\partial x_j}(x_j-z_j) - \sum_{j,k=1}^{n} \frac{\partial^2\varrho(x)}{\partial x_j\partial x_k}(z_j-x_j)(z_k-x_k)$$

defined for $x\in U$ and $z\in\mathbb{C}^n$ is called the Levi polynomial of ϱ.

A direct computation shows (cf., for instance, Lemma 1.4.1 in [H/L]) that the real part of the Levi polynomial together with the Levi form (cf. Sect. 4.1) forms the second-order Taylor polynomial of ϱ, i.e.

$$\varrho(z) = \varrho(x) - \operatorname{Re} F_{\varrho}(z,x) + \sum_{j,k=1}^{n} \frac{\partial^2\varrho(x)}{\partial\bar{x}_j\partial x_k}(\bar{z}_j-\bar{x}_j)(z_k-x_k) + o(|x-z|^2) \tag{7.1}$$

for $z \longrightarrow x\in U$ (Re:= real part of).

7.2. Definition. Let $U \subseteq \mathbb{C}^n$ be an open set. A function $\varrho: U \longrightarrow \mathbb{R}$ will be called normalized q-convex ($1\leq q\leq n$) if it is strictly plurisubharmonic with respect to $z_1,\dots,z_q$ and, moreover, there exist constants $C<\infty$, $\beta>0$ such that

$$\operatorname{Re} F_{\varrho}(z,x) \geq \varrho(x) - \varrho(z) + \beta\sum_{j=1}^{q}|x_j-z_j|^2 - C\sum_{j=q+1}^{n}|x_j-z_j|^2 \tag{7.2}$$

for all $z,x\in U$.

7.3. Lemma. Let X be an n-dimensional complex manifold, and $\varphi: X \longrightarrow \mathbb{R}$ an arbitrary q-convex function ($1\leq q\leq n$). Let $h: U \longrightarrow \mathbb{C}^n$ be holomorphic coordinates in some neighborhood U of a point $y\in X$ such that $\varphi\circ h^{-1}$ is strictly plurisubharmonic with respect to $x_1,\dots,x_q$, where $x_1,\dots,x_n$ are the canonical complex coordinates on $\mathbb{C}^n$ (h exists by Proposition 4.4 (i)). Then there is a neighborhood $V \subseteq U$ of y such that $\varrho \circ h^{-1}$ is normalized q-convex in h(V).

Proof. Since $\varrho \circ h^{-1}$ is strictly plurisubharmonic with respect to $x_1, \ldots, x_q$, for every neighborhood $V \subset\subset U$ of y, we can find $\alpha > 0$ with

$$\sum_{j,k=1}^{q} \frac{\partial^2 \varrho \circ h^{-1}(x)}{\partial \bar{x}_j \partial x_k} \bar{t}_j t_k \geq \alpha \sum_{j=1}^{q} |t_j|^2 \qquad \text{for all } x \in h(V) \text{ and } t \in \mathbb{C}^q.$$

If V is sufficiently small, then, moreover, for some $M < \infty$,

$$\sum_{\max(j,k) \geq q+1} \frac{\partial^2 \varrho \circ h^{-1}(x)}{\partial \bar{x}_j \partial x_k} \bar{t}_j t_k$$

$$\leq M \left| \left(\sum_{j=1}^{q} |t_j|^2 \right)^{1/2} \left(\sum_{j=q+1}^{n} |t_j|^2 \right)^{1/2} + \sum_{j=q+1}^{n} |t_j|^2 \right|$$

if $x \in h(V)$ and $t \in \mathbb{C}^n$. Together this implies that

$$\sum_{j,k=1}^{n} \frac{\partial^2 \varrho \circ h^{-1}(x)}{\partial \bar{x}_j \partial x_k} \bar{t}_j t_k \geq \frac{\alpha}{2} \sum_{j=1}^{q} |t_j|^2 - \left(\frac{2M^2}{\alpha} + M \right) \sum_{j=q+1}^{n} |t_j|^2$$

for all $x \in h(V)$ and $t \in \mathbb{C}^n$. In view of (7.1), after shrinking V, the last estimate yields (7.2) with $\varrho \circ h^{-1}$ and h(V) instead of ϱ and U if we set $\beta = \alpha/4$ and $C = 4M^2/\alpha + 2M$. □

7.4. Definition. $[U, \varrho, \varphi, D]$ will be called a q-convex configuration in $\mathbb{C}^n$ $(0 \leq q \leq n-1)$ if the following conditions (i)-(iv) are fulfilled:

(i) U is a convex domain in $\mathbb{C}^n$, and $\varphi: U \longrightarrow \mathbb{R}$ is a convex C^2 function such that the domain $D_2 := \{z \in U: \varphi(z) < 0\}$ is non-empty and relatively compact in U.

(ii) $\varrho: U \longrightarrow \mathbb{R}$ is a normalized (q+1)-convex function which has at most one critical point, and if such a critical point $y \in U$ exists, then $\varphi(y) < 0$ and y is a non-degenerate critical point of ϱ which is not the point of a local minimum of ϱ.

(iii) $d\varrho(z) \wedge d\varphi(z) \neq 0$ for all $z \in U$ with $\varrho(z) = \varphi(z) = 0$.

(iv) $D = \{z \in U: \varrho(z) < 0, \varphi(z) < 0\}$, and $\emptyset \neq D \neq \{z \in U: \varphi(z) < 0\}$.

If $[U,\varrho,\varphi]$ fulfils conditions (i)-(iii) and the set

$$D:=\{z\in U:\ \varrho(z)<0,\ \varphi(z)<0\}$$

is not empty, then we say $[U,\varrho,\varphi]$ defines the q-convex configuration $[U,\varrho,\varphi,D]$.

We distinguish 4 types of q-convex configurations $[U,\varrho,\varphi,D]$ in $\mathbb{C}^n$:

Type I: $d\varrho(z)\neq 0$ for all $z\in U$.
Type II: ϱ has a critical point $y\in U$ and $\varrho(y)=0$.
Type III: ϱ has a critical point $y\in U$ and $\varrho(y)<0$.
Type IV: ϱ has a critical point $y\in U$ and $\varrho(y)>0$.

A q-convex configuration in $\mathbb{C}^n$ will be called non-critical if it is of type I, and critical otherwise. We say a q-convex configuration in $\mathbb{C}^n$ is smooth if it is of one of the types I, III or IV, and non-smooth if it is of type II.

Remark. If U, φ, D_2 are as in condition (i), then the set $\{z\in U: d\varphi(z)=0\}$ consists precisely of all $z\in U$ with $\varphi(z) = \min_{x\in U}\varphi(x)$. Therefore, then $\partial D_2 = \{z\in U: \varphi(z)=0\}$ and $d\varphi(z)\neq 0$ for all $z\in\partial D_2$.

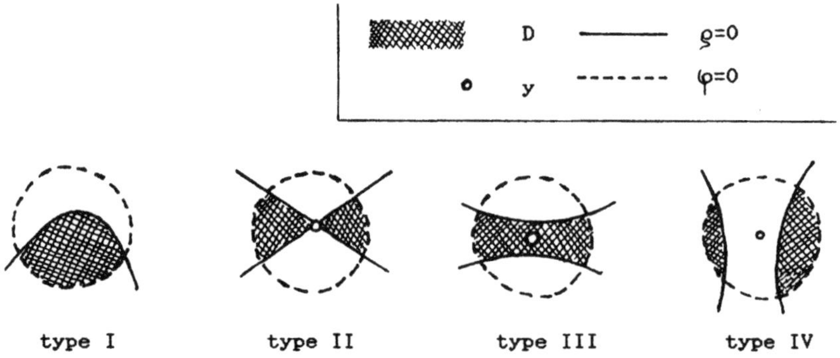

FIGURE 1: q-convex configurations

7.5. Lemma. If $\varrho: X \longrightarrow \mathbb{R}$ is an arbitrary (q+1)-convex function on an n-dimensional complex manifold X ($0\leq q\leq n-1$), then, for any point $y\in X$ which is neither a degenerate critical point of ϱ nor the point of a local minimum of ϱ, there exist holomorphic coordinates $h: V \longrightarrow \mathbb{C}^n$ in some neighborhood V of y such that h(V) is the open unit ball, $h(y) = 0$, and, for all $0<r<1$, $[h(V), \varrho\circ h^{-1}, |z|^2-r^2]$ defines a q-convex configuration, which is of type I if $d\varrho(y)\neq 0$, and of type II if $d\varrho(y)=0$.

Proof. By Lemma 7.3 we can find holomorphic coordinates $h: V \longrightarrow \mathbb{C}^n$ in some neighborhood V of y such that $h(y)=0$, h(V) is the open unit ball, and $\varrho\circ h^{-1}$ is normalized (q+1)-convex in h(V).

If $d\varrho(y)\neq 0$, then, after shrinking V, it can be assumed that moreover

$d\varrho(x) \neq 0$ if $x \in V$ and $d\varrho \circ h^{-1}(z) \wedge d|z|^2 \neq 0$ if $z \in h(V)\setminus\{0\}$. Therefore, then, for any $0<r<1$, $[h(V), \varrho \circ h^{-1}, |z|^2-r^2]$ defines a q-convex configuration of type I.

Now let $d\varrho(y)=0$. Then y is a non-degenerate critical point which is not the point of a local minimum (by hypothesis) and not the point of a local maximum (by Corollary 4.5). So it follows from Proposition 0.6 in Appendix B that, after shrinking V, $d\varrho \circ h^{-1}(z) \wedge d|z|^2 \neq 0$ for all $z \in h(V)\setminus\{0\}$ with $\varrho \circ h^{-1}(z) = 0$. Since non-degenerate critical points are isolated (cf. Proposition 0.1 in Appendix B), after a further shrinking of V, $d\varrho(x) \neq 0$ for all $x \in V\setminus\{y\}$. Thus, $[h(V), \varrho \circ h^{-1}, |z|^2-r^2]$ defines a q-convex configuration of type II if $0<r<1$. □

7.6. The set $\mathrm{Div}(\varrho)$. Let $\varrho: U \longrightarrow \mathbb{R}$ be a normalized (q+1)-convex function defined on a domain $U \subseteq \mathbb{C}^n$ $(0 \leq q \leq n-1)$.

Definition. By $\mathrm{Div}(\varrho)$ we denote the set of all n-tuples $w=(w^1,\dots,w^n)$ of C^1 functions $w^j: C^n \times U \longrightarrow C$ which are obtained by the following

Construction: Let $C<\infty$ and $\beta>0$ be constants such that inequality (7.2) is fulfilled. Choose C^1-functions $a_{j,k}: U \longrightarrow \mathbb{C}$ $(j,k=1,\dots,n)$ such that

$$\left| a_{jk}(x) - \frac{\partial^2 \varrho(x)}{\partial x_j \partial x_k} \right| < \frac{\beta}{2n^2} \tag{7.3}$$

for all $x \in U$. Set

$$w^j(z,x) = \begin{cases} 2\dfrac{\partial \varrho(x)}{\partial x_j} - \displaystyle\sum_{k=1}^{n} a_{jk}(x)(x_k - z_k) & \text{if } 1 \leq j \leq q+1. \\[2ex] 2\dfrac{\partial \varrho(x)}{\partial x_j} - \displaystyle\sum_{k=1}^{n} a_{jk}(x)(x_k - z_k) + (C+\beta)(\bar{x}_j - \bar{z}_j) & \text{if } q+2 \leq j \leq n. \end{cases}$$

Remark. If $w \in \mathrm{Div}(\varrho)$, then $w(z,x)$ is holomorphic with respect to $z_1,\dots,z_{q+1}$, and it follows from (7.3) and (7.2) that

$$\mathrm{Re}\langle w(z,x), x-z \rangle \geq \varrho(x) - \varrho(z) + \frac{\beta}{2}|x-z|^2 \quad \text{for all } x,z \in U. \tag{7.4}$$

7.7. Canonical Leray data and maps. Let $[U, \varrho, \varphi, D]$ be a q-convex configuration in $\mathbb{C}^n$ $(0 \leq q \leq n-1)$. Set

$$Y_1 := \{z \in U: \varrho(z)=0\}, \qquad D_1 := \{z \in U: \varrho(z)<0\},$$
$$Y_2 := \{z \in U: \varphi(z)=0\}, \qquad D_2 := \{z \in U: \varphi(z)<0\},$$

Then $D=D_1\cap D_2$ is a domain with piecewise almost C^1 boundary and (Y_1,Y_2) is a frame for D (cf. Sect. 3.1).[1)]

Now let $w_1\in\mathrm{Div}(\varrho)$ (cf. Sect. 7.6). Then, by (7.4),

$$\langle w_1(z,x),x-z\rangle \neq 0 \qquad \text{for all } x\in Y_1 \text{ and } z\in D_1.$$

Further, we set $w_2(x)=\nabla^{\mathbb{C}}\varphi(x)$ for $x\in U$ (cf. Remark 2.10), and recall that, by Observation 2.11,

$$\langle w_2(x),x-z\rangle \neq 0 \qquad \text{for all } x\in Y_2 \text{ and } z\in D_2.$$

Thus, (w_1,w_2) is a Leray datum for $(D;\ Y_1,Y_2)$ (cf. Sect. 3.4).

Definition. We say (w_1,w_2) is a canonical Leray datum for $[U,\varrho,\ \varphi,D]$ (or for D) if $w_1\in\mathrm{Div}(\varrho)$ and $w_2=\nabla^{\mathbb{C}}\varphi$. We say v is a canonical Leray map for $[U,\varrho,\ \varphi,D]$ (or for D) if v is the canonical combination (cf. Sect. 3.4) of some canonical Leray datum (w_1,w_2) for $[U,\varrho,\ \varphi,D]$, *i.e.*

$$v_j(z,x) = \frac{w_j(z,x)}{\langle w_j(z,x),x-z\rangle} \tag{7.5}$$

for $z\in D$, $x\in Y_j$ $(j=1,2)$, and

$$v_{12}(z,x,t) = (1-t)\frac{w_1(z,x)}{\langle w_1(z,x),x-z\rangle} + t\frac{w_2(x)}{\langle w_2(x),x-z\rangle}. \tag{7.6}$$

for $z\in D$, $x\in Y_1\cap Y_2$ and $0\le t\le 1$.

Remark. If v is a canonical Leray map for a q-convex configuration $[U,\varrho,\ \varphi,D]$, then v_1, v_2, and v_{12} depend holomorphically on $z_1,\dots,z_{q+1}$.

7.8. Theorem. Let $[U,\varrho,\ \varphi,D]$ be a q-convex configuration in $\mathbb{C}^n$ $(0\le q\le n-1)$, v a canonical Leray map for $[U,\varrho,\ \varphi,D]$, and f a continuous $(0,r)$-form on $\bar{D}$ such that $\bar\partial f$ is also continuous on $\bar{D}$. If $r\ge n-q$, then $L^v f=0$ and hence

$$(-1)^r f = \bar\partial T^v f - T^v\bar\partial f \quad \text{in } D. \tag{7.7}$$

Proof. This follows from Corollary 3.13 (ii), because v_1, v_2, and v_{12} are holomorphic in $z_1,\dots,z_{q+1}$ (cf. the remark at the end of Sect. 7.7). □

1) The fact that the volume of the smooth part of Y_1 is finite follows, for instance, from Proposition 0.7 in Appendix B.

8. Local approximation of $\bar\partial$-closed $(0,n-q-1)$-forms on strictly q-convex domains

In this section we prove the following approximation theorem:

8.1. Theorem. If $[U,\varrho,\varphi,D]$ is a q-convex configuration in $\mathbb{C}^n$ ($0\le q\le n-1$), then any $(0,n-q-1)$-form which is continuous and $\bar\partial$-closed in a neighborhood of $\bar D$ can be approximated uniformly on $\bar D$ by $(0,n-q-1)$-forms which are continuous and $\bar\partial$-closed in $\mathbb{C}^n$.

Remark. A stronger result will be obtained in Sect. 10.

8.2. Proof of Theorem 8.1 for $q=n-1$. In this case we have to prove that every holomorphic function f in a neighborhood of $\bar D$ can be approximated uniformly on $\bar D$ by holomorphic functions in $\mathbb{C}^n$. Of course, this is well known from the Oka-Cartan theory (see, for instance, Theorem 2.7.1 in [H/L]). However, to be independent, here we give a direct proof (using the same method as in the proof of Theorem 2.7.1 in [H/L]).

First we prove the following

Statement 1. Let $K \subset\subset D$ be a compact set and f a holomorphic function in a neighborhood of $\bar D$. Then f can be approximated uniformly on K by holomorphic functions in $D_2:=\{z\in U: \varphi(z)<0\}$.

Proof of Statement 1. Let (w_1,w_2) be a canonical Leray data for D (cf. Sect 7.7). Then the corresponding Cauchy-Fantappie formula (3.18) takes the form

$$(2\pi i)^n f(z) = \int\limits_{\substack{\varrho(x)=0\\ \varphi(x)<0}} f(x)\frac{\det(w_1(z,x),\overbrace{\bar d_x w_1(z,x)}^{n-1})}{\langle w_1(z,x),x-z\rangle^n}\wedge\omega(x)$$

$$+\int\limits_{\substack{\varrho(x)<0\\ \varphi(x)=0}} f(x)\frac{\det(w_2(x),\overbrace{\bar d w_2(x)}^{n-1})}{\langle w_2(x),x-z\rangle^n}\wedge\omega(x)$$

$$+\int\limits_{\substack{0\le t\le 1\\ \varrho(x)=\varphi(x)=0}} f(x)\det(v_{12}(z,x,t),\overbrace{\bar d_{x,t}v_{12}(z,x,t)}^{n-1})\wedge\omega(x), \qquad (8.1)$$

where v_{12} is defined by (7.6). Since $w_1(z,x)$ is holomorphic in z, the second two integrals in (8.1) define some holomorphic functions in D_2.

Therefore, we only have to prove that the holomorphic function which is defined by the first integral in (8.1) can be approxmimated uniformly on K by holomorphic functions in D_2. This is a consequence of the following

Statement 2. For any fixed $x_0 \in U$ with $\varrho(x_0)=0$, the function

$$\frac{1}{\langle w_1(z,x_0),x_0-z\rangle} \qquad (z\in D)$$

can be approximated uniformly on K by holomorphic functions in D_2.

Proof of Statement 2. Set $\phi(z,x)=\langle w_1(z,x),x-z\rangle$. Since ϱ does not have local minima and since, by (7.4), $\phi(z,x)\neq 0$ if $z\in D$ and $x\in U\setminus D$, we can find points $x_1,\dots,x_N \in D_2\setminus\bar{D}$ such that $\varphi(x_N)=0$,

$$\varrho(x_N) = \max_{x\in\bar{D}_2} \varrho(x), \tag{8.2}$$

and

$$\max_{z\in K}\left|1-\frac{\phi(z,x_{j-1})}{\phi(z,x_j)}\right| < 1 \qquad \text{for } j=1,\dots,N.$$

Then

$$\frac{1}{\phi(z,x_{j-1})} = \frac{1}{\phi(z,x_j)}\sum_{k=1}^{\infty}\left[1-\frac{\phi(z,x_{j-1})}{\phi(z,x_j)}\right]^k,$$

for $j=1,\dots,N$, where the convergence is uniform on K. Therefore, for any $1\leq j\leq N$, the functions $1/\phi(z,x_{j-1})$ can be approximated uniformly on K by polynomials in $1/\phi(z,x_j)$ and $\phi(z,x_{j-1})$. Hence $1/\phi(z,x_0)$ can be approximated uniformly on K by polynomials in the functions $1/\phi(z,x_N)$ and $\phi(z,x_0),\dots,\phi(z,x_{N-1})$. Since all these functions are holomorphic in D_2 (for $1/\phi(z,x_N)$ this follows from the fact that, by (8.2) and (7.4), $\phi(z,x_0)\neq 0$ for all $z\in D_2$), this completes the proof of Statement 2 (and hence of Statement 1).

End of proof of Theorem 8.1 for q=n-1. Let f be a holomorphic function in a neighborhood W of $\bar{D}$. Take $\varepsilon>0$ sufficiently small, set $\tilde{D}:=\{z\in U: \varrho(z)<\varepsilon,\ \varphi(z)<\varepsilon\}$, and $\tilde{D}_2:=\{z\in U: \varphi(z)<\varepsilon\}$. Then $[U,\varrho-\varepsilon,\varphi-\varepsilon,\tilde{D}]$ is an (n-1)-convex configuration, and $D \subset\subset \tilde{D} \subset\subset W$. Therefore, it follows from Statement 1 that f can be approximated uniformly on $\bar{D}$ by holomorphic functions in $\tilde{D}_2$. Since $\tilde{D}_2$ is convex, this completes the proof of Theorem 8.1 for q=n-1. (Here we used the fact that any holomorphic function in a neighborhood of a convex compact set in $\mathbb{C}^n$ can be approximated uniformly on this set by holomorphic functions in $\mathbb{C}^n$. This can be proved similar as Statement 1.) □

For $0\le q\le n-2$, the proof of Theorem 8.1 will be reduced to the case $q=n-1$, by means of the following

8.3. Lemma. Let $[U,\varrho,\varphi,D]$ be a q-convex configuration in $\mathbb{C}^n$. If $0\le q\le n-2$, then, for each continuous $\bar\partial$-closed $(0,n-q-1)$-form f in some neighborhood U of $\bar D$, there exists a neighborhood $V\subseteq U$ of $\bar D$ such that, over V, f can be written

$$f = h d\bar z_{q+2}\wedge\ldots\wedge d\bar z_n + \bar\partial g, \tag{8.3}$$

where h is a C^∞ function on V which depends holomorphically on $z_1,\ldots,z_{q+1}$, and g is a continuous $(0,n-q-2)$-form on V.

Proof. Notice that, for all sufficiently small $\varepsilon>0$, $[U,\varrho-\varepsilon,\varphi-\varepsilon]$ defines also a q-convex configuration. Therefore it is sufficient to show that f can be written in the form (8.3), over D. Let v be a canonical Leray map for $[U,\varrho,\varphi,D]$ (cf. Sect. 7.7)). Then the piecewise Cauchy-Fantappie formula (Theorem 3.12) yields the representation

$$f = (-1)^{n-q-1} L^v f + \bar\partial g \qquad \text{on } D,$$

where $g=(-1)^{n-q-1}T^v f$. Every canonical Leray map for $[U,\varrho,\varphi,D]$ is of class C^∞, and holomorphic with respect to $z_1,\ldots,z_{q+1}$. Since f is of bidegree $(0,n-q-1)$ and by (3.9) and (3.10), this implies that $L^v f$ is of the form $hd\bar z_{q+2}\ \ldots\ d\bar z_n$ where h is a C^∞ function on D which depends holomorphically on $z_1,\ldots,z_{q+1}$. □

8.4. Proof of Theorem 8.1 if $q<n-1$. Let V,h,g be as in Lemma 8.3. If χ is a C^∞ function on $\mathbb{C}^n$ with supp $\chi\subset\subset V$ and $\chi=1$ in some neighborhood of $\bar D$, then $\bar\partial(\chi g)$ is a continuous $\bar\partial$-closed $(0,n-q-1)$-form on $\mathbb{C}^n$ which is equal to $\bar\partial g$ on $\bar D$. Therefore it is sufficient to prove that h can be approximated uniformly on $\bar D$ by C^∞ functions on $\mathbb{C}^n$ which depend holomorphically on $z_1,\ldots,z_{q+1}$. To do this, first we fix some vector $w=(w_{q+2},\ldots,w_n)\in\mathbb{C}^{n-q-1}$ and set

$$\mathbb{C}^{q+1}_w = \{z\in\mathbb{C}^n:\ z_{q+2}=w_{q+2},\ldots,z_n=w_n\}.$$

Then the set $U\cap\mathbb{C}^{q+1}_w$, the function $\varphi|U\cap\mathbb{C}^{q+1}_w$, and the set $\{z\in U:\varphi(z)<0\}\cap\mathbb{C}^{q+1}_w$ are convex, and

$$\{z\in U: \varphi(z)<0\}\cap \mathbb{C}_w^{q+1} \subset\subset U\cap \mathbb{C}_w^{q+1}.$$

Further, then the restriction of ϱ to $U\cap \mathbb{C}_w^{q+1}$ is normalized (q+1)-convex. Choose $\varepsilon>0$ so small that

$$\{z\in U: \varphi(z)<\varepsilon,\ \varrho(z)<\varepsilon\} \subset\subset V.$$

Then, by Proposition 0.4 in Appendix B, we can find a real-linear map $L:\mathbb{C}^n \longrightarrow \mathbb{R}$ such that

$$D \subset\subset D_{\varepsilon,L} := \{z\in U: \varphi(z)<\varepsilon,\ (\varrho+L)(z)<\varepsilon\} \subset\subset V,$$

and $[U\cap \mathbb{C}_w^{q+1}, \varrho+L-\varepsilon, \varphi-\varepsilon, D_{\varepsilon,L}\cap \mathbb{C}_w^{q+1}]$ is a q-convex configuration in $\mathbb{C}_w^{q+1}$ (or $\bar{D}\cap \mathbb{C}_w^{q+1}$ is empty). Since Theorem 8.1 is already proved for $q=n-1$, this implies that, for each $\delta>0$ and all $w=(w_{q+2},\ldots,w_n)\in \mathbb{C}^{n-q-1}$ with $D\cap \mathbb{C}_w^{q+1} \neq \emptyset$, there exist a neighborhood $W \subseteq \mathbb{C}^n$ of $\bar{D}\cap \mathbb{C}_w^{q+1}$ and a holomorphic function h_w on $\mathbb{C}^n$ which is independent of $z_{q+2},\ldots,z_n$ such that $|h-h_w|<\delta$ on W. Now a partition of unity argument with respect to $z_{q+2},\ldots,z_n$ completes the proof of Theorem 8.1 also in the case $0\leq q\leq n-2$. □

9. Uniform estimates for the local solutions of the $\bar{\partial}$-equation constructed in Sect. 7

In Sect. 7, q-convex configurations were introduced, the notion of a canonical Leray map v for such configurations was defined, and it was observed (Theorem 7.8) that the corresponding Cauchy-Fantappie operator T^v (cf. Sect. 3.10) solves the $\bar{\partial}$-equation for forms of bidegree (0,r) with $r\geq n-q$. Here we prove the following important estimates for T^v:

9.1. Theorem. Let $[U,\varrho,\varphi,D]$ be a q-convex configuration in $\mathbb{C}^n$ $(0\leq q\leq n-1)$, v a canonical Leray map for D, $\varepsilon>0$, and

$$D_\varepsilon := \{z\in D: \varphi(z) < -\varepsilon\}.$$

Then there exists a constant $C<\infty$ such that, for all continuous differential forms f on $\bar{D}$,

$$\| T^v f\|_{1/2,D_\varepsilon} \leq C\|f\|_{0,D} \tag{9.1}$$

(for the definition of these norms, see Sect. 0.11). If $[U,\varrho,\varphi,D]$ is of type II and $y\in U$ is the critical point of ϱ, then, moreover,

$$\|T^v f\|_{0,D_\varepsilon} \le C\Big(\sup_{z\in \operatorname{supp} f}|z-y|\Big)\|f\|_{0,D}, \tag{9.2}$$

and

$$\sup_{\substack{z,x\in D_\varepsilon \\ |z-y|<r \\ |x-y|<r}} \frac{\|T^v f(z) - T^v f(x)\|}{|z-x|^{1/2}} \le C(1 + |\ln r|)r^{1/2}\|f\|_{0,D}. \tag{9.3}$$

for each $r>0$.

In the proof of this theorem we use the following

<u>9.2. Lemma.</u> Let $[U,\varrho,\varphi,D]$ be a q-convex configuration in $\mathbb{C}^n$, $D_2:=\{z\in U: \varphi(z)<0\}$, $w\in \mathrm{Div}(\varrho)$, and $h(z,x):= \operatorname{Im}\langle w(z,x),x-z\rangle$ for $z,x\in U$ (Im:= imaginary part of). Then the following assertions hold true:

(i) $$\Big\|d_x h(z,x)\Big|_{x=z}\Big\| = \|d\varrho(z)\| \qquad \text{for all } z\in U.$$

(ii) $$\|d\varrho(z)\|^2 \le \sqrt{n}\Big\|d\varrho(z)\wedge d_x h(z,x)\Big|_{x=z}\Big\| \qquad \text{for all } z\in U.$$

(iii) If $t_j=t_j(x)$ are the real coordinates of $x\in\mathbb{C}^n$ with $x_j=t_j(x)+ it_{j+n}(x)$, then there is a constant $K<\infty$ such that

$$\left|\frac{\partial h(z,\cdot)}{\partial t_j}(x) - \frac{\partial h(z,\cdot)}{\partial t_j}(z)\right| \le K|x-z| \qquad \text{for all } x,z\in D_2 \ (j=1,\dots 2n).$$

(iv) If $[U,\varrho,\varphi,D]$ is of type II and $y\in U$ is the critical point of ϱ, then there is a constant $K<\infty$ such that

$$|t(z,x)| \le K(|x-y||x-z| + |x-z|^2) \qquad \text{for all } z,x\in D_2.$$

<u>Proof.</u> If $t_j=t_j(x)$ are the real coordinates of $x\in\mathbb{C}^n$ with $x_j=t_j(x)+t_{j+n}(x)$, then, by definition of $\mathrm{Div}(\varrho)$,

$$h(z,x) = \sum_{j=1}^{n} \left(\frac{\partial \varrho}{\partial t_j}(x) t_{j+n}(x-z) - \frac{\partial \varrho}{\partial t_{j+n}}(x) t_j(x-z) \right) + O(|x-z|^2),$$

and

$$d_x h(z,x) = \sum_{j=1}^{n} \left(\frac{\partial \varrho}{\partial t_j}(x) dt_{j+n}(x) - \frac{\partial \varrho}{\partial t_{j+n}}(x) dt_j(x) \right) + O(|x-z|). \quad \square$$

Proof of Theorem 9.1. Recall that

$$T^v = B_D + R_1^v + R_2^v + R_{12}^v.$$

It is well known (cf., for instance, Proposition 1 in Appendix 1 of [H/L]) that there exists a constant $C<\infty$ such that, for all continuous differential forms f on $\bar{D}$,

$$\| B_D f(z) - B_D f(x) \| \leq C(1 + |\ln|x-z||)|x-z| \|f\|_{0,D}$$

for all $z,x \in D$. Hence, in particular, (9.1) and (9.3) hold true with B_D instead of T^v. Further, it is easy to see (cf., for instance, Proposition 5 (i) in Appendix 1 of [H/L]) that (9.2) holds true with B_D instead of T^v. Since $S_2 \cap \bar{D}_\varepsilon = S_{12} \cap \bar{D}_\varepsilon = \emptyset$ (for the definitions of S_2 and S_{12}, see Sect. 3.2), estimates (9.1)-(9.3) hold true also with R_2^v and R_{12}^v instead of T^v.

Therefore, it remains to prove that (9.1)-(9.3) hold true with R_1^v instead of T^v. Recall that (cf. Sect. 3.7)

$$R_1^v f(z) = \int_{S_1 \times \triangle_{01}} f(x) \wedge \hat{R}_1^v(z,x,t), \qquad z \in D,$$

where

$$\hat{R}_1^v(z,x,t) := -\frac{1}{(2\pi i)^n} \det(v_{01}(z,x,t), \overbrace{(\bar{\partial}_{z,x} + d_t) v_{01}(z,x,t)}^{n-1}) \wedge \omega(x)$$

with

$$v_{01}(z,x,t) := t_0 \frac{\bar{x}-\bar{z}}{|x-z|^2} + t_1 \frac{w(z,x)}{\langle w(z,x), x-z \rangle}$$

for some $w=(w^1,\dots,w^n)\in\mathrm{Div}(\varrho)$. Further on (in this proof), we use the abbreviation $\phi = \phi(z,x) = \langle w(z,x), x-z\rangle$. Then

$$(\bar\partial_{z,x}+d_t)V_{01} = \left[\frac{w}{\phi} - \frac{\bar x-\bar z}{|x-z|^2}\right]dt_1 + t_1\left[\frac{\bar\partial_{z,x}w}{\phi} - \frac{w\bar\partial_{z,x}\phi}{\phi^2}\right]$$

$$+ t_0\left[\frac{d\bar x-d\bar z}{|x-z|^2} - \frac{(\bar x-\bar z)\bar\partial_{z,x}|x-z|^2}{|x-z|^4}\right].$$

Therefore, expanding the determinant in the definition of $\hat R_1^v(z,x,t)$ (using the rules collected in Proposition 0.6 and taking into account that only monomials of degree 1 in t contribute to the integral $R_1^v f$) and integrating with respect to t, we obtain

$$R_1^v f(z) = \sum_{s=0}^{n-2} c_s \int_{S_1} f(x)\wedge\tilde R_s(z,x), \qquad z\in D,$$

where

$$\tilde R_s(z,x) = \frac{\det(w,\bar x-\bar z,\overbrace{\bar\partial_{z,x}w}^{n-s-2},\overbrace{d\bar x-d\bar z}^{s})}{\phi^{n-s-1}|x-z|^{2s+2}}\wedge\omega(x)$$

and $c_1,\dots,c_{n-2}$ are complex numbers which are independent of f and z. This implies that the coefficients of $R_1^v f$ are linear combinations of integrals of the form

$$\int_{S_1} \frac{FG}{\phi^{n-s-1}|x-z|^{2s+2}}\, d\bar x_1\wedge\dots\overset{\wedge}{\underset{j}{\dots}}\dots\wedge d\bar x_n\wedge\omega(x), \tag{9.4}$$

where $0\le s\le n-2$, $1\le j\le n$, F is a coefficient of f, and G is the product of some of the functions w^k, $\bar x_k-\bar z_k$, $\partial w^k/\partial\bar z_l$ and $\partial w^k/\partial\bar x_l$. Since

$$w^k(z,x) = \frac{\partial\varrho(x)}{\partial x_k} + O(|x-z|)$$

and the product G contains at least one of the functions w^k and at least one of the functions $\bar{x}_k-\bar{z}_k$, there exists a constant $C_0<\infty$ such that, for the product G in each of the integrals (9.4), we have

$$|G(z,x)| \leq C_0(\|d\varrho(x)\| \, |x-z| + |x-z|^2)$$

and

$$\| d_z G(z,x)\| \leq C_0(\|d\varrho(x)\| + |x-z|)$$

for all z,x $\bar{D}$. Further, by (7.4), there is a constant $\alpha>0$ with

$$|\phi(z,x)| \geq \alpha |x-z|^2$$

for all $x\in S_1$ and $z\in D$. Further, for some constant $C_1<\infty$,

$$\|d_z\phi(z,x)\| \leq C_1(\|d\varrho(x)\| + |x-z|)$$

for all $z,x\in\bar{D}$. Hence there is a constant $C_2<\infty$ such that, for any continuous differential form f on $\bar{D}$ and all $z\in D$,

$$\| R_1^v f(z)\| \leq C_2\|f\|_{0,D}\left[\int\limits_{S_1\cap \operatorname{supp} f} \frac{\|d\varrho\| dS_1}{|\phi|\,|x-z|^{2n-3}} + \int\limits_{S_1\cap \operatorname{supp} f} \frac{dS_1}{|x-z|^{2n-2}} \right], \tag{9.5}$$

where dS_1 is the Euclidean volume form of S_1.

Taking into account the relation

$$d_z \frac{G}{\phi^{n-s-1}|x-z|^{2s+2}} = \frac{d_zG}{\phi^{n-s-1}|x-z|^{2s+2}} - \frac{(n-s-1)Gd_z\phi}{\phi^{n-s}|x-z|^{2s+2}} + \frac{(s+1)Gd_z|x-z|^2}{\phi^{n-s-1}|x-z|^{2s+4}},$$

we can find $C_3<\infty$ such that, for all continuous differential forms f on $\bar{D}$ and all $z\in D$,

$$\|dR_1^v f(z)\| \leq C_3\|f\|_{0,D}\left[\int\limits_{S_1\cap \operatorname{supp} f} \frac{\|d\varrho\|^2 dS_1}{|\phi|^2|x-z|^{2n-3}} + \int\limits_{S_1\cap \operatorname{supp} f} \frac{\|d\varrho\| dS_1}{|\phi|\,|x-z|^{2n-2}} \right.$$

$$+ \int_{S_1 \cap \operatorname{supp} f} \frac{dS_1}{|\phi||x-z|^{2n-3}} \Bigg] . \tag{9.6}$$

Set $h(z,x) = \operatorname{Im} \phi(z,x)$. ($\operatorname{Im}$:= imaginary part of) Then, by (7.4), there is a constant $c>0$ such that

$$|\phi(z,x)| \geq |\varrho(z)| + |h(z,x)| + c|x-z|^2 \tag{9.7}$$

for all $x \in S_1$ aund $z \in D$.

Now we first consider the case that the configuration $[U, \varrho, \varphi, D]$ is smooth. Then S_1 is smooth and it is clear that

$$\sup_{z \in D} \int_{S_1} \frac{dS_1}{|x-z|^{2n-2}} < \infty .$$

Moreover, then it follows from (9.7) and Proposition 1 (ii) in Appendix A (which can be applied in view of assertion (ii) in Lemma 9.2) that

$$\sup_{z \in D} \int_{S_1} \frac{\|d\varrho\| dS_1}{|\phi||x-z|^{2n-3}} < \infty .$$

Therefore and by (9.5), we can find a constant $C_4 < \infty$ such that

$$\|R_1^{\vee} f\|_{0,D} \leq C_4 \|f\|_{0,D} \tag{9.8}$$

for all continuous differential forms f on $\bar{D}$. Further, applying Proposition 1 (i) and (ii) in Appendix A to the integrals on the right hand side of (9.6), we obtain $C_5 < \infty$ such that, for any continuous differential form f on $\bar{D}$ and all $z \in D$,

$$\|dR_1^{\vee} f(z)\| \leq C_5 |\varrho(z)|^{-1/2} \|f\|_{0,D} . \tag{9.9}$$

By Proposition 2 in Appendix A, it follows from (9.8) and (9.9) that, for some $C_6 < \infty$,

$$\|R_1^v f\|_{1/2,D} \le C_6 \|f\|_{0,D}.$$

for any continuous differential form f on $\bar{D}$. This completes the proof of estimate (9.1) in the case that $[U,\varrho,\varphi,D]$ is smooth.

Now we consider the case that $[U,\varrho\ \varphi,D]$ is not smooth, i.e. of type II. Then, without loss of generality, we can assume that the critical point of ϱ is the origin. (So we have the situation considered in Propositions 1-3 in Appendix B.)

In view of assertions (i)-(ii) in Lemma 9.1 and estimate (9.7), we can apply Proposition 3 in Appendix B to the integrals on the right hand side of (9.5), and we obtain a constant $C_7<\infty$ such that, for any continuous differential form f on $\bar{D}$,

$$\|R_1^v f\|_{0,D} \le C_7 \|f\|_{0,D} \tag{9.10}$$

and, moreover,

$$\|R_1^v f\|_{0,D} \le C_7 \Big(\sup_{z \in \operatorname{supp} f} |z|\Big) \|f\|_{0,D}. \tag{9.11}$$

Applying Propositions 2 in Appendix B to the integrals on the right hand side of (9.6), we obtain $C_8<\infty$ such that,

$$\|dR_1^v f(z)\| \le C_8 \Big(|\ln|\varrho(z)|| + \|d\varrho(z)\| \, |\varrho(z)|^{-1/2}\Big) \|f\|_{0,D}. \tag{9.12}$$

for any continuous differential form f on $\bar{D}$ and all $z \in D$.

If $\delta>0$ and $0 \le \chi \le 1$ is a continuous function on $\mathbb{C}^n$ such that $\chi(z) = 0$ for $|z| \le \delta/2$ and $\chi(z) = 1$ for $|z| \ge 1$, then it is clear that $\lim R_1^v(\chi f)(z)$ for $D \ni z \longrightarrow 0$ exists, and it follows from (9.11) (with $(1-\chi)f$ instead of f) that

$$\|R_1^v(1-\chi)f\|_{0,D} \le \delta C_7 \|f\|_{0,D}.$$

Since $R_1^v f = R_1^v \chi f + R_1^v(1-\chi)f$ and $\delta>0$ can be chosen arbitrarily small, so it follows that $\lim R_1^v f(z)$ for $D \ni z \longrightarrow 0$ exists. Therefore, from (9.12) and Propositon 4 in Appendix B we obtain $C_9<\infty$ with

$$\sup_{\substack{z,x\in D\\ |z|,|x|<r}} \frac{\|R_1^v f(z) - R_1^v f(x)\|}{|z-x|^{1/2}} \leq C_9\big(1 + |\ln r|\big) r^{1/2} \|f\|_{0,D}. \qquad (9.13)$$

for all $r>0$ and any continuous differential form f on $\bar{D}$.

It follows from (9.10) and (9.13) (with $r=1$) that (9.1) holds true with R_1^v instead of T^v, i.e. (9.1) is proved also for the case when the configuration $[U,\varrho, \varphi,D]$ in not smooth. Finally, we observe that estimates (9.11) and (9.13) imply that (9.2) and (9.3) hold true with R_1^v instead of T^v. □

10. Local uniform approximation of $\bar\partial$-closed $(0,n-q-1)$-forms on strictly q-convex domains

Here, by means of the estimates obtained in the preceding section, we prove the following strengthening of Theorem 8.1.

10.1. Theorem. Let $[U,\varrho, \varphi,D]$ be a q-convex configuration in $\mathbb{C}^n$ $(0\leq q\leq n-1)$, $\varepsilon>0$, and $D_\varepsilon := \{z\in D: \varphi(z)<-\varepsilon\}$. Then any continuous $(0,r)$-form with $n\geq r\geq n-q-1$ on $\bar{D}$ which is $\bar\partial$-closed in D can be approximated uniformly on $\bar{D}_\varepsilon$ by continuous $\bar\partial$-closed $(0,r)$-forms in $\mathbb{C}^n$.

Proof. Choose a C^∞ function $\psi : \mathbb{R} \longrightarrow \mathbb{R}$ such that $\psi(t)=1$ if $t\leq 1/2$ and $\psi(t)=0$ if $t\geq 1$. We assume, without loss of generality, that $0\in U$ and if the configuration $[U,\varrho, \varphi,D]$ is not of type I, then 0 is the critical point of ϱ. For $k=1,2,\ldots$ and $z\in\mathbb{C}^n$, we set $b_k^0(z) = \psi(k|z|)$ if $[U,\varrho, \varphi,D]$ is not smooth, and $b_k^0(z) = 0$ if $[U,\varrho, \varphi,D]$ is smooth. Then

$$\|db_k^0\|_{0,\mathbb{C}^n} \leq k \max_{0\leq t\leq 1} \left|\frac{d}{dt}\psi(t)\right| \qquad (10.1)$$

and

$$b_k^0(z) = 0 \qquad \text{if } |z| \geq \frac{1}{k}. \qquad (10.2)$$

If $[U,\varrho, \varphi,D]$ is smooth, then the surface $\{z\in U: \varrho(z)=0\}$ is smooth. If $[U,\varrho, \varphi,D]$ is not smooth, i.e. of type II, then the surface $\{z\in U: \varrho(z)=0,\ z\neq 0\}$ is smooth and $b_k^0(z)=1$ for all z with $|z|\leq 1/2k$. Therefore, in both cases, for all $k=1,2,\ldots$, we can find a neighborhood $U_k \subset\subset \{z\in U: \varphi(z)<0\}$ of $\bar{D}_\varepsilon$ as well as real-valued C^∞ functions b_k^j on $\mathbb{C}^n$ and vectors $a_k^j\in\mathbb{C}^n$ $(j=1,\ldots,N(k)<\infty)$ such that

$$\sum_{j=0}^{N(k)} b_k^j = 1 \qquad \text{on } U_k, \tag{10.3}$$

and, for $j=1,\dots,N(k)$,

$$\operatorname{supp} b_k^j \subset\subset \{z\in U:\ \varphi(x)<0\}, \tag{10.4}$$

$$(\bar D \cap \operatorname{supp} b_k^j) + t a_k^j \subseteq D \qquad \text{for all } 0<t\le 1. \tag{10.5}$$

Now let f be some continuous $(0,r)$-form on $\bar D$ which is $\bar\partial$-closed in D, where $n-q-1\le r\le n$.

Take a canonical Leray map v for $[U,\varrho,\varphi,D]$ and set

$$f_k^j = b_k^j f + T^v(\bar\partial b_k^j \wedge f)$$

for $k=1,2,\dots$ and $j=0,1,\dots,N(k)$. Then, by Theorem 9.1,

$$f_k^j \in C^0_{0,r}(\bar D \setminus \{z\in U:\ \varphi(z)=0\}) \tag{10.6}$$

and, by (10.3), we have

$$f = \sum_{j=0}^{N(k)} f_k^j \qquad \text{on } \overline{D\cap U_k}. \tag{10.7}$$

The piecewise Cauchy-Fantappie formula (Theorem 3.12) yields the relation

$$f_k^j = L^v(b_k^j f) + \bar\partial T^v(b_k^j f) \qquad \text{in } D. \tag{10.8}$$

By (10.4).

$$L^v(b_k^j f) = \int_{\varrho(x)=0} b_k^j(x) f(x)\wedge \tilde L_r(\cdot,x), \tag{10.9}$$

where the kernel $\tilde L_r(z,x)$ is of the form

$$\tilde L_r(z,x) = c\,\frac{\det(w(z,x),\overbrace{\bar\partial_z w(z,x)}^{r},\overbrace{\bar\partial_x w(z,x)}^{n-r-1})}{\langle w(z,x),x-z\rangle^n}$$

for some $w \in \mathrm{Div}(\varrho)$ and some constant c (cf. Sect. 3.5 and the proof of Proposition 0.9). In view of (7.4) and (10.4), we can find open sets W_k^j ($j=0,\dots,N(k)$; $k=1,2,\dots$) such that

$$D \cup \left(\{z \in U: \varphi(x)<0, \varrho(z)=0\} \setminus \operatorname{supp} b_k^j\right) \subseteq W_k^j \subseteq U$$

and

$$\langle w(z,x), x-z \rangle \neq 0 \qquad \text{if } x \in \partial D \cap \operatorname{supp} b_k^j \text{ and } z \in W_k^j. \tag{10.10}$$

Therefore, the right hand side of (10.9) defines a C^{∞} form g_k^j on W_k^j with

$$g_k^j = L^v(b_k^j f) \qquad \text{on } D. \tag{10.11}$$

Since $w(z,x)$ depends holomorphically on $z_1,\dots,z_{q+1}$ and since $r \geq n-q-1$, it follows that

$$g_k^j \in Z_{0,r}^0(W_k^j) \tag{10.12}$$

(observe that $g_k^j=0$ if $r \geq n-q$). Moreover, by (10.10) and the definition of T^v we obtain continuous $(0,r-1)$-forms h_k^j on W_k^j which are C^{∞} in $W_k^j \setminus (D \cap \operatorname{supp} b_k^j)$ such that

$$h_k^j = T^v(b_k^j f) \qquad \text{on } D. \tag{10.13}$$

Since h_k^j is C^{∞} in $W_k^j \setminus (D \cap \operatorname{supp} b_k^j)$ and by (10.8), one obtains that

$$\bar{\partial} h_k^j \in Z_{0,r}^0(W_k^j). \tag{10.14}$$

By (10.8), (10.11) and (10.13), we have the relation $f_k^j = g_k^j + \bar{\partial} h_k^j$ on D. Therefore, taking into account (10.12), (10.14), and (10.5), we can find neighborhoods V_k^j of $\bar{D}$ and numbers $c_k^j > 0$ ($j=1,\dots,N(k)$; $k=1,2,\dots$) such such that, by setting

$$F_k^j(z) = (g_k^j + \bar{\partial} h_k^j)(z + a_k^j v_k^j),$$

$z \in V_k^j$, forms $F_k^j \in Z_{0,r}^0(V_k^j)$ with

$$\|F_k^j - f_k^j\|_{0,D_\varepsilon} \leq \frac{1}{k} \tag{10.15}$$

are defined.

By (10.1) and (10.2), it follows from estimate (9.2) in Theorem 9.1 that

$$\lim_{k \to \infty} \|T^v(\bar\partial b_k^0 f)\|_{0,D_\varepsilon} = 0. \tag{10.16}$$

Since f is continuous at 0, without loss of generality we can assume that, moreover, f(0)=0. Then it is clear that also $\lim \|b_k^0 f\|_{0,D_\varepsilon} = 0$ for $k \to \infty$. Together with (10.16) this implies that

$$\lim_{k \to \infty} \|f_k^0\|_{0,D_\varepsilon} = 0. \tag{10.17}$$

Setting $f_k = \sum_{j=0}^{N(k)} F_k^j$, now we obtain a sequence f of $\bar\partial$-closed (0,r)-forms in some neighborhoods V_k of $\bar D$. In view of (10.7), it follows from (10.15) and (10.17) that $\lim \|f_k - f\|_{0,D_\varepsilon} = 0$ for $k \to \infty$. Without loss of generality we may assume that ε is so small that $[U,\varrho, \varphi-\varepsilon, D]$ is also a q-convex configuration. If r=n-q-1, then the proof can be completed by applying Theorem 8.1 to $[U,\varrho, \varphi-\varepsilon, D]$ and the forms f_k. If $r \geq n-q$, then we proceed as follows: For all k, we take a C^∞ function χ_k on $\mathbb{C}^n$ with supp $\chi_k \subset\subset V_k$ and $\chi_k = 1$ on $\bar D$. By Theorem 7.8, after shrinking V_k, we may assume that $f_k = \bar\partial u_k$ for some $u_k \in C^0_{0,r-1}(V_k)$. Therefore, $\tilde f_k := \bar\partial(\chi_k u_k)$ is a continuous $\bar\partial$-closed (0,r)-form on $\mathbb{C}^n$ with $\tilde f_k = f_k$ on $\bar D$. □

11. Finiteness of the Dolbeault cohomology of order r with uniform estimates on strictly q-convex domains with r≥n-q

11.1. Definition. Let $D \subset\subset X$ be a domain in an n-dimensional complex manifold X. Suppose E is a holomorphic vector bundle over X or, more generally, E is a $C^\alpha\mathcal{O}$ vector bundle over $\bar D$ (for the notion of a $C^\alpha\mathcal{O}$ vector bundle, see Sect. 0.12), where $0 \leq \alpha < 1$. Let $r \in \{1,\dots,n\}$.

We say the operator $\bar\partial$ is α-regular at $Z^0_{0,r}(\bar D,E)$ if there exists a linear operator

$$T: E^{\alpha \to 0}_{0,r}(\bar D,E) \longrightarrow C^\alpha_{0,r-1}(\bar D,E)$$

such that the following conditions are fulfilled:

(i) T is bounded as an operator acting from the normed space $E^{\alpha\to 0}_{0,r}(\bar{D},E)$ endowed with norm $\|\cdot\|_{0,D}$ into the Banach space $C^{\alpha}_{0,r-1}(\bar{D},E)$ endowed with the norm $\|\cdot\|_{\alpha,D}$ (for the notations, cf. Sect. 0.11);

(ii) $\bar\partial Tf = f$ for all $f \in E^{\alpha\to 0}_{0,r}(\bar{D},E)$;

(iii) T is compact as an operator with values in the Banach space $C^{0}_{0,r-1}(\bar{D},E)$ (endowed with the norm $\|\cdot\|_{0,D}$).

Remarks to this definition.

I. For $\alpha>0$, condition (iii) follows from (i), because then the embedding of $C^{\alpha}_{0,r-1}(\bar{D},E)$ into $C^{0}_{0,r-1}(\bar{D},E)$ is compact (Ascoli's theorem).

II. If $\bar\partial$ is α-regular at $Z^{0}_{0,r}(\bar{D},E)$, then $E^{\alpha\to 0}_{0,r}(\bar{D},E)$ is a closed subspace of the Banach space $Z^{0}_{0,r}(\bar{D},E)$ (with the norm $\|\cdot\|_{0,D}$). This follows from the fact that $\bar\partial$ is closed as an operator between the Banach spaces $C^{\alpha}_{0,r-1}(\bar{D},E)$ and $Z^{0}_{0,r}(\bar{D},E)$.

III. Simple arguments show (cf., for instance, the proof of Proposition 3 in Appendix 2 of [H/L]) that condition (ii) is equivalent to the following one:

(ii') There exists a compact linear operator K from $E^{\alpha\to 0}_{0,r}(\bar{D},E)$ into $C^{0}_{0,r}(\bar{D},E)$ such that $\bar\partial T - K$ is the identity operator on $E^{\alpha\to 0}_{0,r}(\bar{D},E)$.

IV. If $\bar\partial$ is α-regular at $Z^{0}_{0,r}(\bar{D},E)$ and, moreover,

$$\dim[Z^{0}_{0,r}(\bar{D},E)/E^{\alpha}_{0,r}(\bar{D},E)] < \infty,$$

then $\bar\partial$ is also β-regular at $Z^{0}_{0,r}(\bar{D},E)$ for all $0\le\beta\le\alpha$. This follows from Remark III and the relation $E^{\alpha\to 0}_{0,r}(\bar{D},E) \subseteq E^{\beta\to 0}_{0,r}(\bar{D},E) \subseteq E^{0}_{0,r}(\bar{D},E)$.

11.2. Theorem. Let $D \subset\subset X$ be a non-degenerate strictly q-convex domain in an n-dimensional complex manifold X $(0\le q\le n-1)$, and E a holomorphic vector bundle over X. Then, for all $n-q\le r\le n$, the operator $\bar\partial$ is 1/2-regular at $Z^{0}_{0,r}(\bar{D},E)$, and

$$\dim H^{0,r}_{1/2\to 0}(\bar{D},E) < \infty, \tag{11.1}$$

where

$$H^{0,r}_{1/2\to 0}(D,E) := Z^{0}_{0,r}(\bar{D},E)/E^{1/2\to 0}_{0,r}(\bar{D},E).$$

Moreover, then there exist bounded linear operators

$$T_r : C^0_{0,r}(\bar{D},E) \longrightarrow C^{1/2}_{0,r-1}(\bar{D},E),$$

$$K_r : C^0_{0,r}(\bar{D},E) \longrightarrow C^{1/2}_{0,r}(\bar{D},E)$$

($r=n-q,\dots,n$) such that, for any continuous E-valued $(0,r)$-form f on $\bar{D}$ with $n-q\le r\le n$ and with the property that $\bar{\partial}f$ is also continuous on $\bar{D}$, we have the relation (here $T_{n+1} := 0$)

$$\bar{\partial}T_r f + T_{r+1}\bar{\partial}f = f + K_r f \quad \text{on } D. \tag{11.2}$$

Sometimes the following generalization of this theorem to $C^{\alpha}\mathcal{O}$ vector bundles is useful (see Sect. 0.12 for the definition of such bundles):

<u>11.2'. Theorem.</u> Let $D \subset\subset X$ be a non-degenerate strictly q-convex domain in an n-dimensional complex manifold X ($0\le q\le n-1$), and E a $C^{\alpha}\mathcal{O}$ vector bundle over $\bar{D}$, where $0\le\alpha\le 1/2$. Then, for all $n-q\le r\le n$, the operator $\bar{\partial}$ is α-regular at $Z^0_{0,r}(\bar{D},E)$, and

$$\dim\,[Z^0_{0,r}(\bar{D},E)/E^{\alpha\to 0}_{0,r}(\bar{D},E)] < \infty. \tag{11.3}$$

Moreover, then there exist bounded linear operators

$$T_r : C^0_{0,r}(\bar{D},E) \longrightarrow C^{\alpha}_{0,r-1}(\bar{D},E),$$

$$K_r : C^0_{0,r}(\bar{D},E) \longrightarrow C^{\alpha}_{0,r}(\bar{D},E)$$

($r=n-q,\dots,n$) which are even compact if $\alpha<1/2$ such that, for any continuous E-valued $(0,r)$-form f on $\bar{D}$ with $n-q\le r\le n$ and with the property that $\bar{\partial}f$ is also continuous on $\bar{D}$, we have (here $T_{n+1} := 0$)

$$\bar{\partial}T_r f + T_{r+1}\bar{\partial}f = f + K_r f \quad \text{on } D. \tag{11.4}$$

<u>Proof of Theorems 11.2 and 11.2'</u> (cf. the proof of the Kodaira finiteness Theorem 1.15). Since Theorem 11.2 is a special case of Theorem 11.2', we have to prove only Theorem 11.2'. (Notice that a direct proof of Theorem 11.2 would not give any simplifications.)

So let the hypotheses of Theorem 11.2' be fulfilled.

In view of Lemma 7.5, Theorem 7.8 and Theorem 9.1, we can find open sets $U_j \subset\subset X$ $(j=1,\dots,N<\infty)$ with $\partial D \subseteq U_1\cup\dots\cup U_N$, and bounded linear operators T_r^j from $C^0_{0,r}(D,E)$ into $C^{\alpha}_{0,r-1}(\overline{U_j\cap D},E)$ $(r=1,\dots n)$ which are even compact if $\alpha<1/2$ (since then the embeddings $C^{1/2}_{0,r-1}(\overline{U_j\cap D})\longrightarrow C^{\alpha}_{0,r-1}(\overline{U_j\cap D})$ are compact) such that, for any continuous E-valued $(0,r)$-form f on $\bar{D}$ with $n-q\le r\le n$ and with the property that $\bar{\partial}f$ is also continuous on $\bar{D}$, one has

$$\bar{\partial}T_r^j f + T_{r+1}^j\bar{\partial}f = f \qquad \text{on } \overline{U_j\cap D},$$

where $T_{n+1}:=0$.

By Lemma 1.16, we can find open sets $U_j \subset\subset D$ $(j=N+1,\dots,M<\infty)$ with $D\setminus(U_1\cup\dots\cup U_N) \subseteq U_{N+1}\cup\dots\cup U_M$, as well as compact linear operators T_r^j from $C^0_{0,r}(\bar{D},E)$ into $C^{1/2}_{0,r-1}(\bar{U}_j,E)$ $(r=1,\dots,n)$ and K_r^j from $C^0_{0,r}(\bar{D},E)$ into $C^{1/2}_{0,r}(\bar{U}_j,E)$ $(r=1,\dots,n-1)$ such that, for any continuous E-valued $(0,r)$-form f on $\bar{D}$ with the property that $\bar{\partial}f$ is also continuous on $\bar{D}$,

$$\bar{\partial}T_r^j f + T_{r+1}^j\bar{\partial}f = f + K_r^j f \qquad \text{on } \bar{U}_j,$$

where $K_n^j:=0$ and $T_{n+1}^j:=0$.

Now we take a C^{∞} partition of unity $\{\chi_j\}_{j=1}^M$ subordinated to the covering $\{U_j\}_{j=1}^M$, and define

$$T_r = \sum_{j=1}^{M} \chi_j T^j \qquad \text{and} \qquad K_r = \sum_{j=1}^{M} \bar{\partial}\chi_j\wedge T_r^j + \sum_{j=1}^{M} \chi_j K_r^j$$

for $r=n-q,\dots,n$. It is clear that these operators have the required properties. It remains to prove that, for $n-q\le r\le n$, $\bar{\partial}$ is α-regular at $Z^0_{0,r}(\bar{D},E)$, and (11.3) holds true.

It follows from (11.2) that

$$\bar{\partial}\circ T_r = \text{id} + K_r \qquad \text{on } Z^0_{0,r}(\bar{D},E) \tag{11.5}$$

(id:= identity operator). Since K is compact as an operator with values in $C^0_{0,r}(\bar{D},E)$, by Remark III following Definition 11.1, this implies that $\bar{\partial}$ is α-regular at $Z^0_{0,r}(\bar{D},E)$. Further, this compactness of K_r and relation (11.5) yield that the restriction of $\bar{\partial}\circ T_r$ to $Z^0_{0,r}(\bar{D},E)$ is a Fredholm operator. Since, obviously, the image of this operator is contained in $E^{\alpha\longrightarrow 0}_{0,r}(\bar{D},E)$, it follows (11.3). □

12. Global uniform approximation of $\bar\partial$-closed (0,n-q-1)-forms and invariance of the Dolbeault cohomology of order $\geq$n-q with respect to q-convex extensions, and the Andreotti-Grauert finiteness theorem for the Dolbeault cohomology of order $\geq$n-q on q-convex manifolds

12.1. Definition. Let X be an n-dimensional complex manifold, and $0\leq q\leq n-1$ an integer.

a) If $D \subseteq X$ is a domain, then we say X is a q-convex extension of D if the following two conditions are fulfilled:

(i) ∂D is compact.

(ii) There exists a (q+1)-convex function $\varrho:U \longrightarrow \mathbb{R}$ in some open set $U \subseteq X$ with $X\setminus D \subseteq U$ such that

$$D\cap U = \{z\in U:\ \varrho(z)<0\}$$

and, for any number

$$0 < \alpha < \sup_{z\in U} \varrho(z),$$

the set $\{z\in U:\ 0\leq\varrho(z)\leq\alpha\}$ is compact.

If, in this case, the function ϱ in condition (ii) can be chosen without degenerate critical points in U, then X will be called a non-degenerate q-convex extension of D.

b) If $D \subseteq G \subseteq X$ are two domains, then we say G is a strictly q-convex extension of D in X if G is a q-convex extension of D such that, moreover, the set $G\setminus D$ is relatively compact in X and the following strengthening of condition (ii) is fulfilled:

(ii)' There exists a (q+1)-convex function $\varrho:U \longrightarrow \mathbb{R}$ in some neighborhood $U \subseteq X$ of $\overline{G\setminus D}$ such that

$$D\cap U = \{z\in U:\ \varrho(z)<0\} \qquad \text{and} \qquad G\cap U = \{z\in U:\ \varrho(z)<1\}.$$

If, in this case, the function ϱ in condition (ii)' can be chosen without degenerate critical points in U, then G will be called a non-degenerate strictly q-convex extension of D in X.

c) Instead of "(n-1)-convex extension" we shall say also "pseudoconvex extension"

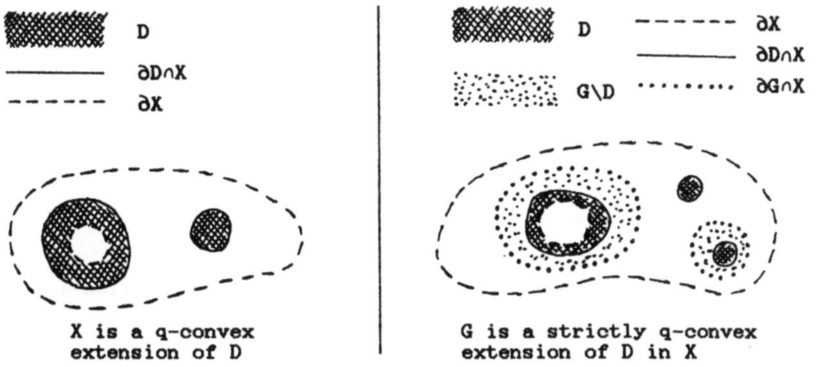

FIGURE 2

Remarks to this definition. In the following remarks, $D \subseteq X$ is a domain in an n-dimensional complex manifold X such that X is a q-convex extension of D ($0 \le q \le n-1$), and $\varrho : U \longrightarrow \mathbb{R}$ is as in condition (ii) in part a) of this definition.

I. For all numbers

$$0 \le \alpha < \beta < \sup_{x \in U} \varrho(x),$$

the domain $\{z \in U: \varrho(z) < \beta\}$ is a strictly q-convex extension in X of the domain $\{z \in U: \varrho(z) < \alpha\}$.

II. D and X need not be q-convex. For instance, if $X = \{z \in \mathbb{C}^n: 1 < |z| < \infty\}$ and $D = \{z \in \mathbb{C}^n: 1 < |z| < 2\}$, then X is a strictly pseudoconvex extension of D, although X and D are not pseudoconvex if $n \ge 2$.

III. Since ∂D is compact, there exists only a finite number of connected components of X which have a non-empty intersection with both D and X\D. If X_0 is some connected component of X such that $X_0 \subseteq X \setminus D$, then the restriction of ϱ to X_0 is a (q+1)-convex exhausting function for X_0. Therefore, then X_0 is completely q-convex.

IV. It is admitted that $\partial D = \emptyset$ or $D = \emptyset$. If $D = \emptyset$, then X is completely q-convex. If $\partial D = \emptyset$, then X\D is completely q-convex.

12.2. Definition. Let X be an n-dimensional complex manifold and $0 \le q \le n-1$ an integer. If A_1, A_2, $V \subseteq X$ are domains in X, then we say $[A_1, A_2, V]$ is a q-convex extension element in X if the following conditions are fulfilled (cf. Def. 7.4 for the definition of a q-convex configuration):

(i) $A_1 \subseteq A_2$.

(ii) $A_2 \setminus A_1 \subset\subset V \subset\subset X$.

(iii) There exist q-convex configurations $[U_j, \varrho_j, \varphi_j, D_j]$ in $\mathbb{C}^n$ ($j=1,2$) with $U:=U_1=U_2$ and $\varphi:=\varphi_1=\varphi_2$ such that, for some biholomorphic map h from U onto a neighborhood of $\bar{V}$,

$$V = h(\{z\in U: \varphi(z)<0\}) \qquad \text{and} \qquad V\cap A_j = h(D_j) \text{ for } j=1,2.$$

If, in this case, the type of $[U, \varrho_j, \varphi, D_j]$ is T_j, then $[A_1, A_2, V]$ will be called of type (T_1, T_2).

A q-convex extension element $[A_1, A_2, V]$ in X will be called <u>smooth</u> if it is of a type (T_1, T_2) with $T_1 \neq$ II and $T_2 \neq$ II, otherwise it will be called <u>non-smooth</u>.

If $A_1 \subseteq A_2$ are domains in X, then we say A_2 <u>can be obtained from A_1 by means of a q-convex extension element in X</u> if there exists a domain $V \subseteq X$ such that $[A_1, A_2, V]$ is a q-convex extension element in X.

Instead of "(n-1)-convex extension element" we shall say also "<u>pseudoconvex extension element</u>".

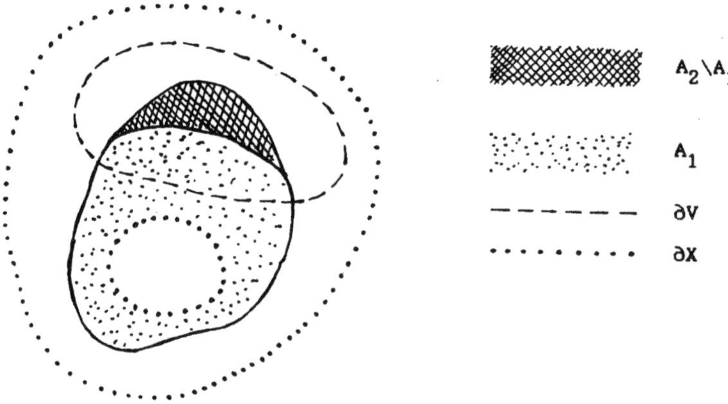

FIGURE 3: q-convex extension element $[A_1, A_2, V]$ in X

<u>Remark to this definition.</u> If $[U, \varrho, \varphi]$ defines a q-convex configuration in $\mathbb{C}^n$ ($0\le q\le n-1$), then for any sufficiently small $\varepsilon>0$, $[U, \varrho, \varphi-\varepsilon]$ and $[U, \varrho, \varphi+\varepsilon]$ also define qconvex configurations in $\mathbb{C}^n$. This yields the following

<u>Proposition.</u> If $[A_1, A_2, V]$ is a q-convex extension element in an n-dimensional complex manifold X ($0\le q\le n-1$), then:

(i) For any neighborhood $W \subseteq X$ of $\bar{V}$, there exists a domain V_+ with $V \subset\subset V_+ \subset\subset W$ such that $[A_1, A_2, V_+]$ is also a q-convex extension element in X.

(ii) For any compact set $K \subset\subset V$, there exists a domain V_- with $K \subset\subset V_- \subset\subset V$ such that $[A_1, A_2, V_-]$ is also a q-convex extension element in X.

12.3. Lemma. Let X be an n-dimensional complex manifold, let $D \subsetneq G \subsetneq X$ be two domains such that G is a non-degenerate strictly q-convex extension of D in X $(0 \le q \le n-1)$, and suppose that the following strengthening of condition (ii)' in Definition 12.1 is fulfilled:

(ii)" There exists a (q+1)-convex function ϱ without degenerate critical points in a neighborhood U of $\overline{G \setminus D}$ such that $D \cap U = \{z \in U: \varrho(z) < 0\}$, $G \cap U = \{z \in U: \varrho(z) < 1\}$, and ϱ does not have local minima in U.

Then there exists a finite number of domains

$$D = A_0 \subsetneq A_1 \subsetneq \dots \subsetneq A_N = G$$

such that, for all $1 \le j \le N$, A_j can be obtained from A_{j-1} by means of a q-convex extension element in X.

Moreover, for any open covering $\{U_i\}_{i \in I}$ of X, these domains A_j can be chosen so that, in addition, for each $1 \le j \le N$, there are an index $i \in I$ and a q-convex extension element $[B_1, B_2, V]$ in X with $B_1 = A_{j-1}$, $B_2 = A_j$, and $V \subset\subset U_i$.

Proof. It is sufficient to prove the following

Statement. Let $\varrho: X \longrightarrow \mathbb{R}$ be a (q+1)-convex function without degenerate critical points on an n-dimensional complex manifold X $(0 \le q \le n-1)$. Set $D_\alpha := \{z \in X: \varrho(z) < \alpha\}$ for $-\infty < \alpha < +\infty$, and suppose that ∂D_0 is compact. Then we can find a number $\varepsilon > 0$ with the following property: For all α, β with $-\varepsilon \le \alpha \le 0 \le \beta \le \varepsilon$, there exists a finite number of domains

$$D_\alpha = A_0 \subseteq A_1 \subseteq \dots \subseteq A_N = D_\beta$$

such that, for each $1 \le j \le N$, there is a q-convex extension element $[B_1, B_2, V]$ in X with $B_1 = A_{j-1}$, $B_2 = A_j$, and $V \subset\subset U_i$ for some $i \in I$.

Proof of the statement. Since ∂D_0 is compact, and non-degenerate critical points are isolated, there exists not more than a finite number of critical points of ϱ which lie on ∂D_0. Denote these points by $y_1, \dots, y_M$ (if exist - otherwise set M=0). By Lemma 7.5, for any $1 \le j \le M$, we can find holomorphic coordinates $h_j: V_j \longrightarrow \mathbb{C}^n$ in some neighborhood V_j of y_j such that $h_j(V_j)$ is the open unit ball in $\mathbb{C}^n$, $h_j(y_j) = 0$ and, for each $0 < r < 1$, $[h_j(V_j), \varrho \circ h_j^{-1}, |z|^2 - r^2]$ defines a q-convex configuration of type II. Choose these neighborhoods V_j so small that, moreover, $V_j \subset\subset U_i$ for some $i = i(j) \in I$, and

$$V_j \cap V_k = \emptyset \qquad \text{if } j \neq k. \tag{12.1}$$

We set

$$W_j^a = h_j^{-1}(\{z\in\mathbb{C}^n:\ |z|<\frac{1}{a+1}\})\qquad \text{for } j=1,\dots,M \text{ and } a=1,2.$$

By Lemma 7.5, for any point $y\in\partial D_0\setminus(W_1^1\cup\dots\cup W_M^1)$, we can find holomorphic coordinates h_y in a neighborhood V_y of y such that $h_y(V_y)$ is the unit ball in C^n, $h_y(y)=0$, and, for all $0<r<1$, $[h_y(V_y),\varrho\circ h_y^{-1},|z|^2-r^2]$ defines a q-convex configuration of type I. We chose these neighborhoods V_y so small that, moreover,

$$V_y \subseteq \partial D_0\setminus(W_1^2\cup\dots\cup W_M^2) \tag{12.2}$$

and $V_y \subset\subset U_i$ for some $i=i(y)\in I$. Set $W_y^1 = h_y^{-1}(\{z\in\mathbb{C}^n:\ |z|<1/2\})$. Since $\partial D_0\setminus(W_1^1\cup\dots\cup W_M^1)$ is compact, we can find points $y_{M+1},\dots,y_N$ in $\partial D_0\setminus(W_1^1\cup\dots\cup W_M^1)$ $(N<\infty)$ such that

$$\partial D_0\setminus(W_1^1\cup\dots\cup W_M^1) \subseteq W_{y_{M+1}}^1\cup\dots\cup W_{y_N}^1.$$

Set $W_j^1 = W_{y_j}^1$ for $j=M+1,\dots,N$. Then we can find $\varepsilon>0$ with

$$D_\varepsilon\setminus D_{-\varepsilon}\subset\subset\bigcup_{j=1}^{N} W_j^1.$$

Choose C^∞ functions χ_j on X $(j=1,\dots,N)$ such that $\chi_1+\dots+\chi_N=1$ on $D_\varepsilon\setminus D_{-\varepsilon}$ and, for any $1\le j\le N$, $\chi_j=0$ in $X\setminus W_j^1$. Then, by (12.1) and (12.2), $\chi_j=1$ on W_j^2 if $1\le j\le M$.

Now let some numbers α,β with $-\varepsilon\le\alpha\le 0\le\beta\le\varepsilon$ be given. Set

$$A_k = \{z\in X:\ \varrho(z)-\alpha<(\beta-\alpha)\sum_{j=1}^{k}\chi_j(z)\}$$

for $k=0,\dots,N$. Then it is clear that $D_\alpha = A_0\subseteq A_1\subseteq\dots\subseteq A_N = D_\beta$, and it is easy to see that all $[A_{j-1},A_j,W_j^1]$ $(j=1,\dots,N)$ are q-convex extension elements in X if ε is sufficiently small. □

Remark to Lemma 12.3. Let us choose the number ε in the proof of this Lemma so small that, moreover, $d\varrho(z)\neq 0$ for all $z\in(\overline{D_\varepsilon\setminus D_{-\varepsilon}})\setminus\{y_1,\dots,y_M\}$. Then the q-convex extension elements $[A_{j-1},A_j,W_j^1]$ constructed in this proof are of type (I,I) if $j=M+1,\dots,N$, and of type (IV,III), (II,III) or (IV,II) according as $\alpha<0<\beta$, $\alpha=0<\beta$ or $\alpha<0=\beta$ if $j=1,\ldots,M$. In particular, if $\alpha<0<\beta$, then all these elements are of one of the types (I,I) or (IV,III). So it is proved that the domains A_j in Lemma 12.3 can be chosen so that, moreover, for each $1\le j\le N$, A_j can be obtained from A_{j-1} by means of a smooth q-convex extension element in X.

12.4. Lemma. Let $[A_1,A_2,V]$ be a q-convex extension element in an n-dimensional complex manifold X ($0\le q\le n-1$), and E a holomorphic vector bundle over X which is holomorphically trivial over some neighborhood $\tilde{V}$ of $\bar{V}$. Then the following assertions hold true:

(i) If $n-q\le r\le n$ and U is a neighborhood of $\overline{A_2\setminus A_1}$, then, for any $f_1\in Z^0_{0,r}(\bar{A}_1,E)$, there exists $f_2\in Z^0_{0,r}(\bar{A}_2,E)$ with $f_1-f_2\ E^{1/2\longrightarrow 0}_{0,r}(\bar{A}_1,E)$ and $f_1 = f_2$ on $\bar{A}_1\setminus U$.

(ii) If A_1 is relatively compact in X, then for each $n-q\le r\le n$, the following statement holds true: If $\bar\partial$ is 1/2-regular at $Z^0_{0,r}(\bar{A}_1,E)$, then $\bar\partial$ is 1/2-regular also at $Z^0_{0,r}(\bar{A}_2,E)$ (for the notion of α-reularity, see Definition 11.1).

(iii) Let $n-q\le r\le n$, where, in the case $r=n-q$, we suppose that A_1 is relatively compact in X and $\bar\partial$ is 1/2-regular at $Z^0_{0,n-q}(\bar{A}_1,E)$. Then

$$E^{1/2\longrightarrow 0}_{0,r}(\bar{A}_2,E) = Z^0_{0,r}(\bar{A}_2,E)\cap E^{1/2\longrightarrow 0}_{0,r}(\bar{A}_1,E).$$

(iv) Let $n-q-1\le r\le n$, and suppose that A_1 is relatively compact in X and (if $r\neq n$) $\bar\partial$ is 1/2-regular at $Z^0_{0,r+1}(\bar{A}_1,E)$. Then, for any $f\in Z^{1/2}_{0,r}(\bar{A}_1,E)$, there exists a sequence $f_k\in Z^{1/2}_{0,r}(\bar{A}_2,E)$ with

$$\lim_k \| f - f_k\|_{0,\bar{A}_1} = 0.$$

We mention also the following generalization of this lemma (for the definition of $C^{\alpha}\mathcal{O}$ vector bundles, see Sect. 0.12):

12.4'. Lemma. Let $[A_1,A_2,V]$ be a q-convex extension element in an n-dimensional complex manifold X ($0\le q\le n-1$), and E a $C^{\alpha}\mathcal{O}$ vector bundle over $\bar{A}_2$ which is $C^{\alpha}\mathcal{O}$ trivial over $\bar{A}_2\cap\tilde{V}$ for some neighborhood $\tilde{V}$ of $\bar{V}$ ($0\le\alpha\le 1/2$). Then the following assertions hold true:

(i) If $n-q\le r\le n$ and U is a neighborhood of $\overline{A_2\setminus A_1}$, then, for any $f_1\in Z^0_{0,r}(\bar{A}_1,E)$, there exists $f_2\in Z^0_{0,r}(\bar{A}_2,E)$ with $f_1-f_2\in E^{\alpha\longrightarrow 0}_{0,r}(\bar{A}_1,E)$ and $f_1 = f_2$ on $\bar{A}_1\setminus U$.

(ii) If A_1 is relatively compact in X, then for each $n-q\le r\le n$, the following statement holds true: If $\bar\partial$ is α-regular at $Z^0_{0,r}(\bar{A}_1,E)$, then

$\bar\partial$ is α-regular also at $Z^0_{0,r}(\bar A_2,E)$.

(iii) Let $n-q\le r\le n$, where, in the case $r=n-q$, we suppose that A_1 is relatively compact in X and $\bar\partial$ is α-regular at $Z^0_{0,n-q}(\bar A_1,E)$. Then

$$E^{\alpha\longrightarrow 0}_{0,r}(\bar A_2,E) = Z^0_{0,r}(\bar A_2,E)\cap E^{\alpha\longrightarrow 0}_{0,r}(\bar A_1,E).$$

(iv) Let $n-q-1\le r\le n$, and suppose that A_1 is relatively compact in X and (if $r=n$) $\bar\partial$ is α-regular at $Z^0_{0,r+1}(\bar A_1,E)$. Then, for any $f\in Z^{\alpha}_{0,r}(\bar A_1,E)$, there exists a sequence $f_k\in Z^{\alpha}_{0,r}(\bar A_2,E)$ with

$$\lim_k \|f - f_k\|_{0,\bar A_1} = 0. \qquad \square$$

In the proof of Lemmas 12.4 and 12.4' we use the following corollary which follows immediately from Theorems 7.8, 9.1 and 10.1:

<u>12.5. Corollary to Theorems 7.8, 9.1 and 10.1.</u> For any q-convex extension element $[A_1,A_2,V]$ in an n-dimensional complex manifold X ($0\le q\le n-1$), the following two assertions hold true:

(i) If $n-q\le r\le n$ and $j=1,2$, then there exists a linear operator

$$T^j_r\colon Z^0_{0,r}(\overline{A_j\cap V}) \longrightarrow C^{1/2}_{0,r-1}(\bar A_j\cap V)$$

which is continuous (with respect to the Banach space topology of $Z^0_{0,r}(\overline{A_j\cap V})$ and the Fréchet space topology of $C^{1/2}_{0,r-1}(\bar A_j\cap V)$ and such that, for all $f\in Z^0_{0,r}(\overline{A_j\cap V})$, $\bar\partial T^j_r f = f$ on $\bar A_j\cap V$.

(ii) If $n-q-1\le r\le n$ and $j=1,2$, then the image of the restriction map

$$Z^0_{0,r}(V) \longrightarrow Z^0_{0,r}(\bar A_j\cap V)$$

is dense in the Fréchet space $Z^0_{0,r}(\bar A_j\cap V)$. $\square$

<u>Proof of Lemmas 12.4 and 12.4'.</u> It is sufficient to prove Lemma 12.4'.

<u>Proof of part (i).</u> Since $\alpha\le 1/2$ and E is $C^{\alpha}\mathcal{O}$ trivial over $\bar A_2\cap\tilde V$, it follows from Corollary 12.5 (i) that $f_1|\bar A_1\cap V = \bar\partial u$ for some $u\in C^{\alpha}_{0,r-1}(\bar A_1\cap V,E)$. Choose a C^∞ function χ on X such that $\chi = 1$ on $\overline{A_2\setminus A_1}$ and supp $\chi \subset\subset U$. Set $f_2 = f_1 - \bar\partial(\chi u)$.

Proof of part (ii). Since $\bar\partial$ is α-regular at $Z^0_{0,r}(\bar A_1,E)$, we have a bounded linear operator

$$T_1 : E^{\alpha\to 0}_{0,r}(\bar A_1,E) \longrightarrow C^{\alpha}_{0,r-1}(\bar A_1,E)$$

which is compact as an operator with values in $C^0_{0,r-1}(\bar A_1,E)$ and such that $\bar\partial\circ T_1$ is the identity operator. Choose a neighborhood V' of $\overline{A_2\setminus A_1}$ with $V' \subset\subset V$. Then from Corollary 12.5 (i) we get a continuous linear operator

$$T_2 : Z^0_{0,r}(\overline{A_2\cap V},E) \longrightarrow C^{\alpha}_{0,r-1}(\overline{A_2\cap V'},E)$$

which is compact as an operator with values in $C^0_{0,r-1}(\overline{A_2\cap V'},E)$ and such that $\bar\partial\circ T_2$ is the restriction operator to $\overline{A_2\cap V'}$.

Denote by B the subspace of all $f\in Z^0_{0,r}(\bar A_2,E)$ with $f|\bar A_1\in E^{\alpha\to 0}_{0,r}(\bar A_1,E)$. Choose a C^{∞} function χ on X such that supp $\chi \subset\subset V'$ and $\chi = 1$ in a neighborhood of $\overline{A_2\setminus A_1}$. Define by $T_B = \chi T_2 + (1-\chi)T_1$ a bounded linear operator T_B from B into $C^{\alpha}_{0,r-1}(A_2,E)$. Then T_B is compact as an operator with values in $C^0_{0,r-1}(\bar A_2,E)$, and $\bar\partial\circ T_B - \bar\partial\chi\wedge(T_2+T_1)$ is the identity operator. Since $\bar\partial\chi\wedge(T_2+T_1)$ is compact in B and $\bar\partial\circ T_B(B)$ is contained in $E^{\alpha\to 0}_{0,r}(\bar A_2,B) \subseteq B$, this implies that $\bar\partial$ is α-regular at $Z^0_{0,r}(\bar A_2,E)$ (cf. Remark III following Definition 11.1).

Proof of part (iii). Let $f\in Z^0_{0,r}(\bar A_2,E)$ with $f|\bar A_1\in E^{\alpha\to 0}_{0,r}(\bar A_1,E)$. Then $f|\bar A_1 = \bar\partial u_1$ for some $u_1\in C^{\alpha}_{0,r-1}(\bar A_1,E)$. Further, by Corollary 12.5 (i), we can find $u_2\in C^{\alpha}_{0,r-1}(\bar A_2\cap V,E)$ with $f|\bar A_2\cap V = \bar\partial u_2$. Now we distinguish the cases $n-q+1\le r\le n$ and $r=n-q$.

First let $n-q+1\le r\le n$. Then, by Corollary 12.5 (i), we can find $v\in C^{\alpha}_{0,r-2}(\bar A_1\cap V,E)$ with $u_1-u_2 = \bar\partial v$ on $\bar A_1\cap V$. Choose a C^{∞} function χ on X such that supp $\chi \subset\subset V$ and $\chi = 1$ on $\overline{A_2\setminus A_1}$. Then, on $\bar A_1\cap V$, we have the relation

$$u_1 - \bar\partial(\chi v) = u_2 + \bar\partial((1-\chi)v),$$

and both sides of this equation together define a form $u\in C^{\alpha}_{0,r-1}(\bar A_2,E)$ with $\bar\partial u=f$.

Now let $r=n-q$. Then, by hypothesis, A_1 is relatively compact in X and $\bar\partial$ is α-regular at $Z^0_{0,n-q}(\bar A_1,E)$. By part (ii), this implies that $\bar\partial$ is α-regular also at $Z^0_{0,n-q}(\bar A_2,E)$. Therefore, it is sufficient to find a sequence $w_k \in C^{\alpha}_{0,n-q-1}(\bar{A}_2, E)$ such that the forms $\bar{\partial} w_k$ are continuous on $\bar{A}_2$ and

$$\lim_k \|f - \bar{\partial} w_k\|_{0,\bar{A}_2} = 0.$$

Choose a C^{∞} function χ on X with supp $\chi \subset\subset V$ and $\chi = 1$ in a neighborhood of $\overline{A_2 \setminus A_1}$. Further, by Corollary 12.5 (ii), we can find a sequence $v_k \in Z^0_{0,n-q-1}(\bar{A}_2 \cap V, E)$ with

$$\lim_k \| v_k + u_2 - u_1\|_{0,\bar{A}_1 \cap \operatorname{supp} \chi} = 0.$$

Then the sequence w_k which is defined if we set $w_k = u_1$ on $\bar{A}_1$ and $w_k = (1-\chi)u_1 + \chi(v_k + u_2)$ on $\bar{A}_2 \cap V$ has the required property.

Proof of part (iv). Let $f \in Z^{\alpha}_{0,r}(\bar{A}_1, E)$. Choose a C^{∞} function χ on X with supp $\chi \subset\subset V$ and $\chi = 1$ in a neighborhood of $\bar{A}_2 \setminus \bar{A}_1$. By Corollary 12.5 (ii), we can find a sequence $v_k \in Z^{\alpha}_{0,r}(\bar{A}_2 \cap V, E)$ such that

$$\lim_k \|f - v_k\|_{0,\bar{A}_1 \cap \operatorname{supp} \chi} = 0$$

and hence

$$\lim_k \|\bar{\partial}\chi \wedge (f - v_k)\|_{0,\bar{A}_2} = 0.$$

Since $\bar{\partial}\chi \wedge (f - v_k) = \bar{\partial}(\chi(f - v_k))$ on $\bar{A}_1$ and since $\bar{\partial}$ is α-regular at $Z^0_{0,r+1}(\bar{A}_1, E)$, it follows from part (iii) that the forms $\bar{\partial}\chi \wedge (f - v_k)$ belong to $E^{\alpha \longrightarrow 0}_{0,r+1}(\bar{A}_2, E)$, i.e. $\bar{\partial}\chi \wedge (f - v_k) = \bar{\partial} w_k$ for some $w_k \in C^{\alpha}_{0,r}(\bar{A}_2, E)$. Since, by part (ii), $\bar{\partial}$ is α-regular at $Z^0_{0,r+1}(\bar{A}_2, E)$, this sequence can be chosen so that, moreover, $\lim_k \|w_k\|_{0,\bar{A}_2} = 0$. Setting $f_k = f - \chi(f - v_k) + w_k$ on $\bar{A}_1$ and $f_k = v_k + w_k$ on $\bar{A}_2 \setminus A_1$, now we obtain a sequence $f_k \in Z^{\alpha}_{0,r}(\bar{A}_2, E)$ with $\lim_k \|f - f_k\|_{0,\bar{A}_2} = 0$. □

12.6. Lemma. Let $\varrho : X \longrightarrow \mathbb{R}$ be a (q+1)-convex function without degenerate critical points on an n-dimensional complex manifold X $(0 \le q \le n-1)$, and let $y \in X$ be the point of a local minimum of ϱ. Then we can find a number $\varepsilon > 0$ such that, for all r=1,...,n, the operator $\bar{\partial}$ is 1/2-regular at $Z^0_{0,r}(\bar{D}_\varepsilon)$, and $E^{1/2 \longrightarrow 0}_{0,r}(\bar{D}_\varepsilon) = Z^0_{0,r}(\bar{D}_\varepsilon)$, where $D_\varepsilon := \{z \in X : \varrho(z) < \varrho(y) + \varepsilon\}$.

Proof. Since ϱ has a local minimum at y and since all critical points of ϱ are non-degenerate, we can find a neighborhood U of y such that, with respect to some holomorphic coordinates $h:U \longrightarrow \mathbb{C}^n$, ϱ is strictly convex in U. Choose $\varepsilon>0$ so small that $\bar{D}_\varepsilon \subseteq U$. Then D_ε is strictly convex with respect to these coordinates, and we could conclude the proof by the remark that the assertion of the lemma now follows from Theorem 2.2.2 in [H/L], for instance. However, to be independent of [H/L], let us give also the following arguments: First observe that $\bar\partial$ is 1/2-regular at $Z^0_{0,r}(\bar{D}_\varepsilon)$, by Theorem 11.2. To prove that $E^{1/2\longrightarrow 0}_{0,r}(\bar{D}_\varepsilon) = Z^0_{0,r}(\bar{D}_\varepsilon)$, we consider an arbitrary form $f\in Z^0_{0,r}(\bar{D}_\varepsilon)$. Then, by Theorem 2.12 (Poincaré $\bar\partial$-lemma), $f = \bar\partial u$ on D_ε for some $u\in C^{1/2}_{0,r-1}(D_\varepsilon)$. In particular, then the restriction of f to $D_{\varepsilon/2} := \{z\in X: \varrho(z)<\varrho(y)+\varepsilon/2\}$ belongs to $E^{1/2\longrightarrow 0}_{0,r}(\bar{D}_{\varepsilon/2})$. Since, by Theorem 11.2, $\bar\partial$ is 1/2-regular also at $Z^0_{0,r}(\bar{D}_{\varepsilon/2})$ and ϱ does not have local minima in $D_\varepsilon\setminus D_{\varepsilon/2}$, it follows from Lemma 12.3 and assertions (ii) and (iii) in Lemma 12.4 that f belongs to $E^{1/2\longrightarrow 0}_{0,r}(\bar{D}_\varepsilon)$. □

12.7. Theorem. let $D \subset\subset X$ be a non-degenerate completely strictly q-convex domain in an n-dimensional complex manifold X ($0\le q\le n-1$), and E a holomorphic vector bundle over X. Then, for all $n-q\le r\le n$, $\bar\partial$ is 1/2-regular at $Z^0_{0,r}(\bar{D},E)$, and $E^{1/2\longrightarrow 0}_{0,r}(\bar{D},E) = Z^0_{0,r}(\bar{D},E)$. In other words, then there exist bounded linear operators T_r from $Z^0_{0,r}(\bar{D},E)$ into $C^{1/2}_{0,r-1}(\bar{D},E)$ ($r=n-q,\dots,n$) with $\bar\partial T f = f$ for all $f\in Z^0_{0,r}(\bar{D},E)$.

Notice also the following generalization of this theorem to $C^{\alpha}\mathcal{O}$ vector bundles (cf. Sect. 0.12):

12.7'. Theorem. let $D \subset\subset X$ be a non-degenerate completely strictly q-convex domain in an n-dimensional complex manifold X ($0\le q\le n-1$), and E a $C^{\alpha}\mathcal{O}$ vector bundle over X ($0\le\alpha\le 1/2$). Then, for all $n-q\le r\le n$, $\bar\partial$ is α-regular at $Z^0_{0,r}(\bar{D},E)$, and $E^{\alpha\longrightarrow 0}_{0,r}(\bar{D},E) = Z^0_{0,r}(\bar{D},E)$.

Proof of Theorems 12.7 and 12.7'. This follows from Lemma 12.6, Lemma 12.3 and assertions (ii) and (iii) in Lemma 12.4'. □

12.8. Lemma. Let X be an n-dimensional complex manifold. Let $D \subseteq G \subseteq X$ be domains in X such that G is a non-degenerate strictly q-convex extension of D in X ($0 \leq q \leq n-1$) and $D \neq \emptyset$ [1]. Further, let E be a holomorphic vector bundle over X. Then the following assertions hold true:

(i) Let $n-q \leq r \leq n$. Then, for any $f_D \in Z^0_{0,r}(\bar{D},E)$ and each neighborhood U of $\partial D \cap G$, there exists $f_G \in Z^0_{0,r}(\bar{G},E)$ with $f_D - f_G \in E^{1/2 \longrightarrow 0}_{0,r}(\bar{D},E)$ and $f_G = f_D$ on $D \setminus U$.

(ii) Let $n-q \leq r \leq n$, and suppose D is relatively compact in X, and $\bar{\partial}$ is 1/2-regular at $Z^0_{0,r}(\bar{D},E)$. Then $\bar{\partial}$ is 1/2-regular also at $Z^0_{0,r}(\bar{G},E)$.

(iii) Let $n-q \leq r \leq n$, where, in the case $r=n-q$, we suppose that D is relatively compact in X, and $\bar{\partial}$ is 1/2-regular at $Z^0_{0,n-q}(\bar{D},E)$. Then

$$E^{1/2 \longrightarrow 0}_{0,r}(\bar{G},E) = Z^0_{0,r}(\bar{G},E) \cap E^{1/2 \longrightarrow 0}_{0,r}(\bar{D},E).$$

(iv) Let $n-q-1 \leq r \leq n-1$, and suppose D is relatively compact in X, and $\bar{\partial}$ is 1/2-regular at $Z^0_{0,r+1}(\bar{D},E)$. Then the image of the restriction

$$Z^{1/2}_{0,r}(\bar{G},E) \longrightarrow Z^{1/2}_{0,r}(\bar{D},E)$$

is dense in $Z^{1/2}_{0,r}(\bar{D},E)$ with respect to the norm $\|\cdot\|_{0,D}$.[2]

Notice also the following generalization of this lemma:

<u>12.8'. Lemma.</u> Let X be an n-dimensional complex manifold, and $D \subseteq G \subseteq X$ domains in X such that G is a non-degenerate strictly q-convex extension of D in X ($0 \leq q \leq n-1$) and $D \neq \emptyset$. Further, let E be a $C^{\alpha}\mathcal{O}$ vector bundle over $\bar{G}$ ($0 \leq \alpha < 1/2$). Then the following assertions hold true:

(i) Let $n-q \leq r \leq n$. Then, for any $f_D \in Z^0_{0,r}(\bar{D},E)$ and each neighborhood U of $\partial D \cap G$, there exists $f_G \in Z^0_{0,r}(\bar{G},E)$ with $f_D - f_G \in E^{\alpha \longrightarrow 0}_{0,r}(\bar{D},E)$ and $f_G = f_D$ on $D \setminus U$.

(ii) Let $n-q \leq r \leq n$, and suppose D is relatively compact in X, and $\bar{\partial}$ is α-regular at $Z^0_{0,r}(\bar{D},E)$. Then $\bar{\partial}$ is α-regular also at $Z^0_{0,r}(\bar{G},E)$.

(iii) Let $n-q \leq r \leq n$, where, in the case $r=n-q$, we suppose that D is relatively compact in X, and $\bar{\partial}$ is α-regular at $Z^0_{0,n-q}(\bar{D},E)$. Then

$$E^{\alpha \longrightarrow 0}_{0,r}(\bar{G},E) = Z^0_{0,r}(\bar{G},E) \cap E^{\alpha \longrightarrow 0}_{0,r}(\bar{D},E).$$

(iv) Let $n-q-1 \leq r \leq n-1$, and suppose D is relatively compact in X, and $\bar{\partial}$ is α-regular at $Z^0_{0,r+1}(\bar{D},E)$. Then the image of the restriction map

[1] For $D = \emptyset$ we have Theorem 12.7.

[2] Of course, this is not true with respect to the norm $\|\cdot\|_{1/2,D}$.

$$Z^{\alpha}_{0,r}(\bar{G},E) \longrightarrow Z^{\alpha}_{0,r}(\bar{D},E)$$

is dense in $Z^{\alpha}_{0,r}(\bar{D},E)$ with respect to the norm $\|\cdot\|_{0,D}$.

Proof of Lemmas 12.8 and 12.8'. This follows immediately from Lemmas 12.6, 12.3 and 12.4'. □

12.9. Proposition.[1] Let $D \subset\subset X$ be a non-degenerate strictly q-convex domain in an n-dimensional complex manifold X ($0 \le q \le n-1$), and E a holomorphic vector bundle over X. Then, for any $n \ge r \ge n-q$,

$$E^{1/2\longrightarrow 0}_{0,r}(\bar{D},E) = Z^{0}_{0,r}(\bar{D},E) \cap E^{0}_{0,r}(D,E).$$

Notice also the following generalization of this proposition:

12.9'. Proposition. Let $D \subset\subset X$ be a non-degenerate strictly q-convex domain in an n-dimensional complex manifold X ($0 \le q \le n-1$), and E a $C^{\alpha}\mathcal{O}$ vector bundle over $\bar{D}$ ($0 \le \alpha \le 1/2$). Then, for all $n \ge r \ge n-q$,

$$E^{\alpha\longrightarrow 0}_{0,r}(\bar{D},E) = Z^{0}_{0,r}(\bar{D},E) \cap E^{0}_{0,r}(D,E).$$

Proof of Propositions 12.9 and 12.9'. It is sufficient to prove Proposition 12.9'. Let $f \in Z^{0}_{0,r}(\bar{D},E) \cap E^{0}_{0,r}(D,E)$. Take a non-degenerate strictly q-convex domain $G \subset\subset D$ such that D is a non-degenerate strictly q-convex extension of G in X. Then it follows from Theorem 1.13 (regularity of $\bar{\partial}$) that $f|G \in E^{\alpha\longrightarrow 0}_{0,r}(\bar{G},E)$. Since, by Theorem 11.2, $\bar{\partial}$ is α-regular at $Z^{0}_{0,r}(\bar{G},E)$, now we may apply Lemma 12.8' (iii) and obtain that f belongs to $E^{\alpha\longrightarrow 0}_{0,r}(\bar{D},E)$. □

12.10. Proposition.[2] Let X be an n-dimensional complex manifold, let $D \subset\subset G \subset\subset X$ be non-degenerate strictly q-convex domains such that G is a non-degenerate strictly q-convex extension of D in X ($0 \le q \le n-1$), and let E be a holomorphic vector bundle over X. Then the following assertions hold true:

(i) If $n-q \le r \le n$, then, for any $f_D \in Z^{0}_{0,r}(\bar{D},E)$ and each neighborhood U of ∂D, there exists a form $f_G \in Z^{0}_{0,r}(\bar{G},E)$ with $f_D - f_G \in E^{1/2\longrightarrow 0}_{0,r}(\bar{D},E)$ and $f_D = f_G$ on $D \setminus U$.

---

1) Below we prove the more complete Theorem 12.15.

2) The assertions of this proposition will be repeated also in the more complete theorems given below: part (i) follows from Theorems 12.13 and 12.15 (i), part (ii) follows from Theorems 12.14 and 12.15 (ii), part (iii) is contained in Theorem 12.11.

(ii) $E^{1/2\longrightarrow 0}_{0,r}(\bar{G},E) = Z^0_{0,r}(\bar{G},E)\cap E^0_{0,r}(D,E)$ for all $n-q\le r\le n$.

(iii) If $n-q-1\le r\le n$, then the image of the restriction map

$$Z^0_{0,r}(\bar{G},E) \longrightarrow Z^0_{0,r}(\bar{D},E)$$

is dense in $Z^0_{0,r}(\bar{D},E)$ (with respect ot the norm $\|\cdot\|_{0,D}$).

Notice also the following generalization of this proposition to $C^{\alpha}\mathcal{O}$ vector bundles (see Sect. 0.12):

<u>12.10'. Proposition.</u> Let X be an n-dimensional complex manifold, let $D \subset\subset G \subset\subset X$ be non-degenerate strictly q-convex domains such that G is a non-degenerate strictly q-convex extension of D in X ($0\le q\le n-1$), and let E be a $C^{\alpha}\mathcal{O}$ vector bundle over $\bar{G}$. Then the following assertions hold true:

(i) If $n-q\le r\le n$, then, for any $f_D\in Z^0_{0,r}(\bar{D},E)$ and each neighborhood U of ∂D, there exists a form $f_G\in Z^0_{0,r}(\bar{G},E)$ with $f_D-f_G\in E^{\alpha\longrightarrow 0}_{0,r}(\bar{D},E)$ and $f_D = f_G$ on $D\setminus U$.

(ii) $E^{\alpha\longrightarrow 0}_{0,r}(\bar{G},E) = Z^0_{0,r}(\bar{G},E)\cap E^0_{0,r}(D,E)$ for all $n-q\le r\le n$.

(iii) If $n-q-1\le r\le n$, then the image of the restriction map

$$Z^0_{0,r}(\bar{G},E) \longrightarrow Z^0_{0,r}(\bar{D},E)$$

is dense in $Z^0_{0,r}(\bar{D},E)$ (with respect ot the norm $\|\cdot\|_{0,D}$).

<u>Proof of Propositions 12.10 and 12.10'.</u> It is sufficient to prove Proposition 12.10'.

Assertion (i) follows from Lemma 12.8' (i) and Proposition 12.9.

To prove part (ii), we consider a form $f\in Z^0_{0,r}(\bar{G},E)\cap E^0_{0,r}(D,E)$. By Proposition 12.9, $f|D\in E^{1/2\longrightarrow 0}_{0,r}(\bar{D},E)$. Since, by Theorem 11.2, $\bar{\partial}$ is 1/2-regular at $Z^0_{0,n-q}(\bar{D},E)$, we may apply Lemma 12.8' (iii) and obtain that f belongs to $E^{\alpha\longrightarrow 0}_{0,r}(\bar{G},E)$.

Assertion (iii) follows from Lemma 12.8' (iv), because, by Theorem 11.2, $\bar{\partial}$ is 0-regular at $Z^0_{0,r+1}(\bar{D},E)$ for $n-q-1\le r\le n-1$ (cf. Remark IV following Definition 11.1). □

<u>12.11. Theorem.</u> Let E be a holomorphic vector bundle over an n-dimensional complex manifold X, and $D \subset\subset X$ a non-degenerate strictly q-convex domain such that X is a non-degenerate q-convex extension of D

$(0 \leq q \leq n-1)$. Then, for all $n-q-1 \leq r \leq n$, the image of the restriction map

$$Z^0_{0,r}(X,E) \longrightarrow Z^0_{0,r}(\bar{D},E)$$

is dense in $Z^0_{0,r}(\bar{D},E)$ (with respect to the norm $\|\cdot\|_{0,D}$).

Proof. Since X is a non-degenerate q-convex extension of D, there is a sequence of non-degenerate strictly q-convex domains $D_k \subset\subset X$ ($k=0,1,\ldots$) such that $D_0 = D$, each D_{k+1} is a non-degenerate strictly q-convex extension of D_k in X, and

$$X = \bigcup_{k=0}^{\infty} D_k.$$

Now let a form $f_0 \in Z^0_{0,r}(\bar{D}_0,E)$ and a number $\varepsilon>0$ be given. Then, by Theorem 12.10 (iii), we can find a sequence $f_k \in Z^0_{0,r}(\bar{D}_k,E)$ with $\|f_k - f_{k-1}\|_{0,D_{k-1}} \leq \varepsilon/2^k$ for $k=1,2,\ldots$. Hence $F := \lim f_k$ exists uniformly on the compact subsets of X, belongs to $Z^0_{0,r}(X,E)$, and satisfies the estimate $\|F-f_0\|_{0,D} \leq \varepsilon$. □

12.12. Corollary. Let E be a holomorphic vector bundle over an n-dimensional complex manifold X, and $D \subset\subset X$ an arbitrary (not necessarily non--degenerate) strictly q-convex domain such that X is a q-convex extension of D. If $n-q-1 \leq r \leq n$, then any E-valued continuous $\bar{\partial}$-closed (0,r)-form in a neighborhood of $\bar{D}$ can be approximated uniformly on $\bar{D}$ by E-valued continuous $\bar{\partial}$-closed (0,r)-forms on X.

Proof. This follows from Theorem 12.11, for $\bar{D}$ has a basis of strictly q-convex neighborhoods with C^∞ boundary (cf. Theorem 5.7 (ii)). □

12.13. Theorem. Let E be a holomorphic vector bundle over an n-dimensional complex manifold X, and $D \subset\subset X$ a non-degenerate strictly q-convex domain such that X is a non-degenerate q-convex extension of D $(0 \leq q \leq n-1)$. Then, for each $n-q \leq r \leq n$, the following two assertions hold true:

(i) For any $f_D \in Z^0_{0,r}(\bar{D},E)$ and each neighborhood U of ∂D, there exists a form $f_X \in Z^0_{0,r}(X,E)$ such that $f_D - f_X \in E^{1/2 \longrightarrow 0}_{0,r}(\bar{D},E)$ and $f_X = f_D$ on $D \setminus U$.

(ii) $E^0_{0,r}(X,E) = Z^0_{0,r}(X,E) \cap E^0_{0,r}(D,E)$. [1)]

Proof. Since X is a non-degenerate q-convex extension of D, there is a sequence of non-degenerate strictly q-convex domains $D_k \subset\subset X$ ($k=0,1,\ldots$) such that $D_0=D$, each D_{k+1} is a non-degenerate strictly

---

1) Part (ii) will be generalized in Theorem 12.14 (injectivity of the restriction map (12.3)).

q-convex extension of D_k in X, and

$$X = \bigcup_{k=0}^{\infty} D_k.$$

Now let a form $f \in Z^0_{0,r}(\bar{D},E)$ and a neighborhood U of ∂D be given. Then from Proposition 12.10 (i) we obtain a sequence $f_k \in Z^0_{0,r}(\bar{D}_k,E)$ (k=0,1,...)such that $f_{k+1}-f_k \in E^{1/2 \to 0}_{0,r}(\bar{D}_k,E)$, $f_1=f_0$ on $D\setminus U$, and $f_{k+1}=f_k$ on D_{k-1} if $k \geq 1$. Then $f_X := \lim f_k$ is a form with the properties required in assertion (i).

To prove (ii), let a form $f \in Z^0_{0,r}(X,E)$ with $f|D \in E^0_{0,r}(D,E)$ be given. Then from Proposition 12.10 (ii) and (iii) we obtain a sequence $u_k \in C^0_{0,r-1}(\bar{D}_k,E)$ such that $\bar{\partial} u_k = f$ on $\bar{D}_k$ and $\|u_{k+1}-u_k\|_{0,D_k} \leq 1/2^k$. Then $u := \lim u_k$ exists uniformly on the compact subsets of X, and solves the equation $\bar{\partial}u=f$ on X. □

12.14. Theorem. Let E be a holomorphic vector bundle over an n-dimensional complex manifold X, and $D \subset\subset X$ a strictly q-convex domain such that X is a q-convex extension of D ($0 \leq q \leq n-1$). Then, for any $n-q \leq r \leq n$, the restriction map

$$H^{0,r}(X,E) \longrightarrow H^{0,r}(D,E) \tag{12.3}$$

is an isomorphism.

Proof. In view of the Dolbeault isomorphism (Theorem 2.14) it is sufficient to prove that the restriction map

$$H^{0,r}_0(X,E) \longrightarrow H^{0,r}_0(D,E) \tag{12.3'}$$

is an isomorphism .

In view of a lemma of M. Morse (cf. Proposition 0.5 in Appendix B) and Observation 4.15, we can find a strictly q-convex domain $G \subset\subset D$ with C^∞ boundary such that both X and D are non-degenerate q-convex extensions of G. Then it follows from Theroem 12.13 (ii) that the restriction map $H^{0,r}_0(X,E) \longrightarrow H^{0,r}_0(G,E)$ in injective. Hence (12.3') is injective.

In order to prove that (12.3') is also surjective, we consider a form $f \in Z^0_{0,r}(\bar{D},E)$. Applying Theorem 12.13 (i) to $f|\bar{G}$, then we obtain a form $f_X \in Z^0_{0,r}(X,E)$ such that $f-f_X$ is $\bar{\partial}$-exact in G, and, by part (ii) of Theorem 12.13, we may conclude that $f-f_X$ is $\bar{\partial}$-exact even on D. □

12.15. Theorem. Let $D \subset\subset X$ be a non-degenerate strictly q-convex domain in an n-dimensional complex manifold ($0 \le q \le n-1$), and E a holomorphic vector bundle over X. Then, for each $n-q \le r \le n$, the following two assertions hold true

(i) For any $f_D \in Z^0_{0,r}(D,E)$ and each neighborhood U of ∂D, there exists a form $f_{\bar{D}} \in Z^0_{0,r}(\bar{D},E)$ such that $f_D - f_{\bar{D}} \in E^0_{0,r}(D,E)$ and $f_{\bar{D}} = f_D$ on $D \setminus U$.

(ii) The restriction map

$$H^{0,r}_{1/2\longrightarrow 0}(\bar{D},E) \longrightarrow H^{0,r}_0(D,E) \qquad (12.4)$$

is an isomorphism, where

$$H^{0,r}_{1/2\longrightarrow 0}(\bar{D},E) := Z^0_{0,r}(\bar{D},E)/E^{1/2\longrightarrow 0}_{0,r}(\bar{D},E).$$

We mention also the following generalization of this theorem to $C^{\alpha}\mathcal{O}$ vector bundles (cf. Sect. 0.12):

12.15'. Theorem. Let $D \subset\subset X$ be a non-degenerate strictly q-convex domain in an n-dimensional complex manifold ($0 \le q \le n-1$), and E a $C^{\alpha}\mathcal{O}$ vector bundle over $\bar{D}$ ($0 \le \alpha \le 1/2$). Then, for each $n-q \le r \le n$, the following two assertions hold true:

(i) For any $f_D \in Z^0_{0,r}(D,E)$ and each neighborhood U of ∂D, there exists a form $f_{\bar{D}} \in Z^0_{0,r}(\bar{D},E)$ such that $f_D - f_{\bar{D}} \in E^0_{0,r}(D,E)$ and $f_{\bar{D}} = f_D$ on $D \setminus U$.

(ii) The restriction map

$$Z^0_{0,r}(\bar{D},E)/E^{\alpha\longrightarrow 0}_{0,r}(\bar{D},E) \longrightarrow H^{0,r}_0(D,E)$$

is an isomorphism.

Proof of Theorems 12.15 and 12.15'. It is sufficient to prove Theorem 12.15'.

First we prove assertion (i). Let $f_D \in Z^0_{0,r}(D,E)$, and let U be a neighborhood of ∂D. Choose a non-degenerate strictly q-convex domain $G \subset\subset D$ such that $D \setminus U \subset\subset G$ and D is a non-degenerate strictly q-convex extension of G in X. Then $f_D|\bar{G}$ belongs to $Z^0_{0,r}(\bar{G},E)$, and, by Proposition 12.10 (i), we get a form $f_{\bar{D}} \in Z^0_{0,r}(\bar{D},E)$ with $f_D - f_{\bar{D}} \in E^{\alpha\longrightarrow 0}_{0,r}(\bar{G},E)$ and $f_{\bar{D}} = f_D$ on $D \setminus U$. By Theorem 12.13 (ii), $f_D - f_{\bar{D}}$ belongs even to $E^0_{0,r}(D,E)$. Hence part (i) is proved.

Assertion (ii) follows from (i) (surjectivity of (12.4)) and Proposition 12.9' (injectivity of (12.4)). □

<u>12.16 Theorem</u> (Andreotti-Grauert finiteness theorem). Let E be a holomorphic vector bundle over an n-dimensional complex manifold X. If X is q-convex ($0 \leq q \leq n-1$), then

$$\dim H^{0,r}(X,E) < \infty \qquad \text{for all } n-q \leq r \leq n. \tag{12.5}$$

If X is <u>completely</u> q-convex ($0 \leq q \leq n-1$), then even

$$H^{0,r}(X,E) = 0 \qquad \text{for all } n-q \leq r \leq n. \tag{12.6}$$

<u>Proof.</u> If X is q-convex, then, by a lemma of M.Morse (cf. Proposition 0.5 in Appendix B) and Observation 4.15, we can find a strictly q-convex domain $D \subset\subset X$ with C^{∞} boundary such that X is a non-degenerate q-convex extension of D. It follows from Theorems 12.14 and 12.15 (ii) that

$$\dim H_0^{0,r}(X,E) = \dim H_0^{0,r}(D,E) = \dim H_{1/2 \longrightarrow 0}^{0,r}(\bar{D},E).$$

By Theorem 11.2, this implies (12.5). If X is <u>completely</u> q-convex, then this domain D can be chosen to be <u>completely</u> strictly q-convex, and, by Theorem 12.7, we obtain (12.6). □

CHAPTER IV. THE CAUCHY-RIEMANN EQUATION ON q-CONCAVE MANIFOLDS

__Summary.__ Sections 13-15 are organized analogously as Chapter II. In Sect. 13 a local Cauchy-Fantappie formula for non-degenerate strictly q-concave domains is constructed, which yields local extension of holomorphic functions (local Hartogs extension phenomenon), as well as local solutions of $\bar{\partial}u = f_{0,r}$ for all $1 \leq r \leq q-1$. In Sect. 14, first we prove 1/2-Hölder estimates for these local solutions, repeating word for word the arguments from Sect. 9. Then, using the same arguments as in Sect. 11, from these estimates we deduce the following version with uniform estimates of the Andreotti-Grauert finiteness theorem: If D is a non-degenerate strictly q-convex domain in a complex manifold X and E is a holomorphic vector bundle over X, then dim $H^{0,r}_{1/2\to 0}(\bar{D},E) < \infty$ for all $0 \leq r \leq q-1$, where $H^{0,r}_{1/2\to 0}(\bar{D},E) := Z^{0}_{0,r}(\bar{D},E)/E^{1/2\to 0}_{0,r}(\bar{D},E)$. In Sect. 15 we introduce the concept of a q-concave extension of a complex manifold, and prove that the Dolbeault cohomology classes of order $0 \leq r \leq q-1$ admit uniquely determined continuations along such extensions (for r=0, this is the global Hartogs extension phenomenon for holomorphic functions). Moreover, corresponding results with uniform estimates are obtained. At the end of Sect. 15 we prove the classical Andreotti-Grauert finiteness theorem: If E is a holomorphic vector bundle over a q-concave manifold X, then dim $H^{0,r}(X,E) < \infty$ for all $0 \leq r \leq q-1$.

Sections 16-19 are devoted to the Dolbeault cohomology of order q of q-concave manifolds. In Sect. 16 we prove that the extensions of such cohomology classes along a q-concave extension are uniquely determined (if exist). Sect. 17 is devoted to the special case of (linearly) concave domains $X \subseteq \mathbb{C}^n$. Here we establish the Martineau isomorphism between $H^{n,n-1}(X)$ and the space of holomorphic functions on the dual domain of X, where the emphasis is on the boundary behavior of this isomorphism. In Sect. 18 we prove the Andreotti-Norguet theorem. The main assertion of this theorem can be formulated as follows: If D is an n-dimensional complex domain which is both strictly q-concave and strictly (n-q-1)-convex, then dim $H^{0,q}(D,E) = \infty$. Further, a version with uniform estimates of this theorem is obtained. In Sect. 19 we prove the following version with uniform estimates of the Andreotti-Vesentini separation theorem: For any non-degenerate strictly q-concave domain D in a compact complex manifold X and all holomorphic vector bundles E over X, the space $E_{0,q}^{1/2\longrightarrow 0}(\bar{D},E)$ is closed with respect to the max-norm. Then, as a consequence, we obtain the classical Andreotti-Vesentini separation theorem.

13. Local solution of $\bar{\partial}u=f_{0,r}$ on strictly q-concave domains with $1\leq r\leq q-1$, and the local Hartogs extension phenomenon

13.1. Definition. $[U,\varrho,\varphi,H,D]$ will be called a q-concave configuration in $\mathbb{C}^n$ $(1\leq q\leq n-1)$ if $[U,-\varrho,\varphi]$ defines a q-convex configuration in $\mathbb{C}^n$ (cf. Definition 7.4) and the following conditions (a)-(d) are fulfilled (here Re:= real part of):

(a) $H = H(z)$, $z\in\mathbb{C}^n$, is a function of the form

$$H(z) = H'(z) + M\sum_{j=q+2}^{n}|z_j|^2, \tag{13.1}$$

where H' is a holomorphic polynomial in $z\in\mathbb{C}^n$ and M is a positive number.

(b) $\varphi(z) < 0$ for all $z\in U$ with $\operatorname{Re} H(z) = \varrho(z) = 0$.

(c) $D = \{z\in U\colon \varrho(z)<0,\ \varphi(z)<0,\ \operatorname{Re} H(z) < 0\} \neq \emptyset$.

(d) $d \operatorname{Re} H(z) \neq 0$ for all $z\in U$ with $\operatorname{Re} H(z) = 0$,
$d \operatorname{Re} H(z)\wedge d\varphi(z) \neq 0$ for all $z\in U$ with $\operatorname{Re} H(z) = \varphi(z) = 0$,
$d \operatorname{Re} H(z)\wedge d\varrho(z) \neq 0$ for all $z\in U$ with $\operatorname{Re} H(z) = \varrho(z) = 0$.

If $[U,\varrho,\varphi,H]$ is so that $[U,-\varrho,\varphi]$ defines a q-convex configuration, conditions (a),(b), and (d) are fulfilled, and the domain

$$D:= \{z\in U\colon \varrho(z)<0,\ \varphi(z)<0,\ \operatorname{Re} H(z) < 0\}$$

is non-empty, then we say $[U,\varrho,\varphi,H]$ defines the q-concave configuration $[U,\varrho,\varphi,H,D]$.

A q-concave configuration $[U,\varrho,\varphi,H,D]$ will be called of type I, II, III or IV according as $[U,-\varrho,\varphi]$ defines a q-convex configuration of type I, II, III of IV.

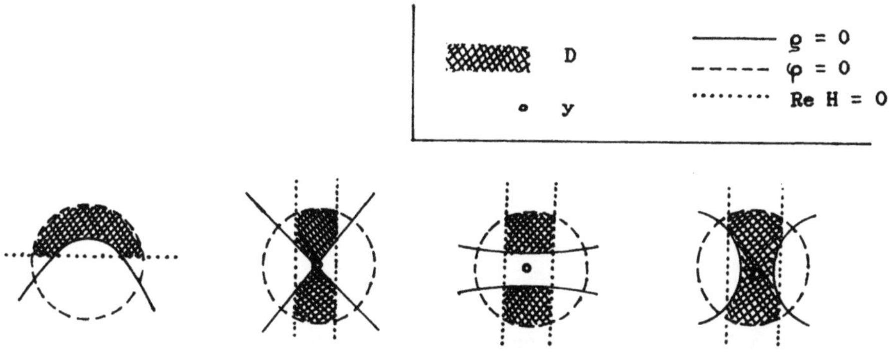

type I type II type III type IV

FIGURE 4: q-concave cofigurations

13.2. Lemma. Let $\varrho: X \longrightarrow \mathbb{R}$ be a (q+1)-concave function without degenerate critical points on an n-dimensional complex manifold ($0 \leq q \leq n-1$), and let $y \in X$ such that $\varrho(y)=0$ and y is not the point of a local maximum of ϱ. Then there exists a biholomorphic map h from some neighborhood V of y onto the unit ball in $\mathbb{C}^n$ such that $h(y)=0$ and the following statement holds true: For any $0<r<1$, we can find a function H such that

$$[h(V),\ \varrho \circ h^{-1},\ |z|^2-r^2,\ H]$$

defines a q-concave configuration in $\mathbb{C}^n$. This configuration is of type I if $d\varrho(y) \neq 0$, and of type II if $d\varrho(y) = 0$.

Proof. By Lemma 7.5, there exist holomorphic coordinates $h: V \longrightarrow \mathbb{C}^n$ in a neighborhood V of y such that h(V) is the unit ball, $h(y) = 0$, and, for any $0<r<1$, $[h(V), -\varrho \circ h^{-1}, |z|^2-r^2]$ defines a q-convex configuration, which is of type I if $d\varrho(y) \neq 0$, and of type II if $d\varrho(y) = 0$. Fix $0<r<1$. It remains to construct the function H.

Since $-\varrho \circ h^{-1}$ is normalized (q+1)-convex (cf. Def. 7.2), there are constants $C<\infty$ and $\beta>0$ with

$$-\operatorname{Re} F_{\varrho \circ h^{-1}}(z,0) \geq \varrho(h^{-1}(z)) + \beta \sum_{j=1}^{q+1} |z_j|^2 - C \sum_{j=q+2}^{n} |z_j|^2$$

for all $z \in h(V)$. Setting

$$\tilde{H}(z) = - F_{\varrho \circ h^{-1}}(z,0) + (C+\beta) \sum_{j=q+2}^{n} |z_j|^2$$

we obtain a polynomial of the form (13.1) such that $\operatorname{Re} \tilde{H}(z) \geq \beta |z|^2$ for all $z\in h(V)$ with $\varrho(h^{-1}(z)) \geq 0$. Hence, $[h(V),\varrho\circ h^{-1},|z|^2-r^2,\tilde{H}-\beta r^2/2]$ fulfils all conditions in order to define a q-concave configuration, except for (possibly) condition (d). By Lemma 0.3 in Appendix B, for almost all complex-linear maps $L:\mathbb{C}^n \longrightarrow \mathbb{C}$ (any real-linear map is the real part of some complex-linear map), the function

$$\operatorname{Re}\left(\tilde{H}(z) - \frac{\beta r^2}{2} + L(z)\right), \qquad z\in\mathbb{C}^n,$$

does not have degenerate critical points. By Proposition 0.4 in Appendix B, the same is true for the restrictions of this function to the surface "$\varphi=0$" as well as to the smooth part of "$\varrho=0$". This implies that, for almost all complex-linear maps $L:\mathbb{C}^n \longrightarrow \mathbb{C}$ and almost all real numbers ε, the function

$$H(z) := \tilde{H}(z) - \frac{\beta r^2}{2} + L(z) + \varepsilon, \qquad z\in\mathbb{C}^n,$$

fulfils condition (d) in Definition 13.1. If, moreover, L and ε are chosen sufficiently small, then H fulfils also the other conditions in this definition. □

13.3. The set Div(H). Let H be of the form (13.1), i.e.

$$H(z) = H'(z) + M\sum_{j=q+2}^{n} |z_j|^2, \qquad z\in\mathbb{C}^n,$$

where H' is a holomorphic polynomial and M is a positive number.

Definition. By Div(H) we denote the set of all n-tuples $v = (v^1,\ldots,v^n)$ of complex-valued C^1 functions $v^j:\mathbb{C}^n\times\mathbb{C}^n \longrightarrow \mathbb{C}$ which are obtained by the following

Construction. Take holomorphic polynomials $v'_j = v'_j(z,x)$ $(j=1,\ldots,n)$ in $(z,x)\in\mathbb{C}^n\times\mathbb{C}^n$ such that

$$H'(x) - H'(z) = \sum_{j=1}^{n} v'_j(z,x)(x_j-z_j),$$

set $v^j = v'_j$ for $j=1,\ldots,q+1$, and $v^j = v'_j + M(\bar{x}_j+\bar{z}_j)$ for $j=q+2,\ldots,n$.

Remark. For any $v \in Div(H)$, we have the relation

$$<v(z,x),x-z> = H'(x) - H'(z) + M \sum_{j=q+2}^{n} (|x_j|^2 - |z_j|^2 + \bar{z}_j x_j - \bar{x}_j z_j)$$

and hence

$$Re<v(z,x),x-z> = Re\, H(x) - Re\, H(z). \tag{13.2}$$

13.4. Canonical Leray data and maps. Let $[U, \varrho, \varphi, H, D]$ be a q-concave configuration in $\mathbb{C}^n$ ($0 \leq q \leq n-1$). Set $\psi_1 = \varrho$, $\psi_2 = \varphi$, $\psi_3 = Re\, H$, and

$$Y_j = \{z \in U: \psi_j(z)=0\}, \qquad D_j = \{z \in U: \psi_j(z)<0\}$$

for $j=1,2,3$. Then $D = D_1 \cap D_2 \cap D_3$ is a domain with piecewise almost C^1 boundary, and (Y_1, Y_2, Y_3) is a frame for D (cf. Sect. 3.1).

Proposition. Let (w_1, w_2, w_3) so that

$$\left.\begin{aligned} &w_1 = w_1(z,x) = - w^*(x,z) \qquad \text{for some } w^* \in Div(-\varrho) \\ &w_2 = \nabla^{\mathbb{C}} \varphi \\ &w_3 \in Div(H) \end{aligned}\right\} \tag{13.3}$$

(cf. Sect. 7.6 for the definition of $Div(-\varrho)$, and Remark 2.10 for the definition of $\nabla^{\mathbb{C}}\varphi$). Then there is a constant $\alpha > 0$ such that

$$Re<w_1(z,x),x-z> \geq \varrho(x) - \varrho(z) + \alpha |x-z|^2 \tag{13.4}$$

for all $z, x \in U$. Moreover, then

$$<w_j(z,x),x-z> \neq 0 \tag{13.5}$$

for all $x \in Y_j$, $z \in D_j$, and $j=1,2,3$. Hence (w_1, w_2, w_3) is a Leray datum for $(D; Y_1, Y_2, Y_3)$.

Proof. (13.4) follows from (7.4). (13.5) follows from (13.4) if $j=1$, and from (13.2) if $j=3$. If $j=2$, then (13.5) follows from the fact that D_2 is convex and, for any $x \in Y_2$, the set $\{z \in \mathbb{C}^n: <w_2(z,x),x-z>=0\}$ is the complex tangent plane of $\partial D_2 = Y_2$ at the point x. □

Definition. We say (w_1, w_2, w_3) is a canonical Leray datum for $[U, \varrho, \varphi, H, D]$ (or for D) if (13.3) holds true, and we say v is a canonical Leray map for $[U, \varrho, \varphi, H, D]$ (or for D) if v is the canonical combination (cf. Sect. 3.4) of some canonical Leray datum for D.

<u>13.5. Lemma.</u> Let $[U,\varrho, \varphi,H,D]$ be a q-concave configuration in $\mathbb{C}^n$, let D_1,D_2,D_3 be as in Sect. 13.4, and let $K:= \overline{D_2\setminus(D_1\cup D_3)}$. Then the following assertions hold true:

(i) The domains D_3 and $D_3\cap D_2$ are pseudoconvex.

(ii) For any neighborhood $W \subseteq U$ of K, there exists a strictly q-convex domain G with $K \subset\subset G \subset\subset W$ such that U is a q-convex extension of G (cf. Def. 12.1).

(iii) If $n-q-1\leq r\leq n$, then any $\bar\partial$-closed continuous $(0,r)$-form in a neighborhood of K can be approximated uniformly on K by $\bar\partial$-closed continuous $(0,r)$-forms on U.

<u>Proof of part (i).</u> This follows from the fact that U and D_2 are convex, and Re H is a continuous plurisubharmonic function (cf., for instance, Sect. 1.4 and 1.5 in [H/L]).

<u>Proof of part (ii).</u> Since $K = \{z\in U: \varphi(z)\leq 0, -\varrho(z)\leq 0, -\operatorname{Re} H(z) \leq 0\}$, there exists $\varepsilon>0$ with

$$K \subset\subset \{z\in U: \varphi(z)<\varepsilon, -\varrho(z)<\varepsilon, - \operatorname{Re} H(z) < \varepsilon\} \subset\subset W,$$

and we can take $\alpha>0$ so small that, for the function

$$\psi(z):= \max\left(\varphi(z)-\varepsilon, -\varrho(z)-\varepsilon, - \operatorname{Re} H(z) - \varepsilon + \alpha|z|^2\right), \qquad z\in U,$$

we have $K \subset\subset \{z\in U: \psi(z)<0\} \subset\subset W$. Since the functions $\varphi(z)-\varepsilon$, $-\varrho(z)-\varepsilon$, and $- \operatorname{Re} H(z) - \varepsilon + \alpha|z|^2$ are strictly plurisubharmonic with respect to $z_1,\dots,z_{q+1}$, it follows from Corollary 4.14 that, for each $\beta>0$, the function

$$\psi_\beta(z):= \max_\beta\left(\varphi(z)-\varepsilon, -\varrho(z)-\varepsilon, - \operatorname{Re} H(z) - \varepsilon + \alpha|z|^2\right), \qquad z\in U,$$

is also strictly plurisubharmonic with respect to $z_1,\dots,z_{q+1}$. In particular, ψ_β is $(q+1)$-convex. Without loss of generality, we can assume that φ is an unbounded exhausting function for U. Then also ψ_β is an unbounded exhausting function for U. Thus, for any sufficiently small $\beta>0$, the domain $G:= \{z\in U: \psi_\beta(z)<0\}$ has the required properties.

<u>Proof of part (iii).</u> This follows from part (ii) and Corollary 12.12. □

<u>13.6. Lemma.</u> Let $[U,\varrho, \varphi,H,D]$ be a q-concave configuration in $\mathbb{C}^n$ $(1\leq q\leq n-1)$, let D_1,D_2,D_3 be as in Sect. 13.4, and let v be a canonical Leray map for D. Then, for any continuous $(0,r)$-form f on $\bar D$, the follow-

ing assertions hold true (for the definition of the operators $L_1^v, L_2^v, L_3^v, L_{13}^v$, see Sect. 3.5):

(i) If $0 \le r \le q-1$, then $L_1^v f = L_3^v f = 0$.

(ii) If $1 \le r \le n$, then $L_2^v f = 0$. If $r=0$, i.e. f is a function, then $L_2^v f$ is holomorphic in D and admits a holomorphic extension into D_2.

(iii) If $0 \le r \le q-2$, then $L_{13}^v f = 0$. If $r=q-1$ and $\bar\partial f = 0$, then also $L_{13}^v f = 0$.

Proof of part (i). Since $w_1(z,x)$ and $w_3(z,x)$ depend holomorphically on $x_1, \dots, x_{q+1}$, this follows from Corollary 3.13 (iii).

Proof of part (ii). Since w_2 is independent of z, the map $v_2(z,x) = w_2(x)/\langle w_2(x), x-z\rangle$ depends holomorphically on $z \in D$, and, by (13.5), it admits a holomorphic extension into D_2. If $r \ge 1$, this implies that $L_2^v f = 0$, by Corollary 3.13 (ii).

Now let $r=0$. Then, by Proposition 3.6,

$$L_2^v f(z) = \frac{1}{(2\pi i)^n} \int_{S_2} f(x) \det(v_2(z,x), \overbrace{\bar\partial_x v_2(z,x)}^{n-1}) \wedge \omega(x), \qquad z \in D,$$

and we see that $L_2^v f$ is holomorphic in D and admits an holomorphic extension into D_2.

Proof of part (iii). If $0 \le r \le q-2$, then it follows from Corollary 3.13 (iii) that $L_{13}^v f = 0$, because both $w_1(z,x)$ and $w_3(z,x)$ depend holomorphically on $x_1, \dots, x_{q+1}$.

Now let $r=q-1$ and $\bar\partial f=0$. Recall that, by Proposition 3.6,

$$L_{13}^v f(z) = \int_{S_{13} \times \triangle_{13}} f(x) \wedge (\hat L_{13}^v)_{q-1}(z,x,t), \qquad z \in D,$$

where, for some constant A,

$$(\hat L_{13}^v)_{q-1} = A \det(v_{13}, d_t v_{13}, \overbrace{\bar\partial_z v_{13}}^{q-1}, \overbrace{\bar\partial_x v_{13}}^{n-q-1}) \wedge \omega(x). \tag{13.6}$$

Set $W_z = \{z \in U: \varrho(x) > \varrho(z), \operatorname{Re} H(x) > \operatorname{Re} H(z)\}$ for all $z \in D$. Then any W_z is a neighborhood of $\overline{D_2 \setminus (D_1 \cup D_2)}$, and it follows from (13.2) and (13.4) that $v_{13}(z,x,t)$ is defined and continuously differentiable for all $z \in D$, $x \in W_z$, and $t \in \triangle_{13}$. Hence the right hand side of (13.6) defines a continuous form $F(z,x,t)$ for $z \in D$, $x \in W_z$ and $t \in \triangle_{13}$. Since $v_{13}(z,x,t)$ depends holomorphically on $x_1, \dots, x_{q+1}$, we have relation

$$\bar\partial_x F(z,x,t) = 0 \tag{13.7}$$

for all $z\in D$, $x\in W_z$, and $t\in\triangle_{13}$. Let P_{q-1} be the set of all strictly increasing collection $I=(i_1,\dots,i_{q-1})$ of integers $1\le i_1<\dots<i_{q-1}\le n$, and let $F_I(z,x,t)$ ($z\in D$, $x\in W_z$, $t\in\triangle_{13}$), $I\in P_{q-1}$, be the forms which are of bidegree $(0,n-q-1)$ in x, and of degree zero in (z,t) such that

$$F(z,x,t) = \sum_{I\in P_{q-1}} F_I(z,x,t)\wedge\omega(x)\wedge dt_1\wedge d\bar z_I,$$

where $d\bar z_I := d\bar z_{i_1}\wedge\dots\wedge d\bar z_{i_{q-1}}$ if $I=(i_1,\dots,i_{q-1})$. Then

$$L^{v}_{13}f(z) = \sum_{I\in P_{q-1}} \left[\int_{\triangle_{13}} dt_1 \int_{S_{13}} f(x)\wedge F_I(z,x,t)\wedge\omega(x)\right]\wedge d\bar z_I$$

for all $z\in D$. Therefore, it is sufficient to prove that, for all $(z,t)\in D\times\triangle_{13}$ and $I\in P_{q-1}$,

$$\int_{S_{13}} f(x)\wedge F_I(z,x,t)\wedge\omega(x) = 0. \tag{13.8}$$

To do this, we fix $(z,t)\in D\times\triangle_{13}$ and $I\in P_{q-1}$. By (13.7), the form $F_I(z,\cdot,t)$ is $\bar\partial$-closed on W_z. Since W_z is a neighborhood of $\overline{D_2\setminus(D_1\cup D_3)}$, and $F_I(z,\cdot,t)$ is of bidegree $n-q-1$, from Lemma 13.5 (iii) we get a sequence F_I^k of $\bar\partial$-closed continuous $(0,n-q-1)$-forms on U such that $F_I^k \longrightarrow F_I(z,\cdot,t)$, uniformly on $\overline{D_2\setminus(D_1\cup D_3)}$. Since $S_{13}\subseteq\overline{D_2\setminus(D_1\cup D_3)}$, it follows that

$$\int_{x\in S_{13}} f\wedge F_I(z,x,t)\wedge\omega(x) = \lim_k \int_{x\in S_{13}} f(x)\wedge F_I^k(x)\wedge\omega(x). \tag{13.9}$$

Since the forms f and F_I^k are $\bar\partial$-closed, the forms $f(x)\wedge F_I^k(x)\wedge\omega(x)$ are closed on S_1. Since S_{13} is the boundary of S_1, this implies, by Stokes' theorem,

$$\int_{S_{13}} f(x)\wedge F_I^k(x)\wedge\omega(x) = 0$$

for all k. Therefore, (13.8) follows from (13.9). □

13.7. Lemma. Let $[U,\varrho,\varphi,H,D]$ be a q-concave configuration in $\mathbb{C}^n$ $(1\le q\le n-1)$, let D_2,D_3 be as in Sect. 13.4, and let v be a canonical Leray map for $[U,\varrho,\varphi,H,D]$. Then, for any $0\le r\le n-2$[1], there exists a linear operator

$$M_r: Z^0_{0,r}(\bar{D}) \longrightarrow Z^0_{0,r}(D_2\cap D_3)$$

which is continuous with respect to the Banach space topology of $Z^0_{0,r}(\bar{D})$ and the Fréchet space topology of $Z^0_{0,r}(D_2\cap D_3)$ such that

$$M_r f\big|_D = L^v_{23} f$$

for all $f\in Z^0_{0,r}(\bar{D})$ (for the definition of L^v_{23}, see Sect. 3.5).

Proof. By (13.5), the map $v_{23}(z,x,t)$ is defined and continuously differentiable for all $z\in D_2\cap D_3$, $x\in S_{23}$ and $t\in\triangle_{23}$. Therefore, by setting

$$\hat{M}_r = \frac{\binom{n-1}{r}}{(2\pi i)^n} \det(v_{23},\overbrace{\bar{\partial}_z v_{23}}^{r},\overbrace{(\bar{\partial}_x+d_t)v_{23}}^{n-r-1})\wedge\omega(x)$$

we obtain a continuous differential form $\hat{M}_r = \hat{M}_r(z,x,t)$ defined for $z\in D_2\cap D_3$, $x\in S_{23}$, and $t\in\triangle_{23}$. Put

$$M_r f = \int_{(x,t)\in S_{23}\times\triangle_{23}} f(x)\wedge\hat{M}_r(\cdot,x,t)$$

for all continuous differential forms f on $\bar{D}$. Since $S_{23}\cap(D_2\cap D_3) = \emptyset$, it is clear that in this way continuous linear operators

$$M_r: C^0_{0,r}(\bar{D}) \longrightarrow C^0_{0,r}(D_2\cap D_3)$$

are defined. By Proposition 3.6, $M_r f|D = L^v_{23} f$ for all $f\in C^0_{0,r}(D)$.

Fix $f\in Z^0_{0,r}(\bar{D})$. It remains to prove that $\bar{\partial} M_r f = 0$. In the same way as

[1] We do not consider the case $n-1\le r\le n$, because then, by Proposition 3.6, $L^v_{23} f = 0$ for each continuous $(0,r)$-form f on $\bar{D}$.

in the case of Proposition 3.9, we see that $\bar\partial_z \hat{M}_r = -\bar\partial_x \hat{M}_{r+1}$, where $\hat{M}_{r+1} := 0$ if $r=n-2$. Since $\hat{M}_{r+1}$ contains the factor $dt_1 \wedge \omega(x)$ and hence $f \wedge \bar\partial_x \hat{M}_{r+1} = d_{x,t}(f \wedge \hat{M}_{r+1})$, this implies that

$$\bar\partial M_r f(z) = -\int_{(x,t)\in S_{23}\times\Delta_{23}} d_{x,t}[f(x)\wedge \hat{M}_{r+1}(z,x,t)]$$

for all $z\in D_2\cap D_3$. Hence, by Stokes' formula ($\partial S_{23} = \emptyset$!),

$$\bar\partial M_r f(z) = \pm \int_{S_{23}\times\partial\Delta_{23}} f(x)\wedge \hat{M}_{r+1}(z,x,t)$$

for all $z\in D_2\cap D_3$. Since $f(x)\wedge \hat{M}_{r+1}(z,x,t)$ is of degree 1 in t whereas $\partial\Delta_{23}$ is of dimension zero, it follows that $\bar\partial M_r f = 0$. $\square$

<u>13.8. Theorem</u> (Hartogs extension). Let $[U,\varrho,\varphi,H,D]$ be a 1-concave configuration in $\mathbb{C}^n$,

$$\tilde{D} := \{z\in U\colon \varphi(z) < 0 \text{ and } \operatorname{Re} H(z) < 0\}.$$

Then any holomorphic function in D admits a holomorphic extension into $\tilde{D}$.

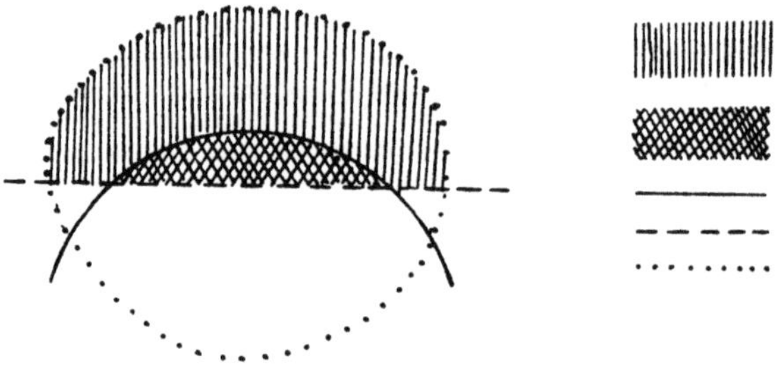

<u>Proof of Theorem 13.8.</u> Since, for any sufficiently small $\varepsilon>0$, $[U,\varrho+\varepsilon,\varphi+\varepsilon,H+\varepsilon]$ also defines a 1-concave configuration, it is sufficient to prove that the theorem holds true for functions which are holomorphic in some neighborhood of $\bar{D}$. Let f be such a function, and let v be a canonical Leray map for $[U,\varrho,\varphi,H,D]$. Then, by Lemma 13.6, $L_1^v f = L_3^v f = L_{13}^v f = 0$. Hence the piecewise Cauchy-Fantappie formula

(3.18) then takes the form $f = L_2^v f + L_{23}^v f$ on D. We complete the proof by the remark that, in view of Lemma 13.6 (ii) and Lemma 13.7, both $L_2^v f$ and $L_{23}^v f$ admit holomorphic extensions into $\tilde{D}$. □

<u>13.9. Corollary.</u> Let $\varrho: X \longrightarrow R$ be a 2-concave function without degenerate critical points on an n-dimensional complex manifold X ($n \geq 2$), and let $D := \{z \in X: \varrho(z) < 0\}$. Then, for any point $y \in \partial D$, there exists a neighborhood U of y such that any holomorphic function in $U \cap D$ admits a holomorphic extension into U.

<u>Proof.</u> This follows from Theorem 13.8 and Lemma 13.2. □

<u>13.10. Theorem.</u> Let $[U, \varrho, \varphi, H, D]$ be a q-concave configuration in $\mathbb{C}^n$ ($1 \leq q \leq n-1$). Then, for each $1 \leq r \leq q-1$, the following two assertions hold true:

(i) For any $\bar\partial$-closed continuous (0,r)-form f in D, there exists a continuous (0,r-1)-form u in D with $\bar\partial u = f$.

(ii) Set

$$\tilde{D} := \{z \in U: \varphi(z) < 0, \mathrm{Re}\, H(z) < 0\}.$$

Let v be a canonical Leray map for D, T^v the corresponding Cauchy--Fantappie operator introduced in Sect. 3.10, and

$$M_r: Z_{0,r}^0(\bar{D}) \longrightarrow Z_{0,r}^0(\tilde{D})$$

the continuous linear operator from Lemma 13.7. Then, for any $f \in Z_{0,r}^0(\bar{D})$, we have the representation

$$f = \bar\partial T^v f + M_r f \qquad \text{in } D\,. \tag{13.10}$$

Moreover, then there exists a continuous (0,r-1)-form g on $\tilde{D}$ with $M_r f = \bar\partial g$ on $\tilde{D}$. Hence $u := T^v f + g$ solves the equation $\bar\partial u = f$ in D.

<u>Proof of part (ii).</u> Let $f \in Z_{0,r}^0(\bar{D})$. Then, by Lemma 13.6, $L_1^v f = L_2^v f = L_3^v f = L_{13}^v f = 0$, and hence the piecewise Cauchy-Fantappie formula (see Theorem 3.12) takes the form $f = \bar\partial T^v f + L_{23}^v f$ in D. Since $L_{23}^v f = M_r f$ on D (cf. Lemma 13.7), this implies (13.10). Since, by Lemma 13.5 (i) and Theorem 5.3, $\tilde{D}$ is completely pseudoconvex, it follows from Theorem 12.16 that $M_r f = \bar\partial g$ for some continuous (0,r-1)-form g on $\tilde{D}$.

<u>Proof of part (i).</u> If f admits a continuous extension onto $\bar{D}$, then assertion (i) is contained in (ii). The general case follows from this special case by means of a modification of the arguments given in the proof of Theorem 12.13 (ii) (the approximation theorem for holomor-

phic functions which we need in the case r=1 follows from the extension Theorem 13.8 and the fact that D is pseudoconvex). □

13.11. Corollary. Let $\varrho: X \longrightarrow \mathbb{R}$ be a q-concave function without degenerate critical points on an n-dimensionsal complex manifold ($1 \le q \le n-1$), $D := \{z \in X: \varrho(z) < 0\}$, and $1 \le r \le n-1$. Then, for any point $y \in \partial D$, there exists a neighborhood U of y such that, for any $\bar\partial$-closed continuous (0,r)-form f on $D \cap U$, there exists a continuous (0,r-1)-form u on $D \cap U$ with $\bar\partial u = f$.

Proof. This follows from Lemma 13.2 and Theorem 13.10 (i). □

14. Uniform estimates for the local solutions of the $\bar\partial$-equation obtained in Sect. 13, and finiteness of the Dolbeault cohomology of order r with uniform estimates on strictly q-concave domains with $1 \le r \le q-1$

In Sect. 13 q-concave configurations were introduced, the notion of a canonical Leray map v for such configurations was defined, and we saw (Theorem 13.10) that the corresponding Cauchy-Fantappie operator T^v (cf. Sect. 3.10) together with the extension operator M_r from Lemma 13.7 solves the $\bar\partial$-equation for forms of bidegree (0,r) with $1 \le r \le q-1$. In the present section we prove uniform estimates for these operators, and give first applications.

14.1. Theorem. Let $[U, \varrho, \varphi, H, D]$ be a q-concave configuration in $\mathbb{C}^n$ ($1 \le q \le n-1$), v a canonical Leray map for D, $\varepsilon > 0$, and

$$D_\varepsilon := \{z \in U: \varphi(z) < -\varepsilon, \ \mathrm{Re}\, H(z) < -\varepsilon\}.$$

Then there exists a constant $C < \infty$ such that: For all continuous differential forms f on $\bar D$,

$$\|T^v f\|_{1/2, D_\varepsilon} \le C \|f\|_{0,D}, \tag{14.1}$$

(for the definition of the norms, see Sect. 0.11). If $[U, \varrho, \varphi, H, D]$ is of type II and $y \in U$ is the critical point of ϱ, then, moreover,

$$\|T^v f\|_{0, D_\varepsilon} \le C \Big(\sup_{z \in \operatorname{supp} f} |z-y| \Big) \|f\|_{0,D}, \tag{14.2}$$

and

$$\sup_{\substack{z,x\in D_\varepsilon \\ |z-y|<r \\ |x-y|<r}} \frac{\|T^v f(z) - T^v f(x)\|}{|z-x|^{1/2}} \leq C\big(1 + |\ln r|\big) r^{1/2} \|f\|_{0,D} \qquad (14.3)$$

for each $r>0$.

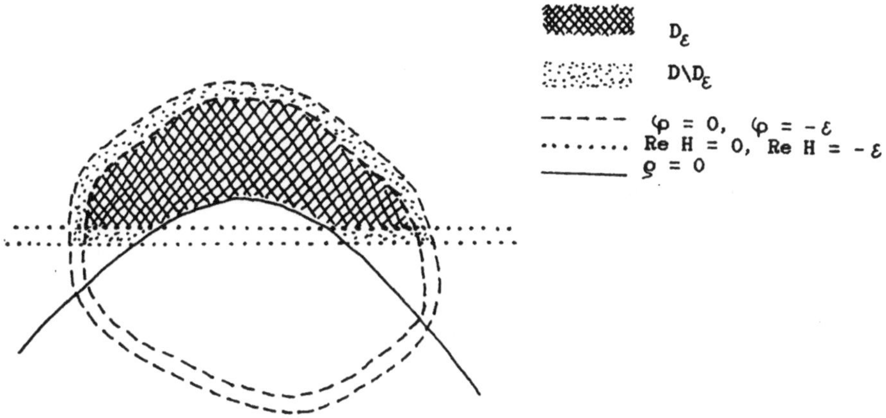

FIGURE 5: The domains D and D_ε in Theorems 14.1 and 14.2

Proof of Theorem 14.1. This is a word for word repetition of the proof of Theorem 9.1, with the following two exceptions:

1) The operator T^v now is of the form

$$T^v = B_D + R_1^v + R_2^v + R_3^v + R_{12}^v + R_{13}^v,$$

and we have to add the remark that since $S_3 \cap \bar{D}_\varepsilon = S_{13} \cap \bar{D}_\varepsilon = \emptyset$, estimates (14.1)-(14.3) hold true also with R_3^v and R_{13}^v instead of T^v.

2) The map $w(z,x)$ now does not belong to $\mathrm{Div}(\varrho)$, but it is of the form

$$w(z,x) = -w^*(x,z) \qquad \text{with } w^* \in \mathrm{Div}(-\varrho), \qquad (14.4)$$

and, instead of Lemma 9.2, we have to use the following

Lemma. If $w(z,x)$ is of the form (14.4) and $h(z,x) = \mathrm{Im}\langle w(z,x), x-z \rangle$, then assertions (i)-(iv) in Lemma 9.2 hold true.

Proof of the latter lemma. If t_j are the real coordinates on $\mathbb{C}^n$ with $x_j=t_j(x)+it_{j+n}(x)$, $x\in\mathbb{C}^n$, then immediately from the definition of Div($-\varrho$) and the relation $h(z,x) = \mathrm{Im}\langle w^*(x,z),z-x\rangle$ one obtains that

$$h(z,x) = \sum_{j=1}^{n}\left(\frac{\partial\varrho(z)}{\partial t_j}t_{j+n}(x-z) - \frac{\partial\varrho(z)}{\partial t_{j+n}}t_j(x-z)\right) + O(|x-z|^2),$$

and

$$d_x h(z,x) = \sum_{j=1}^{n}\left(\frac{\partial\varrho(z)}{\partial t_j}dt_{j+n}(x-z) - \frac{\partial\varrho(z)}{\partial t_{j+n}}dt_j(x-z)\right) + O(|x-z|).$$

It is easy to see that the last two relations yield assertions (i)-(iv) in Lemma 9.2.

14.2. Theorem (cf. Figure 5). Let $[U,\varrho,\varphi,H,D]$ be a q-concave configuration in $\mathbb{C}^n$ ($1\le q\le n-1$), $\varepsilon>0$, and

$$D_\varepsilon := \{z\in U: \varphi(z)<-\varepsilon, \mathrm{Re}\, H(z)< -\varepsilon\}.$$

Then, for all $1\le r\le q-1$, there exists a bounded linear operator

$$R: Z^0_{0,r}(\bar{D}) \longrightarrow C^{1/2}_{0,r-1}(\bar{D}_\varepsilon)$$

such that

$$\bar{\partial}Rf = f\big|_{\bar{D}_\varepsilon} \tag{14.5}$$

for all $f\in Z^0_{0,r}(\bar{D})$.

Proof. Let $\tilde{D}, v, T^v$, and M_r be as in Theorem 13.10. Since $\tilde{D}$ is pseudoconvex (Lemma 13.5 (i)), we can find a strictly pseudoconvex domain G with C^∞ boundary such that $D_\varepsilon \subset\subset G \subset\subset \tilde{D}$ (cf. Theorem 5.7 (i)). Since any strictly pseudoconvex domain in $\mathbb{C}^n$ is completely strictly pseudoconvex (cf. Theorem 5.3), then, by Theorem 12.7, there exists a continuous linear operator $R_G: Z^0_{0,r}(\bar{G}) \longrightarrow C^{1/2}_{0,r-1}(\bar{G})$ such that $\bar{\partial}R_G f = f$ for all

$f\in Z^0_{0,r}(\bar{G})$. Set $R_\varepsilon = T^V + R_G M_r$. Since M_r acts continuously from $Z^0_{0,r}(\bar{D})$ into $Z^0_{0,r}(\tilde{\bar{D}})$, R_G acts continuously from $Z^0_{0,r}(\bar{G})$ into $C^{1/2}_{0,r-1}(\bar{G})$, and $\bar{D}_\varepsilon \subseteq \tilde{G} \subset\subset \tilde{D}$, we see that $R_G M_r$ is bounded as an operator from $Z^0_{0,r}(\bar{D})$ into $C^{1/2}_{0,r-1}(\bar{D}_\varepsilon)$. Together with estimate (14.1) in Theorem 14.1 this implies that R_ε is bounded as an operator from $Z^0_{0,r}(\bar{D})$ into $C^{1/2}_{0,r-1}(\bar{D}_\varepsilon)$. Relation (14.5) follows from (13.10). □

<u>14.3. Definition.</u> Let $D \subset\subset X$ be a domain in an n-dimensional complex manifold X, and $1\le q\le n-1$. We say <u>the boundary of D is strictly q-concave with respect to X</u> if: ∂D is compact, the intersection of ∂D with any connected component of X is non-empty, and there exists a strictly (q+1)-concave function $\varrho: U \longrightarrow \mathbb{R}$ in some neighborhood U of ∂D such that

$$D \cap U = \{z\in U: \varrho(z)<0\}.$$

If, in this case, the function ϱ can be chosen without degenerate critical points in U, then we say the boundary of D is <u>non-degenerate</u> strictly q-concave with respect to X.

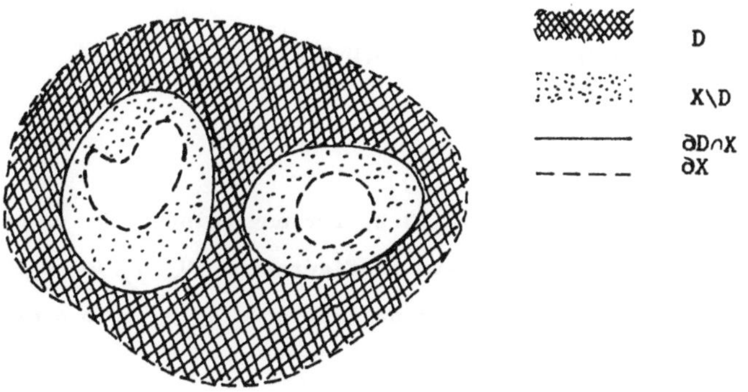

FIGURE 6: The boundary of D is strictly q-concave with respect to X

<u>Remarks to this definition.</u>

<u>I.</u> Of course, the boundary of any strictly q-concave domain in a complex manifold is q-concave with respect to this manifold, because, by definition, q-concave domains are relatively compact. In distinction to this, a domain with q-concave boundary with respect to some manifold

need not be relatively compact in that manifold.

II. In Sect. 12 we did not introduce the notion of a (not necessarily relatively compact) "domain with strictly q-convex boundary", because the theorems which we can prove for such domains are not nice enough, in our opinion.

III. Let X be a complex manifold, and $D \subseteq X$ a domain whose boundary is non-degenerate strictly 1-concave with respect to X. Then any $C^{\alpha}\mathcal{O}$ vector bundle over $\bar{D}$ (for the definition of such bundles, see Sect. 0.12) admits a holomorphic extension into a neighborhood of $\bar{D}$.

This follows from the fact that, in view of Corollary 13.9, the transition functions of such bundles admit holomorphic extensions.

In view of this circumstance, in distinction to Sections 11 and 12, in Sections 14 and 15 we do not formulate results for $C^{\alpha}\mathcal{O}$ vector bundles.

Theorem 14.4 (Hartogs extension). Let X be an n-dimensional complex manifold with $n \geq 2$, and $D \subseteq X$ a domain whose boundary is strictly 1-concave with respect to X. Further, let E be a holomorphic vector bundle over X. Then there exists a neighborhood U of ∂D such that the restriction map

$$\mathcal{O}(D \cup U, E) \longrightarrow \mathcal{O}(D \setminus U, E) \tag{14.6}$$

is an isomorphism.

Proof (cf. Fig. 7). By Lemma 13.2, for any point $y \in \partial D$, we can find a neighborhood G_y of y, a 1-concave configuration $[U_y, \varrho_y, \varphi_y, H_y, W_y]$ in $\mathbb{C}^n$, and a biholomorphic map h_y from a neighborhood of $\bar{U}_y$ onto a neighborhood of $\bar{G}_y$ such that $D\, G_y = h_y(W_y)$ and

$$G_y = h_y(\{z \in U_y : \varphi_y(z) < 0,\ \mathrm{Re}\, H(z) < 0\}). \tag{14.7}$$

For every $y \in \partial D$, we choose a number $\varepsilon_y > 0$ sufficiently small, and set $W'_y := \{z \in W_y : \varrho_y(z) < -\varepsilon_y\}$. Then $[U_y, \varrho_y + \varepsilon_y, \varphi_y, H_y, W'_y]$ is also a 1-concave configuration. If the neighborhoods G_y are chosen so small that E is trivial over each G_y, then it follows from Theorem 13.8 that the restriction maps

$$\mathcal{O}(G_y,E) \longrightarrow \mathcal{O}(h_y(W'_y),E) \tag{14.8}$$

are surjective. Since ∂D is compact, we can find a finite number of points $y_1,\ldots,y_N \in \partial D$ with $\partial D \subseteq G_{y_1} \cup\ldots\cup G_{y_N}$. Since the sets $h_y(W'_y)$ are relatively compact in D, we can find a neighborhood U of ∂D such that

$$h_{y_1}(W'_{y_1}) \cup \ldots \cup h_{y_N}(W'_{y_N}) \subseteq D\setminus U \quad \text{and} \quad D\cup U \subseteq (D\setminus U) \cup G_{y_1} \cup \ldots \cup G_{y_N}.$$

Now it follows from the surjectivity of the restriction maps (14.8) that (14.6) is surjective. Further, we can choose U so that the intersection of $D\setminus U$ with each connected component of $D\cup U$ is non-empty. Then (14.6) is also injective. □

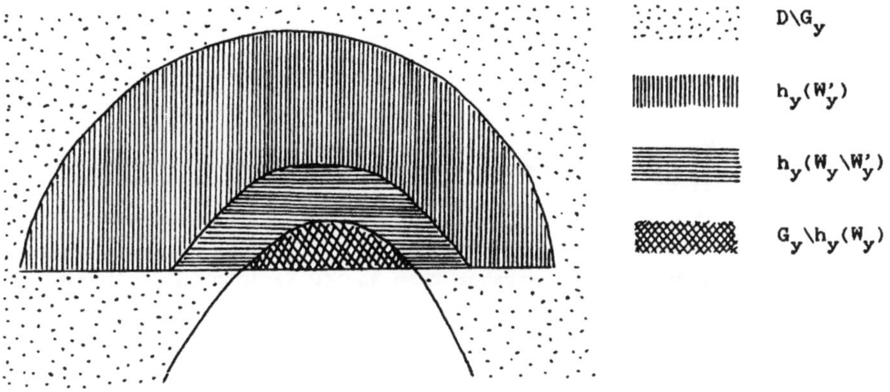

FIGURE 7: To the proof of Theorem 14.4

14.5. Theorem. Let $D \subset\subset X$ be a non-degenerate strictly q-concave domain in an n-dimensional complex manifold X ($1\leq q\leq n-1$), and let E be a holomorphic vector bundle over X. Then, for all $1\leq r\leq q-1$, the operator $\bar\partial$ is 1/2-regular at $Z^0_{0,r}(\bar D,E)$ (cf. Def. 11.1) and, moreover,

$$\dim H^{0,r}_{1/2\longrightarrow 0}(\bar D,E) < \infty,$$

where

$$H^{0,r}_{1/2\to 0}(\bar{D},E) := Z^{0}_{0,r}(\bar{D},E)/E^{1/2\to 0}_{0,r}(\bar{D},E).$$

<u>Proof.</u> By Lemma 13.2 and Theorem 14.2, we can find open sets U_j ($j=1,\dots,N<\infty$) with $\partial D \subseteq U_1\cup\dots\cup U_N$, and bounded linear operators T^j from $Z^0_{0,r}(\bar{D},E)$ into $C^{1/2}_{0,r-1}(\overline{U_j\cap D},E)$ such that $\bar{\partial}T^j f = f$ on $\overline{U_j\cap D}$ for all $f\in Z^0_{0,r}(\bar{D},E)$.

Further, taking into account Lemma 1.16, we can find open sets $U_j \subset\subset D$ ($j=N+1,\dots,M<\infty$) with $D\setminus(U_1\cup\dots\cup U_N) \subseteq U_{N+1}\cup\dots\cup U_M$, as well as bounded linear operators T^j from $Z^0_{0,r}(\bar{D},E)$ into $C^{1/2}_{0,r-1}(\bar{U}_j,E)$ and K^j from $Z^0_{0,r}(\bar{D},E)$ into $C^{1/2}_{0,r}(\bar{U}_j,E)$ such that, for all $f\in Z^0_{0,r}(\bar{D},E)$, $\bar{\partial}T^j f = f + K^j f$ on $\bar{U}_j$.

Choose a C^∞ partition of unity $\{\chi_j\}_1^M$ subordinated to $\{U_j\}_1^M$, and set

$$T = \sum_{j=1}^{M} \chi_j T^j \qquad \text{and} \qquad K = \sum_{j=1}^{M} \bar{\partial}\chi_j \wedge T^j + \sum_{j=N+1}^{M} \chi_j K^j.$$

Then T is a bounded linear operator from $Z^0_{0,r}(\bar{D},E)$ into $C^{1/2}_{0,r-1}(\bar{D},E)$, K is a bounded linear operator from $Z^0_{0,r}(\bar{D},E)$ into $C^{1/2}_{0,r}(\bar{D},E)$, and

$$\bar{\partial}Tf = f + Kf$$

for all $f\in Z^0_{0,r}(\bar{D},E)$. This implies the assertion of the theorem (cf. Remark III following Definition 11.1). □

15. Invariance of the Dolbeault cohomology of order $0 \le r \le q-1$ with respect to q-concave extensions, and the Andreotti-Grauert finiteness theorem for the Dolbeault cohomology of order $0 \le r \le q-1$ on q-concave manifolds

15.1. Definition. Let X be an n-dimensional complex manifold, and $0 \le q \le n-1$ an integer.

a) If $D \subseteq X$ is a domain, then we say X is a q-concave extension of D if the following conditions are fulfilled:

(i) ∂D is compact.

(ii) Every connected component of X has a non-empty intersection with D.

(iii) There exists a (q+1)-concave function $\varrho: U \longrightarrow \mathbb{R}$ in a neighborhood $U \subseteq X$ of $X \setminus D$ such that $D \cap U = \{z \in U: \varrho(z) < 0\}$ and, for any number

$$0 < \alpha < \sup_{z \in U} \varrho(z),$$

the set $\{z \in U: 0 \le \varrho(z) \le \alpha\}$ is compact.

If, in this case, the function ϱ in condition (iii) can be chosen without degenerate critical points in U, then X will be called a non-degenerate q-concave extension of D.

b) If $D \subseteq G$ are two domains in X, then we say G is a strictly q-concave extension of D in X if G is a q-concave extension of D such that, moreover, the set $G \setminus D$ is relatively compact in X and the following strengthening of condition (iii) is fulfilled:

(iii)' There exists a (q+1)-concave function $\varrho: U \longrightarrow \mathbb{R}$ in some neighborhood U of $\overline{G \setminus D}$ such that

$$D \cap U = \{z \in U: \varrho(z) < 0\} \quad \text{and} \quad G \cap U = \{z \in U: \varrho(z) < 1\}.$$

If, in this case, the function ϱ in condition (iii)' can be chosen without degenerate critical points in U, the G will be called a non-degenerate strictly q-concave extension of D in X.

Remarks to this definition.

I. If X is an n-dimensional complex manifold and $D \subseteq G$ are two domains in X, then G is a strictly q-concave extension of D in X if and only if $X\setminus\bar{D}$ is a strictly q-convex extension of $X\setminus\bar{G}$ in X.

II. If $\varrho: U \longrightarrow \mathbb{R}$ is as in condition (iii), then, for all

$$0\leq\alpha<\beta<\sup_{z\in U}\varrho(z),$$

the domain $D\cup\{z\in U:\ \varrho(z)<\beta\}$ is a strictly q-concave extension in X of the domain $D\cup\{z\in U:\ \varrho(z)<\alpha\}$.

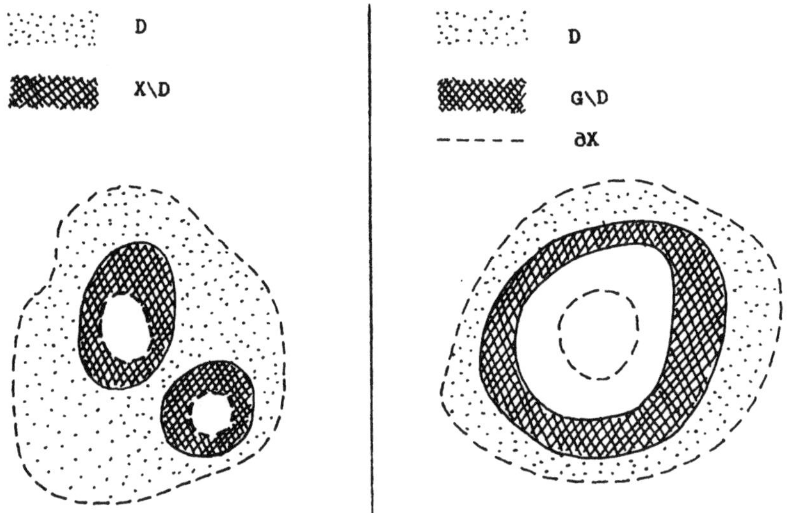

X is a q-concave extension of D

G is a strictly q-concave extension of D with respect to X

FIGURE 8

15.2. Theorem (Hartogs extension). Let E be a holomorphic vector bundle over a complex manifold X, and let $D\subseteq X$ be a domain such that X is a 1-concave extension of D. Then the restriction map

$$\mathcal{O}(X,E) \longrightarrow \mathcal{O}(D,E)$$

is an isomorphism.

Proof. In view of a lemma of Morse (Proposition 0.5 in Appendix B) and Observation 4.15, we can find a sequence D_k (k=0,1,...) of domains in X such that $D_0 \subseteq D$, each D_{k+1} is a non-degenerate strictly 1-concave extension on D_k, and

$$X = \bigcup_{k=0}^{\infty} D_k.$$

If follows from Theorem 14.4 that all restriction maps

$$\mathcal{O}(D_{k+1}, E) \longrightarrow \mathcal{O}(D_k, E)$$

are isomorphisms. Since $D_0 \subseteq D$, this implies that the restriction map

$$\mathcal{O}(X, E) \longrightarrow \mathcal{O}(D, E)$$

is an isomorphism. □

15.3. Corollary. For any 1-concave complex manifold X and all holomorphic vector bundles E over X, we have

$$\dim \mathcal{O}(X, E) < \infty.$$

Proof. Since X is 1-concave, we can find domains $D \subset\subset G \subset\subset X$ such that X is a 1-concave extension of G, and G is a 1-concave extension of D. Then, by Theorem 15.2, the restriction maps

$$\mathcal{O}(X, E) \longrightarrow \mathcal{O}(G, E) \quad \text{and} \quad \mathcal{O}(G, E) \longrightarrow \mathcal{O}(D, E)$$

are isomorphisms. Since D and G are relatively in X, this implies that all sections in $\mathcal{O}(G, E)$ and $\mathcal{O}(D, E)$ are bounded. Hence $\mathcal{O}(G, E)$ and $\mathcal{O}(D, E)$ form Banach spaces with respect to the sup-norm. Since D is relatively compact in G, the restriction map

$$\mathcal{O}(G, E) \longrightarrow \mathcal{O}(D, E)$$

is compact. Since this map is also an isomorphism and compact isomorphisms between Banach spaces exist only in the finite dimensional case, this implies the assertion. □

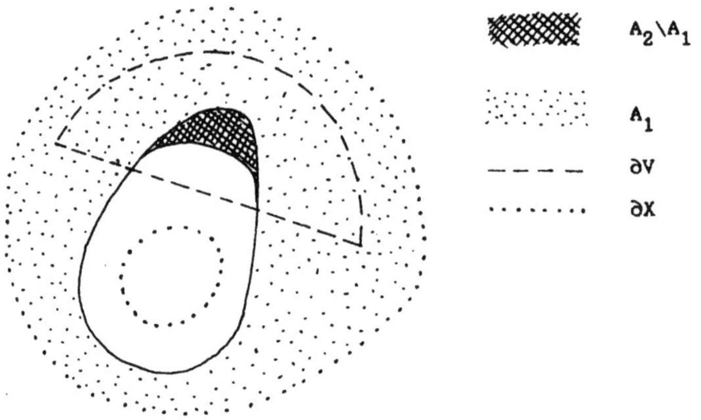

FIGURE 9: q-concave extension element $[A_1, A_2, V]$ in X

<u>15.4. Definition.</u> let X be an n-dimensional complex manifold, and $1 \le q \le n-1$ an integer.

$[A_1, A_2, V]$ will be called a <u>q-concave extension element in X</u> if A_1, A_2, and V are domains in X such that the following conditions are fulfilled (cf. Def. 13.1 for the definition of a q-concave configuration):

(i) $A_1 \subseteq A_2$,

(ii) $A_2 \setminus A_1 \subset\subset V \subset\subset X$.

(iii) There exist q-concave configurations $[U_j, \varrho_j, \varphi_j, H_j, D_j]$ in $\mathbb{C}^n$ $(j=1,2)$ with $U := U_1 = U_2$, $\varphi := \varphi_1 = \varphi_2$, and $H := H_1 = H_2$ such that, for some biholomorphic map h from U onto some neighborhood of $\bar{V}$.

$$V = h(\{z \in U: \varphi(z) < 0, \text{ Re } H(z) < 0\})$$

and

$$V \cap A_j = h(D_j) \qquad \text{for } j=1,2.$$

If, in this case, $[U, \varrho_j, \varphi, H, D_j]$ is of type T_j, then $[A_1, A_2, V]$

will be called of type (T_1, T_2).

If $A_1 \subseteq A_2$ are domains in X, then we say A_2 can be obtained from A_1 by means of a q-concave extension element in X if there exists a domain $V \subseteq X$ such that $[A_1, A_2, V]$ is a q-concave extension element in X.

15.5. Lemma. let X be an n-dimensional complex manifold, and let $D \subseteq G \subseteq X$ be domains such that G is a non-degenerate strictly q-convex extension of D in X ($1 \le q \le n-1$). Then there exists a finite number of domains

$$D = A_0 \subseteq A_1 \subseteq \dots \subseteq A_N = G$$

such that, for all $1 \le j \le N$, A_j can be obtained from A_{j-1} by means of a q-concave extension element in X.

Moreover, for any open covering $\{U_i\}_{i \in I}$ of X, these domains A_j can be chosen so that, in addition, for each $1 \le j \le N$, there exists an index $i \in I$ and a domain $V \subset\subset U_i$ such that $[A_{j-1}, A_j, V]$ is a q-concave extension element in X.

Proof. Since (q+1)-concave functions do not admit local minima, this is a straightforward modification of the proof of Lemma 12.3, using Lemma 13.2 instead of Lemma 7.5. □

15.6. Lemma. Let X be an n-dimensional complex manifold, $[A_1, A_2, V]$ a q-concave extension element in X ($2 \le q \le n-1$), and E a holomorphic vector bundle over X which is holomorphically trivial over some neighborhood $\tilde{V}$ of $\bar{V}$. Then, for all $1 \le r \le q-1$, the following assertions hold true:

(i) For each $f_1 \in Z^0_{0,r}(\bar{A}_1, E)$ and every neighborhood U of $\overline{A_2 \setminus A_1}$, there exists a form $f_2 \in Z^0_{0,r}(\bar{A}_2, E)$ such that $f_1 - f_2 \in E^{1/2 \longrightarrow 0}_{0,r}(\bar{A}_1, E)$ and, moreover, $f_2 = f_1$ on $\bar{A}_1 \setminus U$.

(ii) If A_1 is relatively compact in X, and $\bar{\partial}$ is 1/2-regular at $Z^0_{0,r}(\bar{A}_2, E)$, then $\bar{\partial}$ is 1/2-regular also at $Z^0_{0,r}(\bar{A}_2, E)$.

(iii) $E^{1/2 \longrightarrow 0}_{0,r}(\bar{A}_2, E) = Z^0_{0,r}(\bar{A}_2, E) \cap E^{1/2 \longrightarrow 0}_{0,r}(\bar{A}_1, E)$.

Proof. The proofs of parts (i) and (ii) are repetitions of the proofs of the corresponding assertions in Lemma 12.4, where instead of Corollary 12.5 (i) we have to use Theorem 14.2. The proof of part (iii) is a repetition of the proof of Lemma 12.4 (iii) in the case $n-q+1 \le r \le n$, where again instead of Corollary 12.5 (i) we have to apply Theorem 14.2. □

15.7. Proposition [1]. Let X be an n-dimensional complex manifold, let $D \subsetneq X$ be a domain whose boundary is non-degenerate strictly q-concave with respect to X [2] $(2 \le q \le n-1)$, and let E be a holomorphic vector bundle over X. Then

$$E^{1/2\longrightarrow 0}_{0,r}(\bar{D},E) = Z^{0}_{0,r}(\bar{D},E) \cap E^{0}_{0,r}(D,E) \qquad \text{for all } 1 \le r \le q-1.$$

Proof. Since the boundary of D is non-degenerate strictly q-convex with respect to X, we can find a domain $G \subseteq D$ such that D is a non-degenerate strictly q-concave extension of G in X. Then, by Lemma 15.5 and Lemma 15.6 (iii),

$$E^{1/2\longrightarrow 0}_{0,r}(\bar{D},E) = Z^{0}_{0,r}(\bar{D},E) \cap E^{1/2\longrightarrow 0}_{0,r}(\bar{G},E).$$

Since $\bar{G} \subseteq D$ and, by Theorem 1.13 (regularity of $\bar{\partial}$),

$$E^{0}_{0,r}(D,E) = E^{1/2\longrightarrow 0}_{0,r}(D,E),$$

this implies the assertion. □

15.8. Proposition [3]. Let X be an n-dimensional complex manifold, let $D \subseteq G \subseteq X$ be domains such that G is a non-degenerate strictly q-concave extension of D in X $(2 \le q \le n-1)$, and let E be a holomorphic vector bundle over X. Then, for all $1 \le r \le q-1$, the following assertions hold true:

(i) For any $f \in Z^{0}_{0,r}(\bar{D},E)$ and each neighborhood U of ∂D, there exists a form $F \in Z^{0}_{0,r}(\bar{G},E)$ such that $f - F \in E^{1/2\longrightarrow 0}_{0,r}(\bar{D},E)$ and $F = f$ on $U \setminus D$.

(ii) If D is relatively compact in X and $\bar{\partial}$ is 1/2-regular at $Z^{0}_{0,r}(\bar{D},E)$, then $\bar{\partial}$ is 1/2-regular also at $Z^{0}_{0,r}(\bar{G},E)$.

(iii) $E^{1/2\longrightarrow 0}_{0,r}(\bar{G},E) = Z^{0}_{0,r}(\bar{G},E) \cap E^{0}_{0,r}(D,E)$.

---

1) This proposition is contained also in Theorem 15.12.

2) cf. Def. 14.3

3) Assertion (i) of this proposition is contained also in Theorem 15.9, and part (iii) follows from Theorems 15.11 and 15.12.

Proof. Part (ii) follows immediately from Lemma 15.5 and part (ii) of Lemma 15.6. Taking into account Proposition 15.8, assertions (i) and (iii) follow also from Lemma 15.5 and the corresponding assertions in Lemma 15.6. □

15.9. Theorem. Let E be a holomorphic vector bundle over an n-dimensional complex manifold X, and let $D \subseteq X$ be a domain such that X is a non-degenerate q-concave extension of D ($2 \le q \le n-1$). Then, for any $f \in Z^0_{0,r}(\bar{D},E)$ with $1 \le r \le q-1$ and each neighborhood U of ∂D, there exists a form $F \in Z^0_{0,r}(X,E)$ such that $f-F \in E^{1/2 \longrightarrow 0}_{0,r}(\bar{D},E)$ and $F=f$ on $D \setminus U$.

Proof. Since X is a non-degenerate strictly q-concave extension of D, we can find a sequence D_k ($k=0,1,\dots$) of domains $D_k \subseteq X$ such that $D_0 = D$, each D_{k+1} is a non-degenerate strictly q-concave extension of D_k, and

$$X = \bigcup_{k=0}^{\infty} D_k.$$

Let $f \in Z^0_{0,r}(\bar{D},E)$. Then, by Proposition 15.8 (i), we can find a sequence $f_k \in Z^0_{0,r}(\bar{D}_k,E)$ ($k=1,2,\dots$) such that

$$f_1 - f \in E^{1/2 \longrightarrow 0}_{0,r}(\bar{D},E) \quad \text{and} \quad f_1 = f \text{ on } D \setminus U,$$

$$f_k - f_{k-1} \in E^0_{0,r}(\bar{D}_k,E) \quad \text{and} \quad f_k = f_{k-1} \text{ on } D_{k-2} \text{ if } k \ge 2.$$

Now $F := \lim f_k$ has the required properties. □

15.10. Lemma [1]. Let E be a holomorphic vector bundle over an n-dimensional complex manifold X, and let D be a domain in X such that X is a non-degenerate q-concave extension of D ($2 \le q \le n-1$). Then

$$E^0_{0,r}(X,E) = Z^0_{0,r}(X,E) \cap E^0_{0,r}(D,E) \qquad \text{for all } 1 \le r \le q-1.$$

Proof. Let $f \in Z^0_{0,r}(X,E)$ with $f|D \in E^0_{0,r}(D,E)$ be given, and let D_k be as in the proof of Theorem 15.9. It is sufficient to find a sequence $u_k \in C^0_{0,r-1}(\bar{D}_k,E)$ ($k=0,1,\dots$) such that, for all $k=0,1,\dots$,

[1] The assertion of this lemma is contained also in Theorem 15.11.

$$\bar{\partial} u_k = f \qquad \text{on } \bar{D}_k \tag{15.1}$$

and

$$u_k = u_{k-1} \quad \text{on } \bar{D}_{k-1}, \tag{15.2}$$

where $D_{-1} := \emptyset$.

By Proposition 15.7, we can find $u_0 \in C^0_{0,r-1}(D_0,E)$ with $\bar{\partial} u_0 = f$ on D_0. Assume, for some $m \geq 0$, a collection $u_k \in C^k_{0,r-1}(D_k,E)$ $(k=0,\ldots,m)$ satisfying (15.1) and (15.2) for $k=0,\ldots,m$ is already constructed. Then, by Proposition 15.8 (iii), we can find $g \in C^0_{0,r-1}(\bar{D}_{m+1},E)$ with $\bar{\partial} g = f$ on $\bar{D}_{m+1}$.

If $r=1$, then $g - u_m$ is a holomorphic section on D_m, and, by Theorem 15.2, there exists a holomorphic section $h: X \longrightarrow E$ with $h = g - u_m$ on $\bar{D}_m$. Hence, in this case, $u_{m+1} := g - h$ is a section in $C^0_{0,0}(\bar{D}_{m+1},E)$ with $\bar{\partial} u_{m+1} = f$ on $\bar{D}_{m+1}$ and $u_{m+1} = u_m$ on $\bar{D}_m$.

Now let $r \geq 2$. Then $g - u_m$ $Z^0_{0,r-1}(\bar{D}_m,E)$ and, by Proposition 15.8 (i), we can find a form $v \in Z^0_{0,r-1}(\bar{D}_{m+1},E)$ such that $g - u_m - v = \bar{\partial} w$ for some $w \in C^0_{0,r-2}(\bar{D}_m,E)$. Choose a C^∞ function χ on X with $\chi = 1$ in D_{m-1} and supp $\chi \subset\subset D_m$. Then $u_{m+1} := g - v - \bar{\partial}(\chi w)$ is a form in $C^0_{0,r-1}(\bar{D}_{m+1},E)$ with $\bar{\partial} u_{m+1} = f$ on $\bar{D}_{m+1}$ and $u_{m+1} = u_m$ on $\bar{D}_{m-1}$. □

<u>15.11. Theorem.</u> Let $D \subseteq X$ be a domain in an n-dimensional complex manifold X such that X is a q-concave extension of D $(1 \leq q \leq n-1)$, and let E be a holomorphic vector bundle over X. Then, for all $0 \leq r \leq q-1$, the restriction map

$$H^{0,r}(X,E) \longrightarrow H^{0,r}(D,E)$$

is an isomorphism (for the definition of $H^{0,r}$, see Sect. 1.14).

<u>Proof.</u> In view of the Dolbeault isomorphism (Theorem 2.14) it is sufficient to prove that the restriction map

$$H^{0,r}_0(X,E) \longrightarrow H^{0,r}_0(D,E) \tag{15.3}$$

is an isomorphism. In view of a lemma of Morse (Proposition 0.5 in Appendix B) and Observation 4.15, we can find a domain $G \subseteq D$ such that both X and D are non-degenerate q-concave extensions of G. Since the assertion is already proved for r=0 (Theorem 15.2), we can assume that $1 \le r \le q-1$. Then it follows from Lemma 15.10 that the restriction map $H_0^{0,r}(X,E) \longrightarrow H_0^{0,r}(G,E)$ and hence (15.3) is injective. In order to prove the surjectivity, now let $f \in Z_{0,r}^0(\bar{D},E)$. Then we can apply Theorem 15.9 to $f|\bar{G}$ and obtain $F \in Z_{0,r}^0(X,E)$ with $f-F \in E_{0,r}^0(G,E)$. By Lemma 15.10, it follows that $f-F \in E_{0,r}^0(D,E)$. □

15.12. Theorem. Let X be an n-dimensional complex manifold, and let $D \subseteq X$ be a domain whose boundary is non-degenerate strictly q-concave with respect to X (cf. Def. 14.3). Further, let E be a holomorphic vector bundle over X. Then, for all $1 \le r \le q-1$, the restriction map

$$H_{1/2 \longrightarrow 0}^{0,r}(\bar{D},E) \longrightarrow H_0^{0,r}(D,E)$$

is an isomorphism, where

$$H_{1/2 \longrightarrow 0}^{0,r}(\bar{D},E) := Z_{0,r}^0(\bar{D},E)/E_{0,r}^{1/2 \longrightarrow 0}(\bar{D},E).$$

The assertion on the surjectivity of this restriction map admits the following strengthening: For any $f \in Z_{0,r}^0(D,E)$ and each neighborhood U of of ∂D, there exists $F \in Z_{0,r}^0(\bar{D},E)$ with $f-F \in E_{0,r}^0(D,E)$ such that, moreover, $F|_{(D \setminus U)} = f$.

Proof. By Proposition 15.7, (15.5) is injective.

To complete the proof, let $f \in Z_{0,r}^0(D,E)$ be given. Take a domain $G \subseteq D$ such that $D \setminus U = G \setminus U$ and D is a non-degenerate strictly q-concave extension of G in X. Then $f|\bar{G}$ belongs to $Z_{0,r}^0(\bar{G},E)$ and, by Proposition 15.8 (i), we can find $F \in Z_{0,r}^0(\bar{D},E)$ with $f-F \in E_{0,r}^0(G,E)$ and $F = f$ on $D \setminus U$. By Lemma 15.10, it follows that $f-F$ belongs even to $E_{0,r}^0(D,E)$. □

15.13. Theorem (Andreotti-Grauert finiteness theorem). Let E be a holomorphic vector bundle over a q-concave n-dimensional complex manifold X ($1 \le q \le n-1$). Then

$$\dim H^{0,r}(X,E) < \infty \qquad \text{for all } 0 \le r \le q-1.$$

Proof. In view of a lemma of Morse (Proposition 0.5 in Appendix B) and Observation 4.15, we can find a non-degenerate strictly q-concave domain $D \subset\subset X$ such that X is a non-degenerate q-concave extension of D. Since the assertion is already proved for r=0 (Corollary 15.3), we may assume that $1 \le r \le q-1$. Then, by Theorem 15.11 and the Dolbeault isomorphism (Theorem 2.14),

$$\dim H^{0,r}(X,E) = \dim H^{0,r}(D,E) = \dim H_0^{0,r}(D,E).$$

In view of Theorems 15.12 and 14.5 (ii), this implies that

$$\dim H^{0,r}(X,E) = \dim H^{0,r}_{1/2 \longrightarrow 0}(\bar{D},E) < \infty . \quad \square$$

16. A uniqueness theorem for the Dolbeault cohomology of order q with respect to q-concave extensions

Let X be an n-dimensional complex manifold, and let $D \subseteq X$ be a domain such that X is a q-concave extension of D ($0 \le q \le n-1$).

In the preceding section we saw (Theorem 15.11) that then, for all $0 \le r \le q-1$, the restriction map

$$H^{0,r}(X) \longrightarrow H^{0,r}(D) \tag{16.1}$$

is an isomorphism. This is not true for r=q. For instance, consider the case r=q=0. Then (16.1) takes the form

$$\mathcal{O}(X) \longrightarrow \mathcal{O}(D), \tag{16.2}$$

and it is clear that (16.2) need not be surjective (for instance, if X is the complex plane, and D is the unit disc). However, since, by definition, the intersection of D with any connected component of X is non-empty, (16.2) is injective. In the present section we show that this uniqueness property of holomorphic functions admits the following generalization to cohomology classes:

16.1. Theorem. Let E be a holomorphic vector bundle over an n-dimensional complex manifold X, and let $D \subseteq X$ be a domain such that X is a q-concave extension of D ($0 \le q \le n-1$). Then the restriction map

$$H^{0,q}(X,E) \longrightarrow H^{0,q}(D,E) \tag{16.3}$$

is injective. If $q\geq 1$, then even the following more precise statement holds true: If $f\in Z^0_{0,q}(X,E)$ and $u\in C^0_{0,q-1}(D,E)$ with $\bar\partial u = f$ on D, then, for any neighborhood U of ∂D, there exists a from $v\in C^0_{0,q-1}(X,E)$ such that $\bar\partial v = f$ on X and $v = u$ on $D\setminus U$. □

The main step in the proof of this theorem is the following

<u>16.2. Lemma.</u> Let $[A_1,A_2,V]$ be a q-concave extension element in the n-dimensional complex manifold X ($1\leq q\leq n-1$), and E a holomorphic vector bundle over X. Suppose $f\in Z^0_{0,q}(X,E)$ and $u_1\in C^0_{0,q-1}(\bar A_1,E)$ with $\bar\partial u_1 = f$ on $\bar A_1$. Then, for any neighborhood U of $\overline{A_2\setminus A_1}$, there exists $u_2\in C^0_{0,q-1}(\bar A_2,E)$ such that $\bar\partial u_2 = f$ on $\bar A_2$ and $u_2 = u_1$ on $\bar A_1\setminus U$.

<u>Proof.</u> Since, by Lemma 13.5 (i), V is pseudoconvex and hence (cf. Theorem 5.3) <u>completely</u> pseudoconvex, it follows from Theorem 12.16 (or from Theorem 2.8.1 in [H/L]) that the equation $\bar\partial v = f$ can be solved on V with $v\in C^0_{0,q-1}(V,E)$. Now we distinguish the cases q=1 and $q\geq 2$.

If q=1, then $u_1 - v$ is a holomorphic section of E over $A_1\cap V$, and it follows from Theorem 13.8 that there is a holomorphic section $h:V \longrightarrow E$ such that $h = u_1 - v$ on $\bar A_1\cap V$. Hence, setting $u_2 = u_1$ on $\bar A_1$ and $u_2 = v+h$ on V, we obtain a continuous section $u_2:\bar A_1\cup V \longrightarrow E$ with $\bar\partial u_2 = f$ on $\bar A_1\cup V$ and $u_2 = u_1$ on $\bar A_1$. Since $\bar A_2 \subseteq \bar A_1\cup V$ this completes the proof for for q=1.

Now let $q\geq 2$. We may assume that $U \subset\subset V$. Then, by Theorem 14.2, $u_1 - v = \bar\partial w$ on $\bar A_1\cap U$ for some $w\in C^0_{0,q-2}(\bar A_1\cap U,E)$. Choose a C^∞ function χ on X such that supp $\chi \subset\subset U$ and $\chi = 1$ on $\overline{A_2\setminus A_1}$. Then, on $\bar A_1\cap U$, we have

$$u_1 - \bar\partial(\chi w) = v + \bar\partial((1-\chi)w).$$

The two sides of the last equation together define a form $u_2\in C^0_{0,q-1}(\bar A_1\cup U,E)$ with $\bar\partial u_2 = f$ on $\bar A_1\cup U$ and $u_2 = u_1$ on $\bar A_1\setminus U$. Since $\bar A_2 \subseteq \bar A_1\cup U$, this completes the proof. □

<u>Proof of Theorem 16.1.</u> For q=0 the assertion of the theorem follows from the uniqueness theorem for holomorphic functions. So we can assume that $q\geq 1$. In view of a lemma of Morse (cf. Proposition 0.5 in Appendix B) and Observation 4.15, for every neighborhood U of ∂D, we can find a domain $G \subseteq D$ such that X is a <u>non-degenerate</u> q-concave extension of G, and $G\setminus U = D\setminus U$. Now the assertion of the theorem follows from Lemma 15.5 and Lemma 16.2. □

17. The Martineau theorem on representation of the Dolbeault cohomology of a concave domain by the space of holomorphic functions on the dual domain

17.1. Definition. Let X be a domain of the form

$$X = \mathbb{C}^n \setminus D,$$

where $D \subset\subset \mathbb{C}^n$ is a convex (with respect to the real-linear structure of $\mathbb{C}^n$) domain such that $0 \in D$. Then the domain

$$X^* := \{y \in \mathbb{C}^n : \langle z,y \rangle \neq 1 \quad \text{for all } z \in \bar{D}\} \tag{17.1}$$

will be called the dual domain of X. □

17.2. Propositon. If D, X, and X^* are as in the preceding definiton, then X^* is a bounded domain of holomorphy which contains the origin.

Moreover, for any point $y \in \partial X$, there exists a sequence P_k (k=1,2,...) of (n-1)-dimensional complex planes in $\mathbb{C}^n$ which converges (in the obvious sense) to some (n-1)-dimensional complex plane P_0 in $\mathbb{C}^n$ such that

$$P_k \cap \bar{X}^* = \emptyset \qquad \text{for all } k \geq 1, \qquad \text{but } y \in P_0.$$

If the boundary of D is of class C^1, then these planes can be chosen so that, moreover,

$$P_0 \cap \bar{X}^* = \{y\}.$$

Proof. It is clear that $0 \in X^*$. That X^* is bounded follows from the fact that $0 \in D$. To complete the proof, we fix $y \in \partial X^*$. Then, by (17.1), we can find $z_0 \in \partial D$ with $\langle z_0, y \rangle = 1$. Take a sequence $z_k \in D$ (k=1,2,...) which converges to z_0, and set

$$P_k = \{x \in \mathbb{C}^n : \langle z_k, x \rangle = 1\} \qquad \text{for } k=0,1,\ldots .$$

Then P_k converges to P_0 and $y \in P_0$. Further, by (17.1), $P_k \cap \bar{X}^* = \emptyset$ for $k \geq 1$.

It remains to prove that $P_0 \cap \bar{X}^* = \{y\}$ if ∂D is of class C^1. To do this, we consider an arbitrary point $w \in P_0 \cap \bar{X}^*$. Then $\langle z_0, w\rangle = 1$ and, since $z_0 \in \partial D$, then it follows from (17.1) that $w \in \partial\bar{X}^*$. Since $w \notin X^*$, the plane

$$\{z \in \mathbb{C}^n: \langle z, w\rangle = 1\}$$

does not intersect D. Since $\langle z_0, w\rangle = 1$, this implies that this plane is the complex tangent plane of ∂D at z_0 if ∂D is of class C^1. Since the complex tangent plane is uniquely determined, it follows that $P_0 \cap \bar{X}^*$ consists only of one point. □

Remark. Let D, X, X^*, $y \in \partial X^*$, and P_k be as in Proposition 17.2, and set

$$U := \mathbb{C}^n \setminus \bigcup_{k\ 0}^{\infty} P_k.$$

If ∂D is of class C^1, then $\bar{X}^* \setminus \{y\} \subseteq U$, and there exists a function $h \in C^0\mathcal{O}(\bar{X}^* \cup U)$ (cf. Sect. 0.11 for the definition of $C^0\mathcal{O}$) which is singular at y (cf. Definition 17.5 (i) below).

In fact, take holomorphic functions h_k on $\mathbb{C}^n \setminus P_k$ which have a pole on P_k $(k=1,2,\ldots)$, and set

$$h(z) = \sum_{k=1}^{\infty} \varepsilon_k h_k(z) \qquad (z \in U)$$

for sufficiently small $\varepsilon_k \neq 0$. □

17.3. The Fantappie-Martineau map M_X. Let X be a domain of the form $X = \mathbb{C}^n \setminus \bar{D}$, where $D \subset\subset \mathbb{C}^n$ is a convex domain with $0 \in D$. Then, for any $f \in Z^0_{n,n-1}(X)$, we define a holomorphic function $M_X f$ on X^* as follows: Let $y \in X^*$. Since $\bar{D}$ is compact, we can find a convex domain $G_y \subset\subset \mathbb{C}^n$ with C^∞ boundary such that $D \subset\subset G_y$ and

$$\langle z, y\rangle \neq 1 \qquad \text{for all } z \in \bar{G}_y. \tag{17.2}$$

Define

$$M_X f(y) = -\frac{(n-1)!}{(2\pi i)^n} \int_{\partial G_y} \frac{f(x)}{(1-\langle x, y\rangle)^n}. \tag{17.3}$$

By Stokes' theorem, this definition is independent of the choice of G_y. The continuous linear map

$$M_X : Z^0_{n,n-1}(X) \longrightarrow \mathcal{O}(X^*)$$

obtained in this way will be called the Fantappie-Martineau map for X.

17.4. Proposition [1]. Let X and M_X be as in Definition 17.3. Then

$$\operatorname{Ker} M_X = E^0_{n,n-1}(X),$$

where $\operatorname{Ker} M_X := \{f \in Z^0_{n,n-1}(X) : M_X f = 0\}$.

Proof. Let $f \in E^0_{n,n-1}(X)$. Then $f = \bar{\partial} u$ for some $u \in C^0_{n,n-2}(X)$, and, for any fixed $y \in X^*$, the form

$$\frac{f(x)}{(1-\langle x,y\rangle)^n} = d_x \left[\frac{u(x)}{(1-\langle x,y\rangle)^n} \right]$$

is exact on ∂G_y, where G_y is as in Definition 17.3. Therefore, it follows from Stokes' theorem and (17.3) that $M_X f(y) = 0$.

Now let $f \in \operatorname{Ker} M_X$. In order to prove that $f \in E^0_{n,n-1}(X)$, we take a number $R < \infty$ such that $\bar{D}$ is contained in the ball $B_R := \{z \in \mathbb{C}^n : |z| < R\}$. Consider also the larger ball $B_{R+1} := \{z \in \mathbb{C}^n : |z| < R+1\}$. Then X is an $(n-1)$-concave extension of $\mathbb{C}^n \setminus \bar{B}_R$ and $\mathbb{C}^n \setminus \bar{B}_R$ is an $(n-1)$-convex extension of $B_{R+1} \setminus \bar{B}_R$ (cf. Definitions 15.1 and 12.1). Therefore, by Theorems 16.1 and 12.14, it is sufficient to prove that the restriction of f to $B_{R+1} \setminus \bar{B}_R$ belongs to $E^0_{n,n-1}(B_{R+1} \setminus \bar{B}_R)$. Set

$$w_1(z,x) = (\bar{z}_1, \ldots, \bar{z}_n)$$

for $x \in \partial B_R$ and $z \in B_{R+1} \setminus \bar{B}_R$, and

$$w_2(z,x) = (\bar{x}_1, \ldots, \bar{x}_n)$$

for $x \in \partial B_{R+1}$ and $z \in B_{R+1} \setminus \bar{B}_R$. Then (w_1, w_2) is a Leray data for $(D; -\partial B_R, \partial B_{R+1})$ (cf. Sect. 3.4), and if F is the $(0,n-1)$-form on X with $f(z) = F(z) \wedge dz_1 \wedge \ldots \wedge dz_n$, then the corresponding piecewise Cauchy-Fantappie formula (Theorem 3.12) yields the representation

---

[1] The assertion of this proposition is contained also in Theorem 17.9.

$$F(z) = g(z) - \frac{1}{(2\pi i)^n} \int_{\partial B_R} \frac{F(x) \wedge \det(\bar{z}, \overbrace{d\bar{z}}^{n-1}) \wedge dx_1 \wedge \ldots \wedge dx_n}{\langle \bar{z}, x-z \rangle^n}$$

for all $z \in B_{R+1} \setminus \bar{B}_R$, where g is some form which belongs to $E^0_{n,n-1}(B_{R+1} \setminus \bar{B}_R)$ (the integral over ∂B_{R+1} vanishes, because w_2 is independent of z). Since $F(x) \wedge dx_1 \wedge \ldots \wedge dx_n = f(x)$, therefore it is sufficient to prove that the integral

$$\int_{x \in \partial B_R} \frac{f(x)}{\langle \bar{z}, x-z \rangle^n} \det(\bar{z}, \overbrace{d\bar{z}}^{n-1})$$

vanishes for all $z \in B_{R+1} \setminus \bar{B}_R$. By Proposition 0.9, this integral is equal to

$$\int_{x \in \partial B_R} \frac{f(x)}{\left[\left\langle \frac{\bar{z}}{|z|^2}, x \right\rangle - 1\right]^n} \det\left(\frac{\bar{z}}{|z|^2}, \overbrace{d\frac{\bar{z}}{|z|^2}}^{n-1} \right).$$

Since $\langle \bar{z}/|z|^2, x \rangle = 1$ for all $x \in \bar{B}$ and $z \in B_{R+1} \setminus \bar{B}_R$, it follows that the last integral is equal to $(-1)^n (2\pi i)^n/(n-1)! M_X f(\bar{z}/|z|^2) = 0$. □

<u>17.5. Definition.</u> Let $D \subseteq X$ be a domain in a complex manifold X, and $y \in \partial D$.

(i) If h is a holomorphic function on D, then we say, as usually, h is <u>singular</u> at y (or y is singular for h) if, for each neighborhood U of y, the following condition is fulfilled: If W is a connected component of $U \cap D$ with $y \in W$, then there do not exist a neighborhood V of y and a holomorphic function v on V with $v = h$ on $V \cap W$.

(ii) If f is a $\bar{\partial}$-closed (s,r)-form with $r \geq 1$ on D, then we say f is <u>singular</u> at y if there do not exist a neighborhood U of y and a continuous $(s,r-1)$-form u on $U \cap D$ such that $\bar{\partial} u = f$ on $U \cap D$.

<u>17.6. Definition.</u> Let X be a domain of the form $X = \mathbb{C}^n \setminus D$, where $D \subset\subset \mathbb{C}^n$ is a <u>strictly</u> convex domain with C^2 boundary such that $0 \in D$. Then, clearly, for any $y \in \partial X^*$ there exists a uniquely determined point $z \in \partial X$

with $\langle z,y\rangle=1$. Conversely, for each $z\in\partial X$, there exists a uniquely determined $y\in\partial X^*$ with $\langle z,y\rangle=1$. Thus, by means of the equation

$$\langle z,\phi_X(z)\rangle = 1$$

a bijective map

$$\phi_X:\partial X \longrightarrow \partial X^*$$

is defined. Obviously, this map is continuous [1].

17.7. Theorem. Let X be a set of the form $X = \mathbb{C}^n\setminus D$, where $D \subset\subset \mathbb{C}^n$ is a strictly convex domain with C^2 boundary such that $0\in D$. Then there exists a continuous linear map

$$N: \mathcal{O}(X^*) \longrightarrow Z^0_{n,n-1}(X)$$

such that the following assertions hold true:

(i) If M_X is the Fantappie-Martineau map for X, then $M_XNh = h$ for all $h\in\mathcal{O}(X^*)$.

(ii) Let Y be a subset of ∂X^*. If $h\in\mathcal{O}(X^*)$ admits a continuous extension onto $X^*\cup Y$, then Nh admits a continuous extension onto $\phi_X^{-1}(Y)$. In particular,

$$NC^0\mathcal{O}(\bar{X}^*) \subseteq Z^0_{n,n-1}(\bar{X})$$

(for the definition of $C^0\mathcal{O}$, see Sect. 0.11). Moreover, if $h\in\mathcal{O}(X^*)$ admits a holomorphic extension into some neighborhood of $X^*\cup Y$, then Nh admits a continuous and $\bar{\partial}$-closed extension into some neighborhood of $X\cup\phi_X^{-1}(Y)$.

(iii) A function $h\in\mathcal{O}(X^*)$ is singular at a point $y\in\partial X^*$ if and only if the form Nh is singular at the point $\phi_X^{-1}(y)$.

(iv) $$\dim\,[Z^0_{n,n-1}(\bar{X})/E^0_{n,n-1}(X)] < \infty.$$

Proof. Take a strictly convex C^2 function $\varrho:\mathbb{C}^n\longrightarrow\mathbb{R}$ with the following properties :

---

1) In the proof of Theorem 17.7 below we shall see that ϕ_X is even a C^1 diffeomorphism.

(a) $$D = \{z\in\mathbb{C}^n:\ \varrho(z)<0\};$$

(b) $$\varrho(0) = \min_{z\in\mathbb{C}^n} \varrho(z);$$

(c) there exists $R<\infty$ with $\varrho(z) = |z|^2$ if $|z|\geq R$ or $|z|\leq 1/R$.

Then, for any $z\in\mathbb{C}^n\setminus\{0\}$, $D_z:=\{x\in\mathbb{C}^n:\ \varrho(x)<\varrho(z)\}$ is a strictly convex C^2 domain, and $T_z:=\{x\in\mathbb{C}^n:\ \langle\nabla^{\mathbb{C}}\varrho(z),x-z\rangle=0\}$ is the complex tangent plane of ∂D_z at z (cf. Remark 2.10 for the definition of $\nabla^{\mathbb{C}}\varrho$). Hence

$$\langle\nabla^{\mathbb{C}}\varrho(z),x-z\rangle = 0 \tag{17.4}$$

for all $z,x\in\mathbb{C}^n$ with $\varrho(x)<\varrho(z)$. In particular,

$$\langle\nabla^{\mathbb{C}}\varrho(z),z\rangle = 0 \tag{17.5}$$

for all $z\in\mathbb{C}^n\setminus\{0\}$. Therefore, setting

$$\phi_\varrho(z):=\frac{\nabla^{\mathbb{C}}\varrho(z)}{\langle\nabla^{\mathbb{C}}\varrho(z),z\rangle}$$

for $z\in\mathbb{C}^n\setminus\{0\}$, we obtain a C^1 map ϕ_ϱ from $\mathbb{C}^n\setminus\{0\}$ into itself. It is easy to show that ϕ_ϱ is even an orientation preserving C^1 diffeomorphism from $\mathbb{C}^n\setminus\{0\}$ onto itself. Further, it is clear that

$$\langle z,\phi_\varrho(z)\rangle = 1 \tag{17.6}$$

for all $z\in\mathbb{C}^n\setminus\{0\}$. In particular (cf. Def. 17.6),

$$\phi_\varrho(z) = \phi_X(z). \tag{17.7}$$

Moreover, since $D = \{\varrho<0\}$, it follows from (17.4) that $\langle x,\phi_\varrho(z)\rangle \neq 1$ if $z\in X$ and $x\in\bar{D}$. Hence $\phi_\varrho(X) \subseteq X^*$. Since, on the other hand, by (17.6), $\phi_\varrho(D\setminus\{0\}) \subseteq \mathbb{C}^n\setminus\bar{X}^*$ and ϕ_ϱ is a bijection of $\mathbb{C}^n\setminus\{0\}$, we see that

$$\phi_\varrho(X) = X^* \qquad\text{and}\qquad \phi_\varrho(D\setminus\{0\}) = \mathbb{C}^n\setminus\bar{X}^*. \tag{17.8}$$

Using the diffeomorphism ϕ_ϱ, now we construct the map N. Let $h\in\mathcal{O}(X^*)$ be given. Since h is holomorphic and, by Proposition 0.10, $\omega'(z)\wedge\omega(y)$ is closed on $M:=\{(z,y)\in(\mathbb{C}^n\setminus\{0\})\times X^*:\ \langle z,y\rangle=1\}$, it follows that also the form $h(y)\omega'(z)\wedge\omega(y)$ is closed for all (z,y) in M. Since, by (17.6) and (17.8), the map $X\ni z \longrightarrow \langle z,\phi_\varrho(z)\rangle$ is a C^1 map from X into M, it follows that the form

$$Nh(z):= h(\phi_\varrho(z))\omega'(z)\wedge\omega(\phi_\varrho(z)), \qquad z\in X, \tag{17.9}$$

is closed on X. So a continuous linear map

$$N:\mathcal{O}(X^*) \longrightarrow Z^0_{n,n-1}(X)$$

is defined. It remains to prove assertions (i)-(iv).

First we prove (i). Let $h\in\mathcal{O}(X^*)$ and $y\in X^*\setminus\{0\}$. Further, let G_y be as in Definition 17.3. Then

$$MNh(y) = -\frac{(n-1)!}{(2\pi i)^n}\int\limits_{x\in\partial G_y}\frac{h(\phi_\varrho(x))\omega'(x)\wedge\omega(\phi_\varrho(x))}{(1-\langle x,z\rangle)^n}.$$

In view of (17.6), this yields the relation

$$M_X Nh(y) = -\frac{(n-1)!}{(2\pi i)^n}\int\limits_{z\in\phi_\varrho(\partial G_y)}\frac{h(z)\omega'(\phi_\varrho^{-1}(z))\wedge\omega(z)}{\langle\phi_\varrho^{-1}(z),z-y\rangle^n}. \tag{17.10}$$

Set $W^* = \phi_\varrho(\mathbb{C}^n\setminus\bar{G}_y)\cup\{0\}$. Since $\mathbb{C}^n\setminus\bar{G}_y$ is relatively compact in X, and ϕ_ϱ is an orientation preserving diffeomorphism, it follows from (17.8) that

$$W^* \subset\subset X^* \tag{17.11}$$

and

$$\partial W^* = -\phi_\varrho(\partial G_y). \tag{17.12}$$

Since, by (17.6), $\langle\phi_\varrho^{-1}(y),y\rangle=1$, whereas, by (17.2), $\langle x,y\rangle\neq 1$ for all $x\in\bar{G}_y$, we see that $\phi_y^{-1}(y)\notin\bar{G}_y$, i.e.

$$y\in W^*. \tag{17.13}$$

Further, by (17.2) and (17.6),

$$\langle \phi_\varrho^{-1}(z), z-y\rangle = 1 - \langle \phi_\varrho^{-1}(z), y\rangle = 0 \qquad \text{if } z\in\partial W^*. \tag{17.14}$$

From (17.10)-(17.14), by the Cauchy-Fantappie formula (cf. the supplement following Theorem 2.7), now we obtain that $M_X Nh(y) = h(y)$. This proves assertion (i).

Assertion (ii) follows immediately from (17.9).

Next, we prove assertion (iii). Let $h\in\mathcal{O}(X^*)$ and $y\in\partial X^*$. If h is not singular at y, i.e. if h admits a holomorphic extension into some neighborhood of y, then, by (17.9), Nh is not singular at $\phi_\varrho^{-1}(y)$.

Conversely, assume that Nh is not singular at $\phi_\varrho^{-1}(y)$, i.e. there exists a neighborhood U of $\phi_\varrho^{-1}(y)$ such that $Nh = \bar\partial u$ on $U\cap X$ for some $u\in C^0_{n,n-2}(U\cap X)$. Take a C^∞ function χ on $\mathbb{C}^n$ with supp $\chi \subset\subset U$ and $\chi = 1$ in a neighborhood $V \subset\subset U$ of $\phi_\varrho^{-1}(y)$. Since D is <u>strictly</u> convex, we can find a convex set $D_y \subseteq D$ with $D_y\cap(\mathbb{C}^n\setminus V) = D\cap(\mathbb{C}^n\setminus V)$ and $\phi_\varrho^{-1}(y)\in\mathbb{C}^n\setminus\bar D_y$. Set $X_y = \mathbb{C}^n\setminus\bar D_y$. Since $y\in\partial X^*$ and hence, by (17.7), $\phi_\varrho^{-1}(y) = \phi_X^{-1}(y)$ is the only point $z\in\bar D$ with $\langle z,y\rangle=1$, so we see that $y\in X_y^*$. Now we set $f = Nh - \bar\partial(\chi u)$. Then $f\in Z^0_{n,n-1}(X_y)$ and it follows from Definition 17.3 that

$$M_X(f|X) = M_{X_y}(f)\big|X^*.$$

Since, by Proposition 17.4, $M_X\bar\partial(\chi u) = 0$ and, by assertion (i), $M_X Nh = h$, this yields the relation

$$h = (M_{X_y} f)\big|X^*.$$

Hence h admits a holomorphic extension into $X_y \ni y$, and so h is not singular at y. This proves assertion (iii).

Assertion (iv) follows immediately from (ii) and (iii).

<u>17.8. Corollary.</u> Let X be an n-dimensional complex manifold, $\varrho: X \longrightarrow \mathbb{R}$ an (n-1)-concave function, and $D:= \{x\in X: \varrho(z)<0\}$. Then, for any $y\in\partial D$ with $d\varrho(y) = 0$, there exist arbitrarily small pseudoconvex neighborhoods U of y such that

$$\dim\,[Z^0_{0,n-1}(U\cap\bar D)/E^0_{0,n-1}(U\cap D)] = \infty.$$

More precisely, for any $y\in\partial D$ with $d\varrho(y) \neq 0$, there exist arbitrarily small pseudoconvex neighborhoods U of y such that the following assertion holds true: for each $x \in U \cap \partial D$, we can find an open set $V_x \subseteq X$ with

$$(U \cap \bar{D}) \setminus \{x\} \subseteq V_x \qquad x \notin V_x,$$

as well as a form $f \in Z^0_{0,n-1}(V_x)$ which is singular at x.

Proof. Since $d\varrho(y) \neq 0$ and $-\varrho$ is strictly plurisubharmonic, by an appropriate choice of local holomorphic coordinates in a neighborhood of y, we can achieve that $-\varrho$ is strictly convex (cf., for instance, Theorem 1.4.14 in [H/L]). Hence, we may assume that X is an open subset of $\mathbb{C}^n$ and $-\varrho$ is a strictly convex function on x. Then it is easy to find a strictly convex C^2 domain $G \subset\subset X$ such that $G \subseteq D$ and, for some neighborhood U of y, $\bar{G} \cap U = \bar{D} \cap U$. Without loss of generality, we may assume that $0 \in G$. Set $Y = \mathbb{C}^n \setminus \bar{G}$.

Then, by the remark following the proof of Proposition 17.2, for each point $x \in \partial D \cap U = \partial G \cap U$, we can find a holomorphic function h in a neighborhood of $\bar{Y}^* \setminus \{\phi_Y(x)\}$ which is singular at $\phi_Y(x)$. Now the assertion follows from Theorem 17.7. □

17.9. Theorem (Martineau theorem). Let X be a domain of the form $X = \mathbb{C}^n \setminus \bar{D}$, where $D \subset\subset \mathbb{C}^n$ is a convex [1] domain with $0 \in D$. Then the the Martineau map M_X induces an isomorphism

$$\hat{M}_X : H_0^{n,n-1}(X) \longrightarrow \mathcal{O}(X^*)$$

(for the definition of $H_0^{n,n-1}$, see Sect. 1.14).

Proof. That $\hat{M}_X$ is well-defined and injective we know from Proposition 17.4. In order to prove that $\hat{M}_X$ is surjective, we consider a function $h \in \mathcal{O}(X^*)$. We have to find $f \in Z^0_{n,n-1}(X)$ with $M_X f = h$.

Since D is convex, we can find a sequence of strictly convex open sets $D_k \subset\subset \mathbb{C}^n$ (k=1,2,..) with C^2 boundaries such that

$$D_k \supset\supset D_{k+1} \qquad (k=1,2,\dots) \tag{17.15}$$

and

$$D = \bigcap_{k=1}^{\infty} D_k.$$

[1] We do not assume that D is strictly convex.

Set $X_k = \mathbb{C}^n \setminus \bar{D}_k$. Then $X_k \subset\subset X_{k+1}$, $X_k \supset\supset X^*_{k+1}$ (k=1,2,...), and

$$X = \bigcup_{k=1}^{\infty} X_k, \qquad X^* = \bigcap_{k=1}^{\infty} X^*_k.$$

From Theorem 17.7 (i) we obtain a sequence $f_k \in Z^0_{n,n-1}(X_k)$ with

$$M_{X_k} f_k = h\big|_{X^*_k} \qquad (k=1,2,\ldots).$$

Now we want to modify this sequence so that we obtain a sequence $F_k \in Z^0_{n,n-1}(X_k)$ such that again

$$M_{X_k} F_k = h\big|_{X^*_k} \qquad \text{for} \quad k=1,2,\ldots, \tag{17.16}$$

and, moreover,

$$F_k\big|_{X_{k-2}} = f_{k-1}\big|_{X_{k-2}} \qquad \text{for} \quad k=3,4,\ldots. \tag{17.17}$$

Set $F_1 = f_1$ and assume that, for some $m \geq 1$, a collection $F_k \in Z^0_{n,n-1}(X_k)$ (k=1,...,m) is already costructed such that (17.16) and (17.17) hold true for k=1,...,m. In view of Proposition 17.4, then $f_{m+1} - F_m = \bar{\partial}u$ on X for some $u \in C^0_{n,n-2}(X_m)$. Take a C^∞ function χ on $\mathbb{C}^n$ with supp $\chi \subset\subset X_m$ and $\chi = 1$ on X_{m-1}, and set $F_{m+1} = f_{m+1} - \bar{\partial}(\chi u)$. Then it is clear that (17.17) holds true for k=m+1, and it follows from Proposition 17.4 that also (17.16) is valid for k=m+1. Setting $f = \lim F_k$ we conclude the proof. □

<u>Remark.</u> Under the hypotheses of Theorem 17.9 there exists a continuous linear operator

$$N: \mathcal{O}(X^*) \longrightarrow Z^0_{n,n-1}(X)$$

such that $M_X Nh = h$ for all $h \in \mathcal{O}(X^*)$.

In fact, take a strictly convex C^2 function $\varrho: X \longrightarrow \mathbb{R}$ such that

$$\inf_{z \in X} \varrho(z) = 0, \qquad \sup_{z \in X} \varrho(z) = \infty,$$

and, for any $0 < \alpha < \infty$, the domain $D_\alpha := \{z \in X: \varrho(z) < \alpha\}$ is relatively compact in $\mathbb{C}^n$, and the set $D_\alpha \cup (\mathbb{C}^n \setminus X)$ is open. Then a modification of the arguments given in the proof of Theorem 17.7 (i) shows that the required map N may be defined by setting

$$Nh(z) = h(\phi_\varrho(z))\omega'(z)\wedge\omega(\phi_\varrho(z))$$

for all $z \in X$ and $h \in \mathcal{O}(X^*)$, where $\phi_\varrho(z) := \nabla^{\mathbb{C}}\varrho(z)/\langle\nabla^{\mathbb{C}}\varrho(z), z\rangle$.

17.10. Corollary. Let X be a domain of the form $X = \mathbb{C}^n \setminus \bar{D}$, where $D \subset\subset \mathbb{C}^n$ is a convex domain with $0 \in D$. Then

$$\dim H^{0,n-1}(X) = \infty.$$

Proof. Notice that $H^{0,n-1}(X) \cong H^{n,n-1}(X)$ and, by Theorem 2.14 (smoothing of the Dolbeault cohomology) $H^{n,n-1}(X) \cong H_0^{n,n-1}(X)$. Therefore, the assertion follows from Theorem 17.9 and the fact that, by Proposition 17.2, $\dim \mathcal{O}(X^*) = \infty$. □

Remark. In distinction to the strictly concave case, where we have the local result given by Corollary 17.8, for Corollary 17.10 there does not exist a local version. In fact, under the hypotheses of Corollary 17.10 it is possible that there exist points $y \in \partial X$ such that, for any sufficiently small pseudoconvex neighborhood, $U \cap X$ is also pseudoconvex and hence $H^{0,n-1}(U \cap X) = 0$.

18. Solution of the E. Levi problem for the Dolbeault cohomology: the Andreotti-Norguet theorem on infiniteness of the Dolbeault cohomology of order q on q-concave-(n-q-1)-convex manifolds

If f is a holomorphic function on a domain $D \subseteq \mathbb{C}^n$ such that, for some sequence $y_k \in D$ with $y_k \longrightarrow y \in \partial D$, $\sup f_k(y_k) = \infty$, then f is singular at y. This admits the following generalization to $\bar\partial$-closed (0,r)-forms:

18.1. Lemma. Let $D \subseteq \mathbb{C}^n$ be a domain, f a continuous $\bar\partial$-closed (0,r)-form on D ($0 \le r \le n$), and $y \in \partial D$. Suppose there exist a neighborhood U of y and a sequence $Y_k \subseteq U \cap D$ of r-dimensional closed complex submanifolds of $U \cap D$ which converges to some closed complex submanifold Y of U [1)] such that

---

1) Here "convergence" means that there are holomorphic functions h and h_k in a neighborhood of Y such that $Y = \{h=0\}$, $Y_k = \{h_k = 0\}$, $dh(z) \neq 0$ for all $z \in Y$, and h_k tends to h together with the first-order derivatives, uniformly on the compact sets.

$$Y = \{y\} \cup (Y \cap D),$$

and the following condition is fulfilled: for any neighborhood $V \subset\subset U$ of y, there exists a continuous $\bar\partial$-closed $(r,0)$-form g on $V \cap D$ with

$$\sup_k \left| \int_{Y_k \cap V} f \wedge g \right| = \infty.$$

Then f is singular at y (cf. Def. 17.5).

Proof. Assume f is not singular at y and $r \geq 1$ (for $r=0$ the assertion is trivial). Then there exist a neighborhood V of y and a continuous $(0,r-1)$-form u on $V \cap D$ such that $\bar\partial u = f$ on $V \cap D$. Choose an open ball $B \subset\subset V \cap U$ centered at y such that the intersection $Y \cap \partial B$ is transversal. Then, for any sufficiently large k, the intersection $Y_k \cap \partial B$ is also transversal, and it follows from Stokes' theorem that, for any continuous $\bar\partial$-closed $(r,0)$-form g on $V \cap D$,

$$\sup_k \left| \int_{Y_k \cap B} f \wedge g \right| = \sup_k \left| \int_{Y_k \cap \partial B} u \wedge g \right| = \left| \int_{Y \cap \partial B} u \wedge g \right| < \infty. \qquad \square$$

The main result of this paragraph is the following

18.2. Theorem (Andreotti-Norguet theorem, local version). Let $D \subseteq \mathbb{C}^n$ be a domain with C^2 boundary, let $0 \leq q \leq n-1$, and let $y \in \partial D$ be a point with

$$p^+_{\partial D}(y) = n-q-1 \qquad \text{and} \qquad p^-_{\partial D}(y) = q$$

(for the definition of $p^{\pm}_{\partial D}(y)$, see Sect. 4.10). Then there exists a neighborhood U of y such that, for all neighborhoods $V \subseteq U$ of y,

$$\dim H^{0,q}(D \cap V) = \infty. \tag{18.1}$$

More precisely, then even the following statement holds true: There exists a neighborhood U of y such that, for all $x \in \partial D \cap U$, we can find a neighborhood U_x of $(\bar D \cap U) \setminus \{x\}$ and a form $f_x \in Z^{\infty}_{0,q}(U_x)$ which is singular at x (cf. Def. 17.5).

Proof. We may assume that $q \geq 1$; for $q=0$ the assertion is trivial, because then, with respect to appropriate local holomorphic coordinates,

∂D is strictly convex in a neighborhood of y (cf., for instance, Theorem 1.4.14 in [H/L]).

Since $p^{-}_{\partial D}(y)=q$, using assertion (vi) in Theorem 5.13, we can find a neighborhood U of y such that, for any $x\in\partial D\cap U$, there is a q-dimensional closed complex submanifold Z_x of U which depends continuously on x such that $Z_x = \{x\} \cup \{Z_x \cap D\}$. Since $p^{+}_{\partial D}(y)=n-q-1$, after shrinking U, by Lemma 5.8, there exists an $(n-q)$-convex function in U with $U\cap D = \{x\in U: \varrho(z)<0\}$. Since Z_x depends continuously on x, after a further shrinking of U, for any $x\in\partial D\cap U$, we can find holomorphic coordinates $z^x_1,\dots,z^x_n$ in U which depend continuously on x such that

$$Z_x = \{w\in U:\ z^x_1(w)=\dots=z^x_{n-q}(w)=0\}.$$

By Proposition 4.4 (i), these coordinates can be chosen so that, moreover, ϱ is strictly plurisubharmonic with respect to $z^x_1,\dots,z^x_{n-q}$, and, by Lemma 7.3, we may assume that ϱ is even normalized $(n-q)$-convex with respect to these coordinates. By Proposition 5.5, $d\varrho(y) \neq 0$.

Now it is clear that the assertion of the theorem is a consequence of the following more special

Lemma. Let ϱ be a normalized $(n-q)$-convex function in a neighborhood $U \subseteq \mathbb{C}^n$ of the origin such that $\varrho(0) = 0$ and $d\varrho(0) \neq 0$. Let $D:= \{z\in U: \varrho(z)<0\}$ and $Z:= \{z\in\mathbb{C}^n:\ z_1=\dots=z_{n-q}=0\}$. If $Z\cap U = \{0\}\cup(Z\cap D)$, then, for any sufficiently large number $M<\infty$, the $(0,q)$-form

$$f(z):= \frac{d\bar z_{n-q+1} \cdots d\bar z_n}{[F_\varrho(z,0) + M(|z_{n-q-1}|^2 + \dots + |z_n|^2)]^{n-q-1}} \tag{18.2}$$

is $\bar\partial$-closed in a neighborhood of $\bar D\setminus\{0\}$, but singular at the point 0.

Proof of this lemma. Since Z is contained in the complex tangent plane of ∂D at 0, the restriction of $F_\varrho(\cdot,0)$ (cf. Sect. 7.1 for the definition of the Levi polynomial F_ϱ) to Z is a quadratic homogeneous polynomial. Therefore, we can find $N<\infty$ with

$$|F_\varrho(z,0)| \le N|z|^2 \qquad \text{for all } z\in Z. \tag{18.3}$$

Let C, β be as in estimate (7.2) (in Definition 7.2) with $n-q$ instead of q, and let $M<\infty$ be so large that

$$M \ge C + \beta, \tag{18.4}$$

and

$$\frac{M - N}{M + N} \geq \sin \frac{\pi}{4q}. \tag{18.5}$$

Set

$$h(z) = F_{\varrho}(z,0) + M(|z_{n-q+1}|^2 + \ldots + |z_n|^2).$$

Then it follows from (18.4) and (7.2) that

$$\operatorname{Re} h(z) \geq -\varrho(z) + \beta |z|^2 \tag{18.6}$$

for all $z \in U$. In particular, $h(z) \neq 0$ for all z in some neighborhood of $(\bar{D} \setminus \{0\}) \cap U$. Therefore equation (18.2) defines a continuous $(0,q)$-form f in this neighborhood. Since h is holomorphic with respect to $z_1, \ldots, z_{n-q}$, this form is $\bar{\partial}$-closed.

It remains to prove that f is singular at 0. We want to use Lemma 18.1. From (18.3) and (18.5) we get the estimates $|h(z)| \leq (M+N)|z|^2$ and $\operatorname{Re} h(z) \geq (M-N)|z|^2 \geq |h(z)| \sin \pi/4q$. Hence

$$\operatorname{Re} \frac{1}{(h(z))^q} \geq \frac{\alpha}{|z|^{2q}} \qquad \text{for all } z \in Z, \tag{18.7}$$

where $\alpha := (M+N)^{-1} \sin \pi/4$. Let $t_1, \ldots, t_{2n}$ be the real coordinates in $\mathbb{C}^n$ with $z_j = t_j + it_{j+n}$. Set $g(z) = dz_{n-q+1} \wedge \ldots \wedge dz_n/(2i)^q$ and

$$Z_k = Z - \frac{1}{k}\left(\frac{\partial \varrho}{\partial x_1}(0) , \ldots , \frac{\partial \varrho}{\partial x_n}(0) \right)$$

for $k=1,2,\ldots$. Then the manifolds Z_k converge to Z, and

$$f \wedge g = \frac{dx_{n-q+1} \wedge \ldots \wedge dx_n \wedge dx_{2n-q+1} \wedge \ldots \wedge dx_{2n}}{h^q} .$$

Therefore, by (18.7), for each neighborhood $V \subset\subset U$ of 0,

$$\sup_k \operatorname{Re} \int_{Z_k \cap V} f \wedge g = \infty.$$

So it follows from Lemma 18.1 that f is singular at 0. □

Theorem 18.2 admits several globalizations. For instance, we have

18.3. Theorem (Andreotti-Norguet theorem, a global version). Let E be a holomorphic vector bundle over an n-dimensional complex manifold X, and let $D \subset\subset X$ be a strictly (n-q-1)-convex domain ($0 \leq q \leq n-1$). If there exists at least one point $y \in \partial D$ such that ∂D is of class C^2 in a neighborhood of y and $p^-_{\partial D}(y) = q$, then

$$\dim H^{0,q}(D,E) = \infty.$$

More precisely, then there exists a neighborhood U of y such that, for any $x \in \partial D \cap U$, we can find a neighborhood U_x of $\bar{D} \setminus \{x\}$ and a form $f_x \in Z^{\infty}_{0,q}(U_x,E)$ which is singular at x.

Proof. By Theorem 18.2, we have a neighborhood U of y such that, for any $x \in \partial D \cap U$, we can find a neighborhood W_x of $(\overline{D \cap U}) \setminus \{x\}$ and a form $F_x \in Z^{\infty}_{0,q}(W_x,E)$ which is singular at x.

Fix $x \in \partial D$ and a neighborhood $U_x \subset\subset U$ of x. Since D is strictly (n-q-1)-convex, there is an (n-q)-convex function ϱ in some neighborhood G of ∂D such that $D \cap G = \{z \in G: \varrho(z) < 0\}$. Take $\varepsilon > 0$ sufficiently small. Then $D_\varepsilon := D \cup \{z \in G: \varrho(z) < \varepsilon\}$ is an (n-q-1)-convex extension of $D_{-\varepsilon} := D \setminus \{z \in G: \varrho(z) > -\varepsilon\}$, and we have the relation $D_\varepsilon \cap (U \setminus U_x) \subseteq W_x$. Take a C^∞ function χ on X such that supp $\chi \subset\subset U \setminus D_-$ and $\chi = 1$ in some neighborhood of $U_x \setminus D$. Then, after extending by zero, $\bar{\partial}(\chi F_x)$ is a continuous $\bar{\partial}$-closed (0,q+1)-form on D which vanishes on D_-. Hence, by Theorem 12.14, we can solve the equation $\bar{\partial} u = \bar{\partial}(\chi F\ )$ with $u \in C^{\infty}_{0,q}(D_\varepsilon,E)$. Setting $f_x := \chi F_x - u$ we conclude the proof. □

Notice also the following globalization of Theorem 18.2:

18.4. Theorem (Andreotti-Norguet theroem, a second global version). Let E be a holomorphic vector bundle over an n-dimensional complex manifold X, and let $D \subset\subset X$ be a strictly (n-q-1)-convex domain such that at least one of the following two conditions is fulfilled:

(i) X is a q-concave extension of D;

(ii) X is an (n-q-1)-convex extension of D, and there exists at least one point $y \in \partial D$ such that ∂D is of class C^2 in a neighborhood of y and $p^-_{\partial D}(y) = q$.

Then

$$\dim H^{0,q}(X,E) = \infty.$$

Proof. By Theorem 18.3, this follows from Theorem 16.1 (if condition (i) is fulfilled) and Theorem 12.14 (if condition (ii) is fulfilled). □

The form f_x in Theorem 18.2 which is constructed in the proof of this theorem is unbounded at x. Actually, this form can be chosen to be continuous in x:

18.5. Theorem (Andreotti-Norguet theorem, local version with uniform estimates). Let $D \subseteq \mathbb{C}^n$ be a domain with C^2 boundary, let $0\le q\le n-1$, and let $y\in\partial D$ with $p^+_{\partial D}(y)=n-q-1$ and $p^-_{\partial D}(y)=q$. Then there exists a neighborhood U of y such that, for all neighborhoods $V \subseteq U$ of y,

$$\dim [Z^0_{0,q}(\bar{D}\cap V)/Z^0_{0,q}(\bar{D}\cap V)\cap E^0_{0,q}(D\cap V)] = \infty.$$ [1)]

More precisely, then there exists a neighborhood U of y such that, for each $x\in\partial D\cap U$, there is a neighborhood U_x of $(\bar{D}\cap U)\setminus\{x\}$ and a form $f_x\in Z^0_{0,q}(\bar{U}_x)$ which is singular at x.

Proof. We may assume that $q\ge 1$ (for q=0 the assertion is trivial, because then, with respect to appropriate local holomorphic coordinates, ∂D is strictly convex in a neighborhood of y - cf., for instance, Theorem 1.4.14 in [H/L]).

First we prove the following

Statement A. Let $[U,\varrho, \varphi,D]$ be an (n-q-1)-convex configuration of type I in $\mathbb{C}^n$, and $y\in\partial D$ with $\varphi(y)<0$ and $p^-_{\partial D}(y)=q$ $(0\le q\le n-1)$. Then there exist a neighborhood W_y of the set $\{z\in\bar{D}\colon z\neq y \text{ and } \varphi(z)<0\}$ and a form $F_y\in Z^0_{0,q}(W_y)$ which is singular at y.

Proof of Statement A. By Theorem 18.2, we can find a neighborhood V of y, a neighborhood V_y of $(\bar{D}\cap V)\setminus(y)$ and a form $f_y\in Z^0_{0,q}(V_y)$ which is singular at y. Fix some neighborhood $V' \subset\subset V$ of y, and choose $\varepsilon>0$ sufficiently small. Then $[U,\varrho-\varepsilon, \varphi,D_\varepsilon]$ with $D_\varepsilon:= \{z\in U\colon \varphi(z)<0,\ \varrho(z)<\varepsilon\}$ is also an (n-q-1)-convex configuration, and $D_\varepsilon\cap(V\setminus V') \subset\subset V_y$. Take a C^∞ function χ on $\mathbb{C}^n$ with supp $\chi\subset\subset B$ and $\chi = 1$ on V'. Then, after extending by zero, $\bar\partial(\chi f_y)$ is a continuous and $\bar\partial$-closed (0,q+1)-form on $\bar{D}_\varepsilon$. Hence, by Theorem 7.8, we can solve the equation $\bar\partial u = \bar\partial(\chi f_y)$ with $u\in C^0_{0,q}(D_\varepsilon)$. Setting $W_y = (D_\varepsilon\setminus\bar{V}')\cup V_y$ and $F_y = \chi f_y - u$, we conclude the proof of Statement A.

Next we prove the following

Statement B. Let $[U,\varrho, \varphi,D]$ be an (n-q-1)-convex configuration of type I in $\mathbb{C}^n$ $(0\le q\le n-1)$, and $y\in\partial D$ with $\varphi(y)<0$ and $p^-_{\partial D}(y)=q$. Then, for any neighborhoood V of y, there exist a neighborhood Y of $\{z\in\bar{D}\colon \varphi(z)<0\}$ and a form $f\in Z^0_{0,q}(Y)$ which is "singular at $\partial D\cap V$" in the following sense: there does not exist a neighborhood V' of $\partial D\cap V$ such that the equation $\bar\partial u = f$ admits a continuous solution u on $D\cap V'$.

Proof of Statement B. Fix a neighborhood $V_1 \subset\subset V$ of y. Then it is easy to find an (n-q-1)-convex configuration $[U',\varrho', \varphi',D']$ such that the following conditions (a)-(e) are fulfilled:

---

1) For q=0, this means $\dim C^0\mathcal{O}(\bar{D}\cap V) = \infty$ (cf. Sect. 0.11 for the definition of $C^0\mathcal{O}$).

(a) $U' = U$, $\varphi' = \varphi$.

(b) $\varrho' \leq \varrho$ and hence $D \subseteq D'$.

(c) $\varrho'(y) < 0$ and hence $y \in D'$.

(d) $\varrho' = \varrho$ on $U_1 \setminus V$.

(e) There exists a point $x \in \partial D' \setminus \bar{D}$ with $p^-_{\partial D'}(x) = q$ such that, for each neighborhood Y of the set $\{z \in \bar{D}': \varphi(z)<0,\ z \neq x\}$ and any neighborhood V' of $\partial D \cap V$, there is an increasing sequence of domains $D_k \subseteq Y$ ($k=0,1,\ldots$) such that: $D_0 = V' \cap D$; each D_{k+1} can be obtained from D_k by means of a q-concave extension element in Y (cf. Def. 15.4);

$$D' \setminus (D \setminus V') = \bigcup_{k=0}^{\infty} D_k;$$

and, for all k, there is an integer N(k) with $\bar{D}_k \cap \overline{(D_{N(k)} \setminus D_{N(k)-1})} = \emptyset$.

By statement A, we can find a neighborhood Y of $\{z \in D': z \neq x, \varphi(z)<0\}$ as well as a form $f \in Z^0_{0,q}(Y)$ which is singular at x. Then this form f is "singular at $\partial D \cap V$". In fact, assume that, for some neighborhood V' of $\partial D \cap V$, $f|_{(V' \cap D)} \in E^0_{0,q}(V' \cap D)$. Then, by condition (e) and Lemma 16.2, we obtain the contradiction that f belongs to $E^0_{0,q}(D' \setminus (D \setminus V'))$. So statement B is proved.

Now we go to prove the assertion of the theorem. By Lemma 7.5, we may assume that D belongs to some (n-q-1)concave cofiguration $[W, \varrho, \varphi, D]$ of type I in $\mathbb{C}^n$, where $\varphi(y)<0$. Moreover we may assume that $p^-_{\partial D}(z) = q$ for all $z \in \partial D$ with $\varphi(z)<0$. Set $U = \{z \in W: \varphi(z)<0\}$. Fix $x \in \partial D \cap U$. In order to find the required form f_x, first we take an (n-q-1)-convex configuration $[W', \varrho', \varphi', D']$ such that $W' = W$, $\varphi' = \varphi$, $(U \cap \bar{D}) \setminus \{x\} \subseteq D' \subseteq U \setminus \{x\}$, and $p^-_{\partial D'}(z) = q$ for all $z \in \partial D' \cap U$. It is sufficient to construct a form $f \in Z^0_{0,q}(D')$ which is singular at x.

Take a sequence $x_k \in \partial D' \setminus \{x\}$ which converges to x such that $x_k \neq x_l$ if $k \neq l$. Taking into account Observation 4.15, we can find neighborhoods V_k of x_k such that $V_k \cap V_l = \emptyset$ if $k \neq l$ and, for all k, the domain

$$D_k := D' \cup \bigcup_{\substack{j=1 \\ j \neq k}}^{\infty} V_j$$

belongs to some (n-q-1)-convex configuratiion $[W_k, \varrho_k, \varphi_k, D_k]$ with $W_k = W$ and $\varphi_k = \varphi$, where, moreover, $p^-_{\partial D_k}(z) = q$ for all $z \in \partial D_k \cap U$. Further, we fix open balls $B_k \subset\subset V_k$ centered at x_k. Then, by statement B, we can find a sequence of forms $f_k \in Z^0_{0,q}(D_k)$ such that, for each k, f_k is "singular at $\partial D \cap B_k$". Finally, we choose a sequence of numbers $\varepsilon_k \neq 0$ which are so small that the series $f := \sum_1^\infty \varepsilon_k f_k$ converges uniformly on the compact subsets of D'. So we obtain a form $f \in Z^0_{0,q}(D')$. It remains to show that f is singular at x:

Each of the forms $f - f_k$ admits a continuous and $\bar\partial$-closed extension into the ball B_k. Since the $\bar\partial$-equation is solvable on balls, it follows that, for all k, the form $f - f_k$ is not "singular at $\partial D' \cap B_k$". Since, by construction, each f_k is "singular at $\partial D' \cap B_k$", this implies that, for any k, f is "singular at $\partial D' \cap B_k$". Therefore, f is singular at x. □

Theorem 18.5 admits the following globalization:

18.6. Theorem (Andreotti-Norguet theorem, global version with uniform estimates). Let $D \subset\subset X$ be a strictly (n-q-1)-convex domain in an n-dimensional complex manifold X ($0 \le q \le n-1$), and let E be a holomorphic vector bundle over X. If there exists at least one point $y \in \partial D$ with $p^-_{\partial D}(y) = q$, then

$$\dim [Z^0_{0,q}(\bar D, E)/Z^0_{0,q}(\bar D, E) \cap E^0_{0,q}(D,E)] = \infty.$$

More precisely, then there exists a neighborhood U of y such that, for any $x \in \partial D \cap U$, we can find a form $f_x \in Z^0_{0,q}(\bar D, E)$ which is singular at x, and which admits a continuous and $\bar\partial$-closed extension into some neighborhood of $\bar D \setminus \{x\}$.

Proof. This is a repetition of the proof of Theorem 18.3, where instead of Theorem 18.2 we have to use Theorem18.5. □

18.7. Remark (Supplement to Theorems 18.2 and 18.5). In the next section we shall prove the Andreotti-Vesentini theorem: For any non-degenerate strictly q-concave domain D in a compact complex manifold, the space $E^0_{0,q}(D)$ is closed with respect to the uniform convergence on the compact subsets of D, and the space $E^0_{0,q}(\bar D)$ is closed with respect to uniform convergence on $\bar D$.

This theorem expresses a global property of the domain D, i.e. the corresponding local statement is not true.

In fact, let D, q, y, and U be as in Theorem 18.2. Since $p^+_{\partial D}(y) = n-q-1$, then, by Lemma 7.5 and Theorem 10.1, there exist arbitrarily small pseudoconvex neighborhoods V of y such that the image of the restriction map

$$E^0_{0,q}(V) = Z^0_{0,q}(V) \longrightarrow Z^0_{0,q}(D \cap V)$$

is dense in the Fréchet space $Z^0_{0,q}(D\cap V)$ (one can prove that this is true even for <u>all</u> sufficiently small pseudoconvex neighborhoods V of y). If, moreover, $V \subseteq U$, then, by (18.1) this implies that the space $E^0_{0,q}(D\cap V)$ is not closed in the Fréchet space $Z^0_{0,q}(D\cap V)$.

Under the hypotheses of Theorem 18.5, in the same way we see that the space $Z^0_{0,q}(\bar{D}\cap V)\cap E^0_{0,q}(D\cap V)$ is not closed in $Z^0_{0,q}(\bar{D}\cap V)$.

19. The Andreotti-Vesentini separation theorem for the Dolbeault cohomology of order q on q-concave manifolds

Let Y be an n-dimensional complex manifold, and $X \subset\subset Y$ a domain which is both non-degenerate strictly q-concave and non-degenerate strictly (n-q-1)-convex ($1\leq q\leq n-1$). Then, by the Andreotti-Norguet Theorem 18.3, the complex linear space

$$H^{0,q}(X) \xleftrightarrow[\text{(Theorem 2.14)}]{\text{Dolbeault isomorphism}} H^{0,q}_0(X) = Z^0_{0,q}(X)/E^0_{0,q}(X)$$

is of infinite dimension. So the question arises whether the natural topology of this space, the factor topology of the Fréchet space $Z^0_{0,q}(X)$ with respect to the subspace $E^0_{0,q}(X)$, is separated? In other words: Is $E^0_{0,q}(X)$ closed with respect to uniform convergence on the compact subsets of X? [1] Under certain additional hypotheses the answer is affirmative. For instance, if Y is compact, then one has the following

<u>19.1. Theorem</u> (Andreotti-Vesentini theorem). Let E be a holomorphic vector bundle over an n-dimensional compact complex manifold Y, and let $X \subseteq Y$ be a q-concave domain ($1\leq q\leq n-1$). Then $E^0_{0,q}(X,E)$ is closed with respect to uniform convergence on the compact subsets of X.

In view of the uniqueness Theorem 16.1, this theorem is a consequence of the following

<u>19.2. Theorem</u> (Andreotti-Vesentini theorem with uniform estimates). Let E be a holomorphic vector bundle over an n-dimensional compact complex manifold X, and let $D \subset\subset X$ be a non-degenerate strictly q-concave domain. Then the following two assertions hold true:

(i) For all $0\leq\alpha\leq 1/2$, the operator $\bar{\partial}$ is α-regular at $Z^0_{0,q}(\bar{D},E)$, and hence (cf. Remark II following Def. 11.1) $E^{\alpha\to 0}_{0,q}(\bar{D},E)$ is closed with respect to uniform convergence on $\bar{D}$.

---

[1] For $\dim H^{0,q}(X) < \infty$ this is always the case, by Banach's open mapping theorem.

(ii) Denote by $\hat{Z}^0_{0,q}(\bar{D},E)$ the space of all $f \in Z^0_{0,q}(\bar{D},E)$ such that

$$\int_{\partial D} f \wedge g = 0$$

for any $g \in Z^0_{n,n-q-1}(X \setminus D, E^*)$. Then there exist a bounded linear operator T acting from the Banach space $\hat{Z}^0_{0,q}(\bar{D},E)$ (endowed with the norm $\|\cdot\|_{0,D}$) into the Banach space $C^{1/2}_{0,q-1}(\bar{D},E)$ (with the norm $\|\cdot\|_{1/2,D}$) and a compact linear operator K from $\hat{Z}^0_{0,q}(\bar{D},E)$ into itself such that

$$\bar{\partial} \circ T = id + K \qquad \text{on } \hat{Z}^0_{0,q}(\bar{D},E)$$

(id:= identity operator). Moreover, then

$$E^0_{0,q}(D,E) \cap Z^0_{0,q}(\bar{D},E) \subseteq \hat{Z}^0_{0,q}(\bar{D},E).$$

Hence each of the spaces $E^{\alpha \to 0}_{0,q}(\bar{D},E)$ $(0 \le \alpha \le 1/2)$ and $E^0_{0,q}(D,E) \cap Z^0_{0,q}(\bar{D},E)$ is a closed and finitely codimensional subspace of $Z^0_{0,q}(\bar{D},E)$.

Theorems 19.1 and 19.2 admit the following generalizations:

<u>19.1'. Theorem</u> (Andreotti-Vesentini theorem). Let E be a holomorphic vector bundle over an n-dimensional complex manifold Y which is (n-q)-convex $(1 \le q \le n-1)$, and let $X \subseteq Y$ be a domain such that, for some strictly q-convex domain $G \subset\subset Y$, X is a q-concave extension of $Y \setminus \bar{G}$. Then $E^0_{0,q}(X,E)$ is closed with respect to uniform convergence on the compact subsets of X.

<u>19.2'. Theorem</u> (Andreotti-Vesentini theorem with uniform estimates). Let E be a holomorphic vector bundle over an n-dimensional complex manifold X, and $1 \le q \le n-1$ an integer. Let $G' \subset\subset X$ be a non-degenerate strictly (n-q)-convex domain, $G \subset\subset G'$ a non-degenerate strictly q-convex domain, and $D := G' \setminus \bar{G}$. Then the following two assertions hold true:

(i) For all $0 \le \alpha \le 1/2$, the operator $\bar{\partial}$ is α-regular at $Z^0_{0,q}(\bar{D},E)$, and hence $E^{\alpha \to 0}_{0,q}(\bar{D},E)$ is closed with respect to uniform convergence on $\bar{D}$.

(ii) Denote by $\hat{Z}^0_{0,q}(\bar{D},E)$ the space of all $f \in Z^0_{0,q}(\bar{D},E)$ such that

$$\int_{\partial G} f \wedge g = 0$$

for any $g \in Z^0_{n,n-q-1}(\bar{G},E^*)$. Then there exist a bounded linear operator T from $\hat{Z}^0_{0,q}(\bar{D},E)$ into $C^{1/2}_{0,q-1}(\bar{D},E)$ and a compact linear operator from $\hat{Z}^0_{0,q}(\bar{D},E)$ into itself such that

$$\bar{\partial}\circ T = id + K \qquad \text{on } \hat{Z}^0_{0,q}(\bar{D},E). \tag{19.1}$$

Moreover, then

$$E^0_{0,q}(D,E) \cap Z^0_{0,q}(\bar{D},E) \subseteq \hat{Z}^0_{0,q}(\bar{D},E). \tag{19.2}$$

Hence each of the spaces $E^{\alpha\longrightarrow 0}_{0,q}(\bar{D},E)$ $(0\leq\alpha\leq 1/2)$ and $E^0_{0,q}(D,E)\cap Z^0_{0,q}(\bar{D},E)$ is a closed and finitely codimensional subspace of $\hat{Z}^0_{0,q}(\bar{D},E)$.

Remarks to Theorems 19.2 and 19.2'.

I. The authors believe that, under the hypotheses of Theorem 19.2', even $E^{1/2\longrightarrow 0}_{0,q}(\bar{D},E^*) = E^0_{0,q}(D,E)\cap Z^0_{0,q}(\bar{D},E)$. (By Theorem 19.2' we know only that the difference between these spaces is finitely dimensional.)

II. In Theorems 19.2 and 19.2', assertion (i) follows from assertion (ii) (cf. Remark II following Definition 11.1).

III. The authors believe that, under the hypotheses of Theorem 19.2, the space $E^{1/2\longrightarrow 0}_{0,q}(\bar{D},E)$ is not topologically complemented in the Banach space $Z^0_{0,q}(\bar{D},E)$ (cf. point 6 in Problems at the end of this monograph).

The remainder of the present Section 19 is devoted to the proof of Theorems 19.1' and 19.2' (Theorems 19.1 and 19.2 are special cases of them). By means of Theorems 12.14 and 16.1, Theorem 19.1' can be easily obtained from Theorem 19.2' - the corresponding arguments will be given at the end, in Sect. 19.14. In the following Sects. 19.3-19.13 we prove Theorem 19.2'. In all these sections we assume that the hyptheses of Theorem 19.2' are fulfilled and use the notations from this theorem.

19.3. Construction of the operators W^0_j, W^1_j, W^2_j. Let $\varrho:\Theta \longrightarrow \mathbb{R}$ be a strictly (q+1)-convex function without degenerate critical points in a neighborhood Θ of ∂G such that $G\cap\Theta = \{z\in\Theta: \varrho(z)<0\}$. By lemma 7.5 we can find domains $A^1_j \subset\subset A^2_j \subset\subset A^3_j \subset\subset \Theta$, q-convex configurations $[U_j,\varrho_j,\varphi_j,D_j]$ in $\mathbb{C}^n$, and biholomorphic maps h_j from A^3_j onto U_j $(j=1,\dots,N<\infty)$ such that: $\partial G \subseteq A^1_1\cup\dots\cup A^1_N$; $h_j(A^2_j)=\{z\in U_j: \varphi_j(z)<0\}$; $\varrho_j = \varrho\circ h_j^{-1}$ and hence $h_j(G\cap A^2_j) = D_j$; E is holomorphically trivial over A^3_j.

Further on, let some holomorphic vector bundle isomorphisms

$$P_j:E^*\otimes\Lambda\big|A^3_j \longrightarrow A^3_j\otimes \mathbb{C}^r \qquad (j=1,\dots,N) \tag{19.4}$$

be fixed, where r = rank E, and Λ is the bundle of holomorphic (n,0)-forms on X. We fix also some canonical Leray maps v_j for $[U_j, \varrho_j, \varphi_j, D_j]$ (cf. Sect. 7.7). Then, by Theorems 9.1 and 7.8, the corresponding Cauchy-Fantappie operators T^{v_j} define bounded linear operators

$$T^{v_j}: C^0_{0,r}(\bar{D}_j) \longrightarrow C^{1/2}_{0,r-1}(\overline{h_j(G\cap A^1_j)}) \qquad (1\le r\le n) \qquad (19.5)$$

such that

$$(-1)^{n-q} f = \bar\partial T^{v_j}_{n-q} f + T^{v_j}_{n-q+1}\bar\partial f \qquad \text{on } \overline{h_j(G\cap A^1_j)} \qquad (19.6)$$

for all $f\in C^0_{0,n-q}(\bar{D}_j)$ with the property that $\bar\partial f$ is also continuous on $\bar{D}_j$.

Let

$$\tilde{W}^j_r: C^0_{n,r}(\overline{G\cap A^2_j}, E^*) \longrightarrow C^{1/2}_{n,r-1}(\overline{G\cap A^1_j}, E^*) \qquad (1\le r\le n)$$

be the operators which are defined, via the biholomorphic maps h_j and the isomorphisms (19.4), by the operators (19.5) (here we identify E^*-valued (n,r)-forms with $E^*\otimes\Lambda$-valued (0,r)-forms). Then (19.6) takes the form

$$(-1)^{n-q} f = \bar\partial \tilde{W}^j_{n-q} f + \tilde{W}^j_{n-q+1}\bar\partial f \qquad \text{on } G\cap A^1_j \qquad (j=1,\dots N) \qquad (19.7)$$

for all $f\in C^0_{n,n-q}(\overline{G\cap A^2_j}, E^*)$ such that $\bar\partial f$ is also continuous on $\overline{G\cap A^2_j}$.

Further, we choose some open sets $A^1_j \subset\subset A^2_j \subset\subset G$ ($j=N+1,\dots,M<\infty$) such that

$$\bar{G} \subseteq A^1_1\cup\dots\cup A^1_N\cup A^1_{N+1}\cup\dots\cup A^1_M, \qquad (19.8)$$

and, for any $N+1\le j\le M$, A^2_j is a ball with respect to approp r iate holomorphic coordinates in a neighborhood of $\bar{A}^2_j$. From Observation 2.11 and Proposition 1.4 (i) then we obtain bounded linear operators

$$\tilde{W}^j_r: C^0_{n,r}(\overline{A^2_j}, E^*) \longrightarrow C^{1/2}_{n,n-1}(\overline{A^1_j}, E^*) \qquad (1\le r\le n)$$

such that, for all $f\in C^0_{n,n-q}(\bar{A}^2_j, E^*)$ with the property that $\bar\partial f$ is also continuous on $\bar{A}^2_j$,

$$(-1)^{n-q}f = \bar\partial\tilde W^j_{n-q}f + \tilde W^j_{n-q+1}\bar\partial f \qquad \text{on } A^1_j \qquad (j=N+1,\dots,M). \tag{19.9}$$

By (19.5) we can find C^∞ functions $\chi_1,\dots,\chi_M$ on X such that supp $\chi_j \subset\subset A^1_j$ and

$$\chi_1^2 + \dots + \chi_M^2 = 1 \tag{19.10}$$

in some neighborhood of $\bar G$. Now, setting

$$W^0_j f = (-1)^{n-q}\chi_j\tilde W^j_{n-q})\chi_j f),$$

$$W^1_j f = (-1)^{n-q}\bar\partial\chi_j\wedge\tilde W^j_{n-q}(\chi_j f),$$

$$W^2_j f = (-1)^{n-q+1}\chi_j\tilde W^j_{n-q+1}(\bar\partial\chi_j\wedge f)$$

for $f\in C^0_{n,n-q}(\bar G,E^*)$, we obtain bounded linear operators

$$W^0_j: C^0_{n,n-q}(\bar G,E^*) \longrightarrow C^{1/2}_{n,n-q-1}(\bar G,E^*)$$

and

$$W^1_j, W^2_j: C^0_{n,n-q}(\bar G,E^*) \longrightarrow C^{1/2}_{n,n-q}(\bar G,E^*).$$

From (19.7) and (19.9) it follows that

$$\bar\partial W^0_j f = \chi_j^2 f + W^1_j f + W^2_j f \tag{19.11}$$

for all $f\in Z^0_{n,n-q}(\bar G,E^*)$ and $j=1,\dots,M$.

<u>19.4. Lemma.</u> There exist a bounded linear operator

$$S: C^0_{n,n-q}(\bar G,E^*) \longrightarrow C^0_{n,n-q-1}(\bar G,E^*)$$

and a topologically closed and finitely codimensional subspace $\tilde Z^0_{n,n-q}(\bar G,E^*)$ of $Z^0_{n,n-q}(\bar G,E^*)$ such that (id:= identity operator)

$$\bar\partial\circ\sum_{j=1}^{M}(W^0_j+S(W^1_j+W^2_j)) = \text{id} \qquad \text{on } \tilde Z^0_{n,n-q}(\bar G,E^*). \tag{19.12}$$

19.5. Proof of Lemma 19.4. Set

$$W = \sum_{j=1}^{M} W_j^0 \qquad \text{and} \qquad V = \sum_{j=1}^{M} (W_j^1 + W_j^2).$$

Then, by (19.11) and (19.10),

$$\bar{\partial} \circ W = \mathrm{id} + V \qquad \text{on } Z^0_{n,n-q}(\bar{G}, E^*) \tag{19.13}$$

and hence

$$V(Z^0_{n,n-q}(\bar{G}, E^*)) \subseteq Z^0_{n,n-q}(\bar{G}, E^*). \tag{19.14}$$

Since the embedding operator $C^{1/2}_{n,n-q}(\bar{G}, E^*) \longrightarrow C^0_{n,n-q}(\bar{G}, E^*)$ can be approximated with respect to the uniform operator norm by finitely dimensional linear operators, the operator V considered as an operator with values in $C^0_{n,n-q}(\bar{G}, E^*)$ has the same property. Therefore, we can find a finitely dimensional linear operator F_1 from $C^0_{n,n-q}(\bar{G}, E^*)$ into $C^0_{n,n-q}(\bar{G}, E^*)$ such that the uniform operator norm of $V - F_1$ is $\leq 1/2$. Then $\mathrm{id} + V - F_1$ is invertible in $C^0_{n,n-q}(\bar{G}, E^*)$, and, setting $F_2 = F_1(\mathrm{id} + V - F_1)^{-1}$, we obtain a finitely dimensional linear operator F_2 such that, by (19.12),

$$\bar{\partial} W(\mathrm{id} + V - F_1)^{-1} f = f + F_2 f \tag{19.15}$$

for all $f \in (\mathrm{id} + V - F_1)(Z^0_{n,n-q}(\bar{G}, E^*))$.

We set

$$\tilde{Z}^0_{n,n-q}(\bar{G}, E^*) = \mathrm{Ker}\, F_1 \cap \mathrm{Ker}\, F_2 \cap (\mathrm{id} + V)\big(Z^0_{n,n-q}(\bar{G}, E^*) \cap \mathrm{Ker}\, F_1\big),$$

where $\mathrm{Ker}\, F_j$ is the space of all $f \in C^0_{n,n-q}(\bar{G}, E^*)$ with $F_j f = 0$. Further, we set

$$S = -W(\mathrm{id} + V - F_1)^{-1}.$$

Since V restricted to $Z^0_{n,n-q}(\bar{G}, E^*)$ is a compact linear operator from $Z^0_{n,n-q}(\bar{G}, E^*)$ into itself, and since $\mathrm{Ker}\, F_1$ and $\mathrm{Ker}\, F_2$ are closed and finitely codimensional subspaces of $C^0_{n,n-q}(\bar{G}, E^*)$, then $\tilde{Z}^0_{n,n-q}(\bar{G}, E^*)$ is a closed and finitely codimensional subspace of $Z^0_{n,n-q}(\bar{G}, E^*)$. Further, since

$$W+SV = W[\mathrm{id}-(\mathrm{id}+V-F_1)^{-1}V] = W(\mathrm{id}+V-F_1)^{-1}(\mathrm{id}-F_1),$$

it follows from (19.15) that $\bar\partial(W+SV)f = f$ for all $f\in\tilde Z^0_{n,n-q}(\bar G,E^*)$, i.e. (19.12) holds true.

<u>19.6. Lemma.</u> Denote by $F^0_{n,n-q-1}(\bar G,E^*)$ the space of all $f\in C^0_{n,n-q-1}(\bar G,E^*)$ such that $\bar\partial f$ is continuous on $\bar G$. Set

$$Af = f = \sum_{j=1}^{M}(W^0_j+S(W^1_j+W^2_j))\bar\partial f$$

for $f\in F^0_{n,n-q-1}(\bar G,E^*)$, where W^0_j,W^1_j,W^2_j are the operators from Construction 19.3, and S is the operator from Lemma 19.4. Then $Z^0_{n,n-q-1}(\bar G,E^*)$ is a finitely codimensional subspace of $\mathrm{Im}\,A := A(F^0_{n,n-q-1}(\bar G,E^*))$.

<u>19.7. Proof of Lemma 19.6.</u> Let $\tilde F^0_{n,n-q-1}(\bar G,E^*)$ be the space of all $f\in F^0_{n,n-q-1}(\bar G,E^*)$ with $\bar\partial f\ \tilde Z^0_{n,n-q}(\bar G,E^*)$, where $\tilde Z^0_{n,n-q}(\bar G,E^*)$ is the finitely codimensional subspace of $Z^0_{n,n-q}(\bar G,E^*)$ from Lemma 19.4. Then it is clear that $\tilde F^0_{n,n-q-1}(\bar G,E^*)$ is of finite codimension in $F^0_{n,n-q-1}(\bar G,E^*)$. Henc $A(\tilde F^0_{n,n-q-1}(\bar G,E^*))$ is of finite codimension in Im A. This completes the proof, because, by (19.12) in Lemma 19.4,

$$A(\tilde F^0_{n,n-q-1}(\bar G,E^*)) \subseteq Z^0_{n,n-q-1}(\bar G,E^*) \subseteq \mathrm{Im}\,A.$$

<u>19.8. Construction of the operators $\tilde T,\tilde K$, and L.</u> Let Θ and ϱ be as at the beginning of Construction 19.3. Then, by Lemma 13.3, we can find open sets $B^1_k \subset\subset B^2_k \subset\subset B^3_k \subset\subset \Theta$, q-concave configurations $[V_k,\sigma_k,\psi_k,H_k,G_k]$ in $\mathbb{C}^n$, and biholomorphic maps $g_k:B^3_k\longrightarrow V_k$ $(k=1,\dots,M'<\infty)$ such that: $\partial G \subseteq B^1_k\cup\dots\cup B^1_k$; $g_k(B^2_k) = \{z\in V_k:\ \psi_k(z)<0, \mathrm{Re}\ H_k(z)<0\}$; $\sigma_k = -\varrho\circ g_k^{-1}$ and hence $g_k(D\cap B^2_k) = G_k$; E is holomorphically trivial over B^3_k.

Further on, let some holomorphic vector bundle isomorphisms

$$Q_k:E\,|\,B^3_k \longrightarrow B^3_k\otimes \mathbb{C}^r \qquad (1\le k\le M') \tag{19.16}$$

be fixed. Moreover, we fix some canonical Leray maps u_k for $[V_k,\sigma_k,\psi_k,H_k,G_k]$ (cf. Sect. 13.4).

Let

$$T_r^k : C^0_{0,r}(\overline{D\cap B_k^2}, E) \longrightarrow C^0_{0,r-1}(D\cap B_k^2, E)$$

and

$$L_r^k : C^0_{0,r}(\overline{D\cap B_k^2}, E) \longrightarrow C^0_{0,r}(D\cap B_k^2, E)$$

be the operators which are induced, via the isomorphisms (19.16) and the biholomorphic maps g_k, by the operators $(-1)^r T^{u_k}$ and $(-1)^r L^{u_k}$, respectively (for the definition of L^{u_k} and T^{u_k}, see Sects. 3.5 and 3.7). Choose C^{∞} function $\lambda_1, \ldots, \lambda_{M'}$ on X such that supp $\lambda_k \subset\subset B_k^1$ and

$$\lambda_1^2 + \ldots + \lambda_{M'}^2 = 1 \qquad (19.17)$$

in some neighborhood of ∂G. Then, by the piecewise Cauchy-Fantappie formula (Theorem 3.12),

$$\lambda_k f = \bar\partial T_q^k(\lambda_k f) + T_{q+1}^k(\bar\partial\lambda_k \wedge f) + L_q^k(\lambda_k f) \qquad \text{on } D \qquad (19.18)$$

for all $f \in Z^0_{0,q}(\bar D, E)$ and $1 \le k \le M'$.

By (19.17) we can find a C^{∞} function λ_0 on X with supp $\lambda_0 \subset\subset X\setminus\bar G$ and

$$\lambda_0^2 + \lambda_1^2 + \ldots + \lambda_{M'}^2 = 1 \qquad (19.19)$$

in some neighborhood of $\bar D$. Since G' is non-degenerate strictly $(n-q)$-convex, by Theorem 11.2, for $n-q \le r \le n$, there are bounded linear operators T_r^0 from $C^0_{0,r}(\bar G', E)$ into $C^{1/2}_{0,r-1}(\bar G', E)$ and K_r^0 from $C^0_{0,r}(\bar G', E)$ into $C^{1/2}_{0,r}(\bar G', E)$ such that

$$\lambda_0 f = \bar\partial T_q^0(\lambda_0 f) + T_{q+1}^0(\bar\partial\lambda_0\ f) - K_q^0(\lambda_0 f) \qquad \text{on } \bar D \qquad (19.20)$$

for all $f \in Z^0_{0,q}(\bar D, E)$.

Now, for any $f \in Z^0_{0,q}(\bar D, E)$, we define

$$\tilde{T}f = \sum_{k=0}^{M'} \lambda_k T_q^k(\lambda_k f),$$

$$\tilde{K}f = \sum_{k=0}^{M'} [\bar{\partial}\lambda_k \wedge T_q^k(\lambda_k f) - \lambda_k T_{q+1}^k(\bar{\partial}\lambda_k \wedge f)] + \lambda_0 K_q^0(\lambda_0 f),$$

$$Lf = -\sum_{k=1}^{M'} \lambda_k L_q^k(\lambda_k f). \tag{19.21}$$

Since T_r^0 and K_r^0 are bounded linear operators from $C_{0,r}^0(\bar{G}',E)$ into $C_{0,r-1}^{1/2}(\bar{G}',E)$ resp. $C_{0,r}^{1/2}(\bar{G}',E)$, and since, by Theorem 14.1, the operators T_r^k $(1\le k\le M',\ 1\le r\le n)$ are bounded as operators acting from $C_{0,r}^0(\overline{D\cap B_k^2},E)$ into $C_{0,r-1}^{1/2}(\overline{D\cap B_k^1},E)$, the following statement holds true:

$$\left.\begin{array}{l} \tilde{T} \text{ is a bounded linear operator from } Z_{0,q}^0(\bar{D},E) \text{ into} \\ C_{0,q-1}^{1/2}(\bar{D},E), \text{ and } \tilde{K} \text{ is a bounded linear operator from} \\ Z_{0,q}^0(\bar{D},E) \text{ into } C_{0,q}^{1/2}(\bar{D},E). \end{array}\right\} \tag{19.22}$$

Further, from (19.18)-(19.20) it follows that

$$\bar{\partial}(\tilde{T}f) = f + \tilde{K}f + Lf \qquad \text{on } \bar{D}\setminus\partial G \tag{19.23}$$

for all $f\in Z_{0,q}^0(\bar{D},E)$.

By (19.22), $\tilde{K}$ is compact as an operator from $Z_{0,q}^0(\bar{D},E)$ into $C_{0,q}^0(\bar{D},E)$. This is not true for the operator L. However, we have the following lemma, which is the key to the proof of Theorem 19.2':

<u>19.9. Lemma.</u> Let A and Im A be as in Lemma 19.6, and let $(\text{Im } A)^{\perp}$ be the space of all $f\in Z_{0,q}^0(\bar{D},E)$ with

$$\int_{\partial G} f\wedge g = 0 \qquad \text{for any } g \in \text{Im } A.$$

Then, for each $f\in(\text{Im } A)^{\perp}$, the form Lf (which is defined by (19.21) immediately only on $\bar{D}\setminus\partial G$) admits a continuous extension onto $\bar{D}$. Therefore, a linear operator $L:(\text{Im } A)^{\perp} \longrightarrow C_{0,q}^0(\bar{D},E)$ is well-defined. This operator is compact (here both $(\text{Im } A)^{\perp}$ and $C_{0,q}^0(\bar{D},E)$ are considered as Banach spaces endowed with the norm $\|\cdot\|_{0,D}$).

We divide the proof of this lemma into three parts. The first part is a reduction to some estimates for certain local integrals. Then these estimations will be proved in parts 2 and 3. In the following proof we meet many constants. Therefore, all "large" constants will be denoted by C, i.e. if we write "$f(y)\leq Cg(y)$ for all y Γ", then we mean: there exists a constant $C<\infty$ such that $f(y)\leq Cg(y)$ for all $y\in\Gamma$. Analogously, the "small" constants will be denoted by c.

<u>19.10. Part 1 of the proof of Lemma 19.9: reduction to local estimates.</u> By definition of L, it is sufficient to prove that, for all $1\leq k\leq M'$, the assertion of Lemma 19.9 holds true if L is replaced by the operator

$$f \longrightarrow \lambda_k L_q^k(\lambda_k f).$$

Therefore, let us fix $1\leq k\leq M'$ (for the rest of the proof of Lemma 19.9). Let $[V_k,\sigma_k,\psi_k,H_k,G_k]$ and u_k be as in Sect. 19.8, and let (w_1,w_2,w_3) be the canonical Leray data for $[V_k,\sigma_k,\psi_k,H_k,G_k]$ such that u_k is the canonical combination of (w_1,w_2,w_3). Set

$$\lambda = \lambda_k \circ g_k^{-1}.$$

Since $\operatorname{supp}\lambda \subset\subset \{t\in V_k:\ \psi_k(t)<0, \operatorname{Re} H_k(t)<0\}$, then $L^{u_k}(\lambda f) = L_1^{u_k}(\lambda f)$ for all $f\in C^0_{0,q}(\bar{G}_k)$. Therefore, by definition of $L_1^{u_k}$,

$$\Big((-1)^q \lambda L^{u_k}(\lambda f)\Big)(t) = \int\limits_{g_k(\partial G\cap B_k^1)} f(s)\wedge \tilde{L}(t,s)\wedge\omega(s) \tag{19.24}$$

for all $f\in C^0_{0,q}(\bar{G}_k)$ and $t\in G_k$, where

$$\tilde{L}(t,s) := \frac{(-1)^{q+1}\binom{n-1}{q}}{(2\pi i)^n}\lambda(t)\lambda(s)\frac{\det(w_1,\overbrace{\bar{\partial}_s w_1}^{n-q-1},\overbrace{\bar{\partial}_t w_1}^{q})}{\langle w_1(s,t),s-t\rangle^n} \tag{19.25}$$

Let $p_z,p_x: X\times X \longrightarrow X$ be the projections defined by $p_z(z,x)=z$ and $p_x(z,x)=x$, and let p_z^*E and p_x^*E be the pull-backs of E onto $X\times X$ with respect to p_z and p_x. Then, for any $\operatorname{Hom}(p_x^*E,p_z^*E)$-valued continuous differential form $Y(z,x)$ defined for $(z,x)\in D\times\partial G$, and for each E-valued continuous differential form f on ∂G, the integral

$$\int_{x\in\partial G} f(x)\wedge Y(z,x), \qquad z\in D,$$

is well-defined, and the result of this integration is an E-valued differential form on D.

Recall that, by (13.4),

$$\operatorname{Re} \langle w_1(t,s), s-t\rangle \geq \sigma_k(s) - \sigma_k(t) + c|s-t|^2 \tag{19.26}$$

for all $s,t\in V_k$. Therefore the form $\tilde{L}(t,s)$ is defined and continuous for all $t,s\in V_k$ with $\sigma_k(s)\geq\sigma_k(t)$ and $s\neq t$. Since $\varrho = -\sigma_k\circ g_k$, it follows that the pull-back of $\tilde{L}(t,s)$ with respect to $g_k\times g_k$ is defined and continuous for all $z,x\in B_k^3$ with $\varrho(z)\geq\varrho(x)$ and $x\neq z$. We denote this pull-back by $\tilde{L}_X(z,x)$. Since $\operatorname{supp}\lambda_k \subset\subset B_k^1$ and hence $\tilde{L}_X(z,x) = 0$ if $(z,x)\notin B_k^1\times B_k^1$, after extending by zero, $\tilde{L}_X(z,x)$ is defined and continuous for all $(z,x)\in\bar{D}\times\bar{G}$ with $x\neq z$.

By $\hat{L}_X(z,x)$ we denote the continuous $\operatorname{Hom}(p_x^*E, p_z^*E)$-valued differential form which is defined for all $(z,x)\in\bar{D}\times\bar{G}$ with $x\neq z$ by the following two conditions:

1) $\hat{L}_X(z,x) = 0$ if $(z,x)\notin B_k^1\times B_k^1$;
2) with respect to the isomorphism (19.16), $\hat{L}_X(z,x)$ is represented by the matrix $\tilde{L}_X(z,x)I_r$, where I_r is the $r\times r$ unit matrix.

Let $g_k^1,\dots,g_k^n$ be the components of the biholomorphic map $g_k: B_k^3\longrightarrow V_k$ and $\omega_k = dg_k^1\wedge\dots\wedge dg_k^n$. Then, by definition of the operator L_q^k and by (19.24), one has

$$[\lambda_k L_q^k(\lambda_k f)](z) = \int_{x\in\partial G} f(x)\wedge\hat{L}_X(z,x)\wedge\omega_k(x) \tag{19.27}$$

for all $f\in Z_{0,q}^0(\bar{D},E)$ and $z\in\bar{D}\setminus\partial G$.

For fixed $z\in\bar{D}\setminus\partial G$, the form $\hat{L}_X(z,x)\wedge\omega_k(x)$ may be interpreted as a vector of continuous E^*-valued differential forms on $\bar{G}$, and it follows from (19.25) that these forms belong to the space $F_{n,n-q-1}^0(\bar{G},E^*)$ defined in Lemma 19.6. Therefore

$$[A(\hat{L}_X(z,\cdot)\wedge\omega_k)](x) \qquad (z\in\bar{D}\setminus\partial G,\ x\in\bar{G})$$

is a well-defined $\operatorname{Hom}(p_x^*E, p_z^*E)$-valued differential form on $(\bar{D}\setminus\partial G)\times\bar{G}$, which is of bidegree $(0,q)$ in z, and of bidegree $(n,n-q-1)$ in x. If

$f\in(\mathrm{Im}\ A)^{\perp}$, then (19.27) may be written

$$[\lambda_k L_q^k(\lambda_k f)](z) = \int_{x\in\partial G} f(x)\wedge\Big(\hat{L}_X(z,x)\wedge\omega_k(x) - [A(\hat{L}_X(z,\cdot)\wedge\omega_k)](x)\Big), \qquad (19.28)$$

$z\in\bar{D}\setminus\partial G$. By definition of A, one has

$$\hat{L}_X(z,x)\wedge\omega_k(x) - [A(\hat{L}_X(z,\cdot)\wedge\omega_k)](x) = \sum_{j=1}^{M} K_j(z,x),$$

where $K_j(z,x)$ are the $\mathrm{Hom}(p_x^*E,p_z^*E)$-valued differential forms defined by

$$K_j(z,x) = [\big(W_j^0+S(W_j^1+W_j^2)\big)\big(\bar{\partial}\hat{L}_X(z,\cdot)\wedge\omega_k\big)](x),$$

$z\in\bar{D}\setminus\partial G$, $x\in\bar{G}$ $(1\le j\le M)$. In view of (19.28), this implies that

$$[\lambda_k L_q^k(\lambda_k f)](z) = \sum_{j=1}^{M}\int_{x\in\partial G} f(x)\wedge K_j(z,x) \qquad (19.29)$$

for all $f\in(\mathrm{Im}\ A)^{\perp}$ and $z\in\bar{D}\setminus\partial G$. Therefore it is sufficient to show that, for all $1\le j\le M$, the map

$$f \longrightarrow \int_{x\in\partial G} f(x)\wedge K_j(\cdot,x) \qquad (19.30)$$

defines a compact operator from $C^0_{0,q}(\bar{D},E)$ into itself. So, let us fix also $1\le j\le M$ (for the rest of the proof of Lemma 19.9).

Denote by $\mathrm{dist}(\cdot,\cdot)$ a metric on X which is locally defined as the Euclidean metric with respect to appropriate C^∞ coordinates. For any $\varepsilon>0$, we choose a C^∞ function $d_\varepsilon : X\times X \longrightarrow [0,1]$ such that $d_\varepsilon(z,x)=0$ if $\mathrm{dist}(z,x)\ge\varepsilon$, and $d_\varepsilon(z,x)=1$ if $\mathrm{dist}(z,x)\le\varepsilon/2$. Set

$$K^{\varepsilon}(z,x) = [\big(W_j^0+S(W_j^1+W_j^2)\big)\big(d_\varepsilon(z,\cdot)\bar{\partial}\hat{L}_X(z,\cdot)\wedge\omega_k\big)](x)$$

for $z\in\bar{D}\setminus\partial G$ and $x\in\bar{G}$. Since $1-d_\varepsilon(z,x) = 0$ if $\mathrm{dist}(z,x)\le\varepsilon/2$ and $\bar{\partial}\hat{L}_X(z,x)\wedge\omega_k(x)$ is continuous for all $(z,x)\in\bar{D}\times\bar{G}$ with $x\neq z$, we see that, for any $\varepsilon>0$, the assignment

$$\bar{D} \ni z \longrightarrow (1-d_\varepsilon(z,\cdot))\bar{\partial}\hat{L}_X(z,\cdot)\wedge\omega_k$$

may be interpreted, locally, as a vector of continuous maps with values in the Banach space $C^0_{n,n-q}(G,E^*)$. Since the operator $W^0_j+S(W^1_j+W^2_j)$ is continuous with respect to $\|\cdot\|_{0,G}$, this implies that, for each $\varepsilon>0$, the form $K_j(z,x)-K^\varepsilon(z,x)$ is continuous for all $(z,x)\in\bar{D}\times\bar{G}$. Hence, for any $\varepsilon>0$, the map

$$f \longrightarrow \int_{x\in\partial G} f(x)\wedge[K_j(\cdot,x)-K\ (\cdot,x)],$$

defines a compact linear operator from $C^0_{0,q}(\bar{D},E)$ into itself.

In order to prove that (19.30) defines a compact operator from $C^0_{0,q}(\bar{D},E)$ into itself, therefore it is sufficient to show that

$$\lim_{\varepsilon\to 0}\ \sup_{z\in D} \int_{x\in\partial G} \|K^\varepsilon(z,x)\| d\sigma = 0,$$

where $d\sigma$ is a measure on ∂G which is locally defined to be the $(2n-1)$-dimensional Euclidean volume with respect to appropriate C^∞ coordinates in X. Since the operator S is bounded with respect to $\|\cdot\|_{0,G}$, therefore, it remains to show (in order to prove Lemma 19.9) that

$$\lim_{\varepsilon\to 0}\ \sup_{z\in D} \int_{x\in\partial G} \left\|[W^l_j(d_\varepsilon(z,\cdot)\bar{\partial}\hat{L}_X(z,\cdot)\wedge\omega_k)](x)\right\| d\sigma = 0 \qquad (19.31)$$

for $l=1,2,3$. If $N+1\le j\le M$, this is trivial, because then supp $\chi_j \subset\subset G$ (cf. Construction 19.3) and hence, by definition of W^1_j,

$$[W^1_j(d_\varepsilon(z,\cdot)\bar{\partial}\hat{L}_X(z,\cdot)\wedge\omega_k)](x) = 0 \qquad \text{if } x\in\partial G.$$

So we may assume that $1\le j\le N$.

Let (p_1,p_2) be the canonical Leray data for $[U_j,\varrho_j,\varphi_j,D_j]$ such that v_j is the canonical combination of (p_1,p_2). Set

$$\chi = \chi_j \circ h_j^{-1}.$$

Since supp $\chi \subset\subset \{t \in U_j : \varphi_j(t)<0\}$, then, for all $f \in C^0_{0,n-q}(\bar{D}_j)$, we have

$$\chi T^{\overset{v}{j}}(\chi f) = \chi B_{D_j}(\chi f) + \chi R_1^{\overset{v}{j}}(\chi f), \tag{19.32$_0$}$$

$$\bar{\partial}\chi \wedge T^{\overset{v}{j}}(\chi f) = \bar{\partial}\chi \wedge B_{D_j}(\chi f) + \bar{\partial}\chi \wedge R_1^{\overset{v}{j}}(\chi f), \tag{19.32$_1$}$$

$$\chi T^{\overset{v}{j}}(\bar{\partial}\chi \wedge f) = \chi B_{D_j}(\bar{\partial}\chi \wedge f) + \chi R_1^{\overset{v}{j}}(\bar{\partial}\chi \wedge f). \tag{19.32$_2$}$$

We set

$$\tilde{B}^0(t,s) = \chi(t)\chi(s)\,\frac{\det(\bar{s}-\bar{t}, \overbrace{d\bar{s},}^{q-1}\ \overbrace{d\bar{t}}^{n-q-1})}{|s-t|^{2n}} \wedge \omega(s), \tag{19.33$_0$}$$

$$\tilde{B}^1(t,s) = \bar{\partial}\chi(t)\ \chi(s)\,\frac{\det(\bar{s}-\bar{t}, \overbrace{d\bar{s},}^{q-1}\ \overbrace{d\bar{t}}^{n-q-1})}{|s-t|^{2n}} \wedge \omega(s), \tag{19.33$_1$}$$

$$\tilde{B}^2(t,s) = \chi(t)\bar{\partial}\chi(s)\,\frac{\det(\bar{s}-\bar{t}, \overbrace{d\bar{s},}^{q-1} \overbrace{d\bar{t}}^{n-q})}{|s-t|^{2n}} \wedge \omega(s). \tag{19.33$_2$}$$

Further, we introduce the abbreviation

$$r = r(t,s,y) = y\frac{\bar{s}-\bar{t}}{|s-t|^2} + (1-y)\frac{p_1(t,s)}{\langle p_1(t,s), s-t\rangle}, \tag{19.34}$$

and set

$$\tilde{R}^0(t,s,y) = \chi(t)\chi(s)\ \det(r, d_y r, \overbrace{\bar{\partial}_s r,}^{q-1}\ \overbrace{\bar{\partial}_t r}^{n-q-1}) \wedge \omega(s), \tag{19.35$_0$}$$

$$\tilde{R}^1(t,s,y) = \bar{\partial}\chi(t) \wedge \chi(s)\ \det(r, d_y r, \overbrace{\bar{\partial}_s r,}^{q-1}\ \overbrace{\bar{\partial}_t r}^{n-q-1}) \wedge \omega(s), \tag{19.35$_1$}$$

$$\tilde{R}^2(t,s,y) = \chi(t)\bar{\partial}\chi(s) \wedge \det(r, d_y r, \overbrace{\bar{\partial}_s r,}^{q-2} \overbrace{\bar{\partial}_t r}^{n-q}) \wedge \omega(s). \tag{19.35$_2$}$$

Then, by definition of B_{D_j} and $R_1^{\overset{v}{j}}$ (cf. Sects. 1.3 and 3.7), for any $f \in C^0_{0,n-q}(\bar{D}_j)$ and all $t \in D_j$, we have

$$[\chi B_{D_j}(\chi f)](t) = b_0 \int_{s\in D_j} f(x)\wedge \tilde{B}^0(t,s), \tag{19.36$_0$}$$

$$[\bar\partial\chi\wedge B_{D_j}(\chi f)](t) = b_1 \int_{s\in D_j} f(x)\wedge \tilde{B}^1(t,s), \tag{19.36$_1$}$$

$$[\chi B_{D_j}(\bar\partial\chi\wedge f)](t) = b_2 \int_{s\ D_j} f(x)\wedge \tilde{B}^2(t,s), \tag{19.36$_2$}$$

$$[\chi R_1^{v_j}(\chi f)](t) = a_0 \int_{\substack{s\in h_j(\partial G\cap A_j^2)\\ 0\le y\le 1}} f(s)\wedge \tilde{R}^0(t,s,y), \tag{19.37$_0$}$$

$$[\bar\partial\chi\wedge R_1^{v_j}(\chi f)](t) = a_1 \int_{\substack{s\in h_j(\partial G\cap A_j^2)\\ 0\le y\le 1}} f(s)\wedge \tilde{R}^1(t,s,y), \tag{19.37$_1$}$$

$$[\chi R_1^{v_j}(\bar\partial\chi\wedge f)](t) = a_2 \int_{\substack{s\in h_j(\partial G\cap A_j^2)\\ 0\le y\le 1}} f(s)\wedge \tilde{R}^2(t,s,y), \tag{19.37$_2$}$$

where b_l and a_l are some constants which depend only on n and q (similar as in (19.25)).

Now, for all $x,u\in A_j^3$ with $x\neq u$, by $\tilde{B}_X^l(x,u)$ we denote the pull-back of $\tilde{B}^l(t,s)$ with respect to $h_j\times h_j$ ($l=0,1,2$). Since $\chi_j = 0$ outside A_j^1, then $\tilde{B}_X^l(x,u)=0$ if $(x,u)\notin A_j^1\times A_j^1$. Therefore, after extending by zero, we may assume that the forms $\tilde{B}_X^l(x,u)$ are defined and continuous for all $x,u\in X$ with $x\neq u$. Analogously, by $\tilde{R}_X^l(x,u,y)$ ($x\in G$, $u\in\partial G$, $0\le y\le 1$) we denote the form which is obtained after extending by zero from the pull-back of $\tilde{R}^l(t,s,y)$ ($t\in D_j$, $s\in h_j(\partial G\cap A_j^1)$, $0\le y\le 1$) with respect to $h_j\times h_j$ ($l=0,1,2$).

By $\hat{B}_X^l(x,u)$ and $\hat{R}_X^l(x,u,y)$ we denote the $\mathrm{Hom}(p_u^*E,p_x^*E)$-valued differential forms on $X\times X$ resp. $X\times X\times[0,1]$ which are defined by the following two conditions:

1) $\hat{B}_X^l(x,u)$ and $\hat{R}_X^l(x,u,y)$ vanish if $(x,u)\notin A_j^1\times A_j^1$;

2) with repect to the isomorphism (19.4), $\hat{B}_X^l(x,u)$ is represented by the matrix $\tilde{B}_X^l(x,u)I_r$, and $\hat{R}_X^l(x,u,y)$ is represented by the matrix $\tilde{R}_X^l(x,u,y)I_r$, where I_r is the $r\times r$ unit matrix.

Then it follows from $(19.32)_1$, $(19.36)_1$, and the definition of the operators W_j^l (cf. Construction 19.3) that

$$\left\| [W_j^l(d_\varepsilon(z,\cdot)\bar\partial\hat L_X(z,\cdot)\wedge\omega_k)](w)\right\| \le C\left[\left\|\int_{u\in G} d_\varepsilon(z,u)\bar\partial_u\hat L_X(z,u)\wedge\hat B_X^l(w,u)\right\| + \left\|\int_{\substack{u\in\partial G\\0\le y\le 1}} d_\varepsilon(z,u)\bar\partial_u\hat L_X(z,u)\wedge\hat R_X^l(w,u,y)\right\|\right]$$

for all $z\in D$, $\varepsilon>0$, $w\in G$, and $l=0,1,2$. Thus, for the proof of (19.31), it is sufficient to show that

$$\lim_{\varepsilon\to 0}\sup_{z\in D}\int_{x\in\partial G}\lim_{G\ni w\to x}\left\|\int_{u\in G} d_\varepsilon(z,u)\bar\partial_u\hat L_X(z,u)\wedge\hat B_X^l(w,u)\right\| d\sigma = 0 \qquad (19.38)$$

and

$$\lim_{\varepsilon\to 0}\sup_{z\in D}\int_{x\in\partial G}\lim_{G\ni w\to x}\left\|\int_{\substack{u\in G\\0\le y\le 1}} d_\varepsilon(z,u)\bar\partial_u\hat L_X(z,u)\wedge\hat R_X^l(w,u,y)\right\| d\sigma = 0 \qquad (19.39)$$

for $l=0,1,2$.

Now we want to express (19.38) and (19.39) in the coordinates h_j. First notice that, by definition of $\hat B^l$, $\hat R^l$, $\hat L$, and d_ε, for all sufficiently small $\varepsilon>0$, we have

$$d_\varepsilon(z,u)\bar\partial_u\hat L_X(z,u)\wedge\hat B_X^l(w,u) = 0 \qquad (19.40)$$

and

$$d_\varepsilon(z,u)\bar\partial_u\hat L_X(z,u)\wedge\hat R_X^l(w,u,y) = 0 \qquad (19.41)$$

if $u\notin B_k^l\cap A_j^l$ or $w\notin A_j^l$ or $z\notin B_k^l\cap A_j^2$ $(l=0,1,2)$. Further, by definition of $\hat L$, $\hat B^l$, and $\hat R^l$, there is a holomorphic map H from $B_k^3\times A_j^3$ into the group of invertible complex $r\times r$ matrices such that

$$\bar\partial_u \hat L_X(z,u)\wedge \hat B^1_X(w,u) = H(z,u)\bar\partial_u \tilde L_X(z,u)\wedge \tilde B^1_X(w,u) \tag{19.42}$$

for all $z\in D\cap B^3_k\cap A^3_j$, $w\in G\cap A^3_j$, $u\in\partial G\cap B^3_k\cap A^3_j$ $(l=0,1,2)$, and

$$\bar\partial_u \hat L_X(z,u)\wedge \hat R^1_X(w,u,y) = H(z,u)\bar\partial_u \tilde L_X(z,u)\wedge \tilde R^1_X(w,u,y) \tag{19.43}$$

for all $z\in D\cap B^3_k\cap A^3_j$, $w\in G\cap A^3_j$, $u\in\partial G\cap B^3_k\cap A^3_j$, $0\le y\le 1$ $(l=0,1,2)$.

We introduce also the following abbreviations:

$Y^l := h_j(B^l_k\cap A^l_j)$	for $l=0,1,2$;
$\beta := g_k\circ h_j^{-1}$	on Y^3;
$a(t,s) := w_1(\beta(t),\beta(s))$	for $s,t\in Y^3$;
$A(t,s) := \langle a(t,s),\beta(t)-\beta(s)\rangle$	for $s,t\in Y^3$;
$D^- := D_j\cap Y^2 = \{t\in Y^2 :\ \varrho_j(t)<0\}$;	
$D^+ := \{t\in Y^2 :\ \varrho_j(t)>0\}$;	
$S := \{t\in Y^2 :\ \varrho_j(t)=0\}$;	
$G_\varepsilon(t,s) := d_\varepsilon(h_j^{-1}(t),h_j^{-1}(s))H(h_j^{-1}(t),h_j^{-1}(s))$	for $s,t\in Y^3$.

Then $\varrho_j = -\delta_k\circ\beta$ on Y^3 and hence, by (19.26),

$$\mathrm{Re}\, A(t,s) \ge \varrho_j(t) - \varrho_j(s) + c|s-t|^2 \tag{19.44}$$

for all $s,t\in Y^2$. Therefore, setting $b=\lambda_k\circ h_j^{-1}$ and

$$P(t,s) = \frac{(-1)^{q+1}\binom{n-1}{q}}{(2\pi i)^n}\, b(t)b(s)\, \frac{\det(a, \overbrace{\bar\partial_s a}^{n-q-1}, \overbrace{\bar\partial_t a}^{q})}{A^n}, \tag{19.45}$$

we obtain a continuous differential form defined for all $s,t\in Y^2$ with $\varrho_j(t)>\varrho_j(s)$. By (19.25), $P(t,s)$ is the pull back of $\tilde L_X(z,u)$ with respect to $h_j^{-1}\times h_j^{-1}$. Therefore, it follows from (19.40)-(19.43) that

$$\left\| \int_{u\in G} d_\varepsilon(z,u)\bar\partial_u \hat L_X(z,u)\wedge \hat B^1_X(w,u) \right\|$$

$$\leq C \left\| \int_{s\in D^-} G_\varepsilon(h_j(z),s)\bar\partial_s P(h_j(z),s)\wedge \hat{B}^l(h_j(w),s) \right\|$$

and

$$\left\| \int_{\substack{u\in\partial G\\ 0\leq y\leq 1}} d_\varepsilon(z,u)\bar\partial_u \hat{L}_X(z,u)\wedge \hat{R}^l_X(w,u,y) \right\|$$

$$\leq C \left\| \int_{\substack{s\in S\\ 0\leq y\leq 1}} G_\varepsilon(h_j(z),s)\bar\partial_s P(h_j(z),s)\wedge \hat{R}^l(h_j(w),s,y) \right\|$$

for all $z\in D\cap B^2_k\cap A^2_j$ and $w\in G\cap A^2_j$ ($l=0,1,2$). Taking into account again (19.40) and (19.41), so we see: in order to prove (19.38) and (19.39) it is sufficient to show that, for $l=0,1,2$,

$$\lim_{\varepsilon\to 0}\ \sup_{t\in D^+} I(\hat{B}^l,\varepsilon,t) = 0$$

and

$$\lim_{\varepsilon\to 0}\ \sup_{t\in D^+} I(\hat{R}^l,\varepsilon,t) = 0,$$

where ($dS:=$ $(2n-1)$-dimensional Euclidean volume form of S)

$$I(\hat{B}^l,\varepsilon,t) := \int_{x\in S} \lim_{D^-\ni z\to x} \left\| \int_{s\in D^-} G_\varepsilon(t,s)\bar\partial_s P(t,s)\wedge \hat{B}^l(z,s) \right\| dS$$

and

$$I(\hat{R}^l,\varepsilon,t) := \int_{x\in S} \lim_{D^-\ni z\to x} \left\| \int_{\substack{s\in S\\ 0\leq y\leq 1}} G_\varepsilon(t,s)\bar\partial_s P(t,s)\wedge \hat{R}^l(z,s,y) \right\| dS.$$

This will be done in Sects. 19.11 and 19.12. Actually, we shall obtain even the following more precise estimates: If $\delta>0$ is an arbitrarily small (fixed) number, then

$$I(\hat{B}^{l},\varepsilon,t) \leq C\varepsilon^{1/2-\delta} \tag{19.46}$$

and

$$I(\hat{R}^{l},\varepsilon,t) \leq C\varepsilon^{1/2-\delta} \tag{19.47}$$

for all $t\in D^{+}$, $\varepsilon>0$ and $l=0,1,2$.

19.11. Part 2 of the proof of Lemma 19.9: proof of estimate (19.46).
From the definition of $\tilde{B}^{l}$ (cf. $(19.33)_{l}$) it is clear that in the definition of $I(\tilde{B}^{l},\varepsilon,z)$ the limes may be interchanged with the integration over S. Hence

$$I(\hat{B}^{l},\varepsilon,z) \leq C \int\limits_{x\in S} dS \int\limits_{u\in D^{-}} \left\|G_{\varepsilon}(z,u)\bar{\partial}_{u}P(z,u)\right\| \left\|\hat{B}^{l}(x,u)\right\| d\sigma$$

for all $z\in D^{+}$, $\varepsilon>0$, and $l=0,1,2$, where $d\sigma$ is the 2n-dimensional Euclidean volume form in $\mathbb{C}^{n}$. Since h_{j}^{-1} is C^{1} in a neighborhood of $\bar{D}^{-}$, we can find $K<\infty$ such that $G_{\varepsilon}(z,u)=0$ for all $z\in D^{+}$, $u\in D^{-}$, $\varepsilon>0$ with $|z-u|\geq K\varepsilon$. Therefore, it follows that

$$I(\hat{B}^{l},\varepsilon,z) \leq C \int\limits_{x\in S} dS \int\limits_{\substack{u\in D^{-}\\ |u-z|\leq K\varepsilon}} \left\|\bar{\partial}_{u}P(z,u)\right\| \left\|\hat{B}^{l}(x,u)\right\| d\sigma \tag{19.48}$$

for all $z\in D^{+}$, $\varepsilon>0$, and $l=0,1,2$.

Recall that, by the definition of canonical Leray maps for q-concave configurations (cf. Sect. 13.4), w_{1} is of the form $w_{1}(z,u) = -w_{1}^{*}(u,z)$ for some $w_{1}^{*}\in \mathrm{Div}(-\sigma_{k})$. Since $-\sigma_{k}$ is normalized (q+1)-convex, this implies that $w_{1}(z,u)$ depends holomorphically on $u_{1},\ldots,u_{q+1}$ (cf. Sect. 7.6). Taking into account that $a(z,u)=w_{1}(\beta(z),\beta(u))$ and β is biholomorphic, so we see that

$$\bar{\partial}_{u}\frac{\det(a,\overbrace{\bar{\partial}_{u}a}^{n-q-1},\overbrace{\bar{\partial}_{z}a}^{q})}{A^{n}} = 0$$

and hence, by (19.45),

$$\bar\partial_u P(z,u) = \frac{(-1)^{q+1}\binom{n-1}{q}}{(2\pi i)^n} b(z)\bar\partial b(u)\wedge \frac{\det(a, \overbrace{\bar\partial_u a}^{n-q-1}, \overbrace{\bar\partial_z a}^{q})}{A^n}.$$

Since $\sigma_k = -\varrho_j \circ \beta^{-1}$, moreover, it follows from (19.49) that $|a(z,u)| \le C(\|d\varrho_j(u)\| + |u-z|)$ for all $z,u\in Y^2$. Hence

$$\left\| \bar\partial_u \; P(z,u)\right\| \le C\frac{\|d\varrho_j\| + |u-z|}{|A(z,u)|^n} \tag{19.50}$$

for all $z\in D^+$ and $u\in D^-$. Together with (19.48) this yields the estimate

$$I(\tilde B^l, \varepsilon, z) \le C\int\limits_{x\in S} dS \int\limits_{\substack{u\in D^- \\ |u-z|\le K\varepsilon}} \frac{\|d\varrho_j(u)\| \, \|\tilde B^l(x,u)\|}{|A(z,u)|^n} d\sigma + C\int\limits_{x\in S} dS \int\limits_{\substack{u\in D^- \\ |u-z|\le K}} \frac{\tilde B^l(x,u)}{|A(z,u)|^n |u-z|^{-1}} d\sigma$$

for all $z\in D^+$, $\varepsilon>0$, and $l=0,1,2$. Since $\|\tilde B^l(x,u)\| \le C|x-u|^{-2n+1}$ and $|A(z,u)| \ge c|u-z|^2$ (cf. (19.33) and (19.44)), this implies that

$$I(\tilde B^l, \varepsilon, z) \le C\int\limits_{\substack{u\in D^- \\ |u-z|\le K\varepsilon}} \frac{\|d\varrho_j(u)\| d\sigma}{|A(z,u)|^n} \int\limits_{x\in S} \frac{dS}{|u-x|^{2n-1}} + C\int\limits_{\substack{u\in D^- \\ |u-z|\le K\varepsilon}} \frac{d\sigma}{|z-u|^{2n-1}} \int\limits_{x\, S} \frac{dS}{|u-x|^{2n-1}} \tag{19.51}$$

for all $z\in D^+$, $\varepsilon>0$, and $l=0,1,2$.

By Proposition 3 (i) in Appendix B, we have

$$\int\limits_{x\in S} \frac{dS}{|u-x|^{2n-1}} \le \frac{1}{[\text{dist}(u,S]^{1/4}} \int\limits_{x\in S} \frac{dS}{|u-x|^{2n-1-1/4}} \le \frac{C}{[\text{dist}(u,S)]^{1/4}}$$

By Proposition 5 (i) in Appendix B, this implies

$$\int\limits_{\substack{u\in D^- \\ |u-z|\le K\varepsilon}} \frac{d\sigma}{|z-u|^{2n-1}} \int\limits_{x\in S} \frac{dS}{|u-x|^{2n-1}} \le C\varepsilon^{1/2} \tag{19.52}$$

for all $z\in D^+$ and $\varepsilon>0$. Since $\varrho_j=0$ on S, we have $|\varrho_j(u)| \leq C|u-x|$ for all $u\in D^-$ and $x\in S$. Therefore

$$\int\limits_{x\in S} \frac{dS}{|u-x|^{2n-1}} \leq C \int\limits_{x\in S} \frac{dS}{(|\varrho_j(u)|^2 + |u-x|^2)|u-x|^{2n-3}}$$

for all $u\in D^-$. By Proposition 2 (i) in Appendix B, this implies that

$$\int\limits_{x\in S} \frac{dS}{|u-x|^{2n-1}} \leq C(1 + |\ln|\varrho_j(u)||)$$

for all $u\in D^-$. Hence

$$\int\limits_{\substack{u\in D^- \\ |u-z|\leq K\varepsilon}} \frac{\|d\varrho_j(u)\|d\sigma}{|A(z,u)|^n} \int\limits_{x\in S} \frac{dS}{|u-x|^{2n-1}} \leq C \int\limits_{\substack{u\in \bar{D} \\ |u-z|\leq K\varepsilon}} \frac{\|d\varrho_j(u)\|(1 + |\ln|\varrho_j(u)||)}{|A(z,u)|^n} d\sigma \quad (19.53)$$

for all $z\in D^+$ and $\varepsilon>0$. By (19.44), $|A(z,u)|^n \geq (|\varrho_j(u)|+|u-z|^2)^2|u-z|^{2n-4}$ for all $z\in D^+$ and $u\in D^-$. Since $\varrho_j=0$ on S, we can find a constant $K'<\infty$ such that $|\varrho_j(u)|\leq K'\varepsilon$ for all $u\in D^-$ with $\mathrm{dist}(u,D^+)\leq K\varepsilon$. Therefore, the integral on the right hand side of (19.53) can be estimated by

$$C \int\limits_{\substack{u\in D^- \\ |\varrho_j(u)|\leq K'\varepsilon}} \frac{\|d\varrho_j(u)\|(1 + |\ln|\varrho_j(u)||)}{(|\varrho_j(u)|+|u-z|^2)^2|u-z|^{2n-4}} d\sigma.$$

In view of Proposition 5 (ii) in Appendix B, this implies that, for every fixed $\delta>0$, the right hand side of (19.53) is bounded by $C\varepsilon^{1/2-\delta}$ Together with (19.51) and (19.52) this implies (19.46).

<u>19.12. Part 3 (and end) of the proof of Lemma 19.9: proof of estimate (19.47).</u> Set

$$\phi_- = \phi_-(w,u) = \langle p_1(w,u),u-w\rangle - 2\varrho_j(u)$$

for $w,u\in U_j$, where $p_1\in \mathrm{Div}(\varrho_j)$ is as in Sect. 19.10. Then, by (7.4),

$$\mathrm{Re}\,\phi_-(w,u) \geq c(-\varrho_j(u) - \varrho_j(w) + |u-w|^2) \tag{19.54}$$

for all $w,u\in U_j$. In particular, $\phi_-(w,u)\neq 0$ for all $w,u \in S\cup D^-$, except for the case $u=w\in S$. Therefore, the map

$$v_- := v_-(w,u,t) := t\frac{\bar{u}-\bar{w}}{|u-w|^2} + (1-t)\frac{p_1(w,u)}{\phi_-(w,u)}$$

is defined and of class C^1 for all $0\leq t\leq 1$ and $w,u \in S\cup D^-$ with $w\neq u$. Therefore

$$\tilde{R}^0_-(w,u,t) := \chi(w)\chi(u)\ \det(v_-,d_t v_-,\overbrace{\bar\partial_u v_-}^{q-1},\overbrace{\bar\partial_w v_-}^{n-q-1})\wedge\omega(u),$$

$$\tilde{R}^1_-(w,u,t) := \bar\partial\chi(w)\wedge\chi(u)\ \det(v_-,d_t v_-,\overbrace{\bar\partial_u v_-}^{q-1},\overbrace{\bar\partial_w v_-}^{n-q-1})\wedge\omega(u),$$

$$\tilde{R}^2_-(w,u,t) := \chi(w)\bar\partial\chi(u)\wedge\det(v_-,d_t v_-,\overbrace{\bar\partial_u v_-}^{q-2},\overbrace{\bar\partial_w v_-}^{n-q})\wedge\omega(u)$$

are continuous differential forms for all $0\leq t\leq 1$ and $w,u\in S\cup D^-$ with $w\neq u$. We have (cf. $(19.35)_1$)

$$\tilde{R}^l_-(w,u,t) = \tilde{R}^l(w,u,t) \qquad \text{if } u\in S\ (l=0,1,2). \tag{19.55}$$

Since

$$\det(\bar{u}-\bar{w},\bar{u}-\bar{w},a_1,\dots,a_{n-2}) = \det(p_1,p_1,a_1,\dots,a_{n-2}) = 0 \tag{19.56}$$

for any collection $a_1,\dots,a_{n-2}$ of vectors of differential forms, it follows that

$$\tilde{R}^0_-(w,u,t) := \frac{\chi(w)\chi(u)(2t-1)dt}{\phi_-(w,u)|u-w|^2}\det(v_-,d_t v_-,\overbrace{\bar\partial_u v_-}^{q-1},\overbrace{\bar\partial_w v_-}^{n-q-1})\wedge\omega(u),$$

$$\tilde{R}^1_-(w,u,t) := \frac{\bar\partial\chi(w)\wedge\chi(u)(2t-1)dt}{\phi_-(w,u)|u-w|^2}\det(v_-,d_t v_-,\overbrace{\bar\partial_u v_-}^{q-1},\overbrace{\bar\partial_w v_-}^{n-q-1})\wedge\omega(u),$$

$$\tilde{R}^2_-(w,u,t) := \frac{\chi(w)\bar\partial\chi(u)\wedge(2t-1)dt}{\phi_-(w,u)|u-w|^2}\det(v_-,d_t v_-,\overbrace{\bar\partial_u v_-}^{q-2},\overbrace{\bar\partial_w v_-}^{n-q})\wedge\omega(u),$$

for all $0 \le t \le 1$ and $w,u \in S \cup D^-$ with $w \ne u$. Since

$$\bar\partial_u v_- = t\frac{d\bar u}{|u-w|^2} + (1-t)\frac{\bar\partial_u p_1}{\phi_-} - \frac{t\bar\partial_u |u-w|^2}{|u-w|^4}(\bar u-\bar w) - \frac{(1-t)p_1\bar\partial_u\phi_-}{\phi_-^2}p_1$$

and

$$\bar\partial_w v_- = t\frac{-d\bar w}{|u-w|^2} + (1-t)\frac{\bar\partial_w p_1}{\phi_-} - \frac{t\bar\partial_w |u-w|^2}{|u-w|^4}(\bar u-\bar w) - \frac{(1-t)p_1\bar\partial_w\phi_-}{\phi_-^2}p_1,$$

this implies (taking into account again (19.56)) that

$$\tilde R^0_-(w,u,t) = \chi(w)\chi(u)\sum_{r=0}^{q-1}\sum_{s=0}^{n-q-1} p_{rs}dt\wedge R_{rs}, \tag{19.57$_0$}$$

$$\tilde R^1_-(w,u,t) = \bar\partial\chi(w)\wedge\chi(u)\sum_{r=0}^{q-1}\sum_{s=0}^{n-q-1} p_{rs}dt\wedge R_{rs}, \tag{19.57$_1$}$$

$$\tilde R^2_-(w,u,t) = \chi(w)\bar\partial\chi(u)\wedge\sum_{r=0}^{q-2}\sum_{s=0}^{n-q} p'_{rs}dt\wedge R'_{rs} \tag{19.57$_2$}$$

for all $0 \le t \le 1$ and $w,u \in S \cup D^-$ with $w \ne u$, where p_{rs} and p'_{rs} are some polynomials in t the coefficients of which depend only on n and q,

$$R_{rs} = \frac{\det(p_1,\bar u-\bar w,\overbrace{d\bar u}^{r},\overbrace{\bar\partial_u p_1}^{q-1-r},\overbrace{d\bar w}^{s},\overbrace{\bar\partial_w p_1}^{n-q-1-s})}{(\phi_-(w,u))^{n-1-r-s}|u-w|^{2(r+s+1)}}\wedge\omega(u)$$

and

$$R'_{rs} = \frac{\det(p_1,\bar u-\bar w,\overbrace{d\bar u}^{r},\overbrace{\bar\partial_u p_1}^{q-2-r},\overbrace{d\bar w}^{s},\overbrace{\bar\partial_w p_1}^{n-q-s})}{(\phi_-(w,u))^{n-1-r-s}|u-w|^{2(r+s+1)}}\wedge\omega(u).$$

Since $\phi_-(w,u) \ne 0$ if $w \in D^-$ and $u \in S \cup D^-$ (cf. (19.54)), it follows from (19.57)$_1$ that the forms $\tilde R^1_-(w,u,t)$ have a singularity of order $\le 2n-3$ at $u=w \in D^-$, and the forms $\bar\partial_u \tilde R^1(w,u,t)$ are continuous for all $0 \le t \le 1$ and $w,u \in S \cup D^-$ with $w \ne u$ $(1=0,1,2)$. Therefore we may apply Stokes' formula, and so, taking into account (19.55) and the relation $\bar\partial_u P(z,u)\wedge\tilde R^1_-(w,u,t) = 0$ for $u \notin Y^1$, we obtain that

$$\int_{\substack{u\in S\\ 0\le t\le 1}} G_\varepsilon(z,u)\bar\partial_u P(z,u)\wedge\tilde R^l(w,u,t) = \int_{\substack{u\in D^-\\ 0\le t\le 1}} \bar\partial_u G_\varepsilon(z,u)\wedge\bar\partial_u P(z,u)\wedge\tilde R^l_-(w,u,t)$$

$$+ (-1)^n \int_{\substack{u\in D^-\\ 0\le t\le 1}} G_\varepsilon(z,u)\wedge\bar\partial_u P(z,u)\wedge\bar\partial_u\tilde R^l_-(w,u,t) \qquad (19.58)$$

for all $z\in D^+$, $w\in D^-$, $\varepsilon>0$, and $l=0,1,2$.

Recall that, by (19.54), $|u-w|^2\le C|\phi_-(w,u)|$ for all $w,u\in S\cup D^-$. Further, it is clear that

$$\Big\|\det(p_1,\bar u-\bar w,\overbrace{d\bar u}^{m_1},\overbrace{\bar\partial_u p_1}^{m_2},\overbrace{d\bar w}^{m_3},\overbrace{\bar\partial_w p_1}^{m_4})\Big\| \le C|u-w|\,|p_1(w,u)|$$

and

$$\Big\|\bar\partial_u[\det(p_1,\bar u-\bar w,\overbrace{d\bar u}^{m_1},\overbrace{\bar\partial_u p_1}^{m_2},\overbrace{d\bar w}^{m_3},\overbrace{\bar\partial_w p_1}^{m_4})]\Big\| \le C\big(|u-w| + |p_1(w,u)|\big)$$

for all $w,u\in S\cup D^-$ if $m_1,\dots,m_4\ge 0$ is an arbitrary collection of integers with $m_1+\dots+m_4=n-2$. Therefore, it follows from $(19.57)_1$ that

$$\Big\|\bar\partial_u\tilde R^l_-(w,u,t)\Big\| \le C\left(\frac{|u-w| + |p_1(w,u)|}{|\phi_-(w,u)|\,|u-w|^{2n-2}} + \frac{|p_1(w,u)|\,\|\bar\partial_u\phi_-(w,u)\|}{|\phi_-(w,u)|^2\,|u-w|^{2n-3}}\right.$$

$$\left. + \frac{|p_1(w,u)|\,\|\bar\partial_u|u-w|\|}{|\phi_-(w,u)|\,|u-w|^{2n-2}}\right) \qquad (19.59)$$

for all $0\le t\le 1$, $w,u\in S\cup D^-$, and $l=0,1,2$. Since $p_1\in \mathrm{Div}(\varrho_j)$ (cf. Sects. 7.6 and 7.7), one has the estimate

$$|p_1(w,u)| \le C\big(\|d\varrho_j(w)\| + |u-w|\big) \qquad (19.60)$$

for all $w,u\in S\cup D^-$. In view of the definition of $\phi_-(w,u)$, this implies that also $\|\bar\partial_u\phi_-(w,u)\| \le C(\|d\varrho_j(w)\| + |u-w|)$ for all $w,u\in S\cup D^-$. Therefore, it follows from (19.59) that

$$\left\| \bar\partial_u \tilde{R}^l_-(w,u,t) \right\| \le C\left(\frac{1}{|\phi_-(w,u)||u-w|^{2n-3}} + \frac{\|d\varrho_j(w)\|}{|\phi_-(w,u)||u-w|^{2n-2}} + \frac{\|d\varrho_j(w)\|^2}{|\phi_-(w,u)|^2|u-w|^{2n-3}} \right) \qquad (19.61)$$

for all $0\le t\le 1$, $w,u\in S\cup D^-$ and $l=0,1,2$. Further, by $(19.57)_1$ and (19.60), we have

$$\left\| \tilde{R}^l_-(w,u,t) \right\| \le C\left(\frac{1}{|\phi_-(w,u)||u-w|^{2n-4}} + \frac{\|d\varrho_j(w)\|}{|\phi_-(w,u)||u-w|^{2n-3}} \right) \qquad (19.62)$$

for all $0\le t\le 1$, $w,u\in S\cup D^-$, and $l=0,1,2$.

Now, for $m=1,2,\dots$, we set $S_m = \{w\in D^- : \varrho_j(w)=-1/m\}$. Since ϱ_j does not have degenerate critical points, for all sufficiently large m, S_m is smooth and, for any continuous function f on $S\cup D^-$, we have

$$\int_{S\cap Y^1} f\, dS = \lim_{m\longrightarrow\infty} \int_{S_m\cap Y^1} f\, dS_m, \qquad (19.63)$$

where dS_m is the Euclidean volume form of S_m; this can be proved by similar arguments as in point 6 in Appendix B. Recall that, for any continuous differential form g on $\bar{D}_1$, the form $\check{R}_1{}^j f$ admits a continuous extension onto $S\cup D^-$ (cf. the beginning of the proof of Theorem 9.1). In view of $(19.37)_1$ and since $G_\varepsilon(z,\cdot)\bar\partial P(z,\cdot)$ is continuous on $\bar{D}_j$ and vanishes on $h_j(\partial G\cap(A^2_j\setminus B^2_k)) = h_j(\partial G\cap A^2_j)\setminus S$ for all fixed $\varepsilon>0$ and $z\in D^+$, this implies that, for all fixed $\varepsilon>0$, $z\in D^+$, and $l=0,1,2$, the function

$$f^l_{\varepsilon,z}(w) := \left\| \int_{\substack{u\in S\\ 0\le t\le 1}} G_\varepsilon(z,u)\bar\partial_u P(z,u)\wedge \tilde{R}^l(w,u,t) \right\|, \qquad w\in D^-,$$

admits a continuous extension onto $S\cup D^-$ which will be denoted by $\hat{f}^l_{\varepsilon,z}$. Since, for all sufficiently small $\varepsilon>0$, supp $f^l_{\varepsilon,z} \subset\subset Y^1$, therefore it follows from (19.63) that

$$I(\tilde{R}^l, \varepsilon, z) = \int_S \hat{f}^l_{\varepsilon, z} dS = \lim_{m \longrightarrow \infty} \int_{S_m} f^l_{\varepsilon, z} dS_m \tag{19.64}$$

for all $z \in D^+$ and all sufficiently small $\varepsilon > 0$ $(l=0,1,2)$. Take a number $K < \infty$ such that $G_\varepsilon(z,u)=0$ if $|z-u| \geq K\varepsilon$ (cf. the beginning of Sect. 19.11). Then it follows from (19.58) that

$$f^l_{\varepsilon, z}(w) \leq C \int_{\substack{u \in D^- \\ |u-z| \leq K\varepsilon \\ 0 \leq t \leq 1}} \| \bar{\partial}_u P(z,u) \| \| \tilde{R}^l_-(w,u,t) \| d\sigma + C \int_{\substack{u \in D^- \\ |u-z| \leq K\varepsilon \\ 0 \leq t \leq 1}} \| \bar{\partial}_u P(z,u) \| \| \bar{\partial}_u \tilde{R}^l_-(w,u,t) \| d\sigma$$

for all $z \in D^+$, $w \in D^-$, $\varepsilon > 0$, and $l=0,1,2$, where $d\sigma$ is the Euclidean volume form in $\mathbb{C}^n$. In view of (19.64), this implies that

$$I(\tilde{R}^l, \varepsilon, z) \leq C \sup_m \int_{w \in S_m} dS_m \int_{\substack{u \in D^- \\ |u-z| \leq K\varepsilon \\ 0 \leq t \leq 1}} \| \bar{\partial}_u P(z,u) \| \| \tilde{R}^l_-(w,u,t) \| d\sigma$$

$$+ C \sup_m \int_{w \in S_m} dS_m \int_{\substack{u \in D^- \\ |u-z| \leq K\varepsilon \\ 0 \leq t \leq 1}} \| \bar{\partial}_u P(z,u) \| \| \bar{\partial}_u \tilde{R}^l_-(w,u,t) \| d\sigma$$

for all $z \in D^+$, all sufficiently small $\varepsilon > 0$, and $l=0,1,2$. In view of (19.61) and (19.62), this implies that

$$I(R^l, \varepsilon, z) \leq C \sup_m \int_{\substack{u \in D^- \\ |u-z| \leq K\varepsilon}} \| \bar{\partial}_u P(z,u) \| \left(I^1_m(u) + I^2_m(u) + I^3_m(u) \right) d\sigma$$

for all $z \in D^+$, all sufficiently small $\varepsilon > 0$, and $l=0,1,2$, where

$$I^1_m(u) := \int_{w \in S_m} \frac{1}{|\phi_-(w,u)| \, |u-w|^{2n-3}} dS_m,$$

$$I_m^2(u) := \int\limits_{w\in S_m} \frac{\|d\varrho_j(w)\|}{|\phi_-(w,u)|\,|u-w|^{2n-2}}\,dS_m,$$

$$I_m^3(u) := \int\limits_{w\in S_m} \frac{\|d\varrho_j(w)\|^2}{|\phi_-(w,u)|^2\,|u-w|^{2n-3}}\,dS_m.$$

Taking into account (19.50), so we see that

$$I(\tilde{R}^l,\varepsilon,z) \le C \sup_m \int\limits_{\substack{u\in D^-\\ |u-z|\le K}} \frac{(|u-z|+\|d\varrho_j(u)\|)}{|\phi_-(z,u)|^n}(I_m^1(u)+I_m^2(u)+I_m^3(u))\,d\sigma \quad (19.65)$$

for all $z\in D^+$, all sufficiently small $\varepsilon>0$, and $l=0,1,2$.

Now we go to estimate the integrals $I_m^s(u)$. By (19.54), we have

$$I_m^1(u) \le C\int\limits_{w\in S_m} \frac{dS_m}{(|\varrho_j(u)|+|u-w|^2)\,|u-w|^{2n-3}},$$

$$I_m^2(u) \le C\int\limits_{w\in S_m} \frac{\|d\varrho_j(w)\|\,dS_m}{(|\varrho_j(u)|+|u-w|^2)\,|u-w|^{2n-2}},$$

$$I_m^3(u) \le C\int\limits_{w\in S_m} \frac{\|d\varrho_j(w)\|^2 dS_m}{(|\varrho_j(u)|+|t(w,u)|+|u-w|^2)^2\,|u-w|^{2n-3}}$$

for all $u\in D^-$ and m, where $t(w,u) := \operatorname{Im}\phi_-(w,u) = \operatorname{Im}\langle p_1(w,u),u-w\rangle$. Therefore and by Lemma 9.2, we can apply Proposition 2 in Appendix B to the integrals $I_m^s(u)$, and obtain the estimate

$$I_m^1(u)+I_m^2(u)+I_m^3(u) \le C\left(1 + |\ln|\varrho_j(u)|| + \frac{\|d\varrho_j(u)\|}{|\varrho_j(u)|^{1/2}}\right) \quad (19.66)$$

for all $u\in D^-$ and $m=1,2,\dots$.

Since $\|d\varrho_j(w)\| \le C(\|d\varrho_j(u)\| + |u-w|)$, we see that

$$I_m^2(u) \le C\|d\varrho_j(u)\| \int\limits_{w\in S_m} \frac{dS_m}{|\phi_-(w,u)|\,|u-w|^{2n-2}} + C \int\limits_{w\in S_m} \frac{dS_m}{|\phi_-(w,u)|\,|u-w|^{2n-3}}$$

and

$$I_m^3(u) \le C\|d\varrho_j(u)\| \int\limits_{w\in S_m} \frac{dS_m}{|\phi_-(w,u)|^2|u-w|^{2n-4}} + C \int\limits_{w\in S_m} \frac{dS_m}{|\phi_-(w,u)|^2|u-w|^{2n-5}}$$

$$+ C\|d\varrho_j(u)\|^2 \int\limits_{w\in S_m} \frac{dS_m}{|\phi_-(w,u)|^2|u-w|^{2n-3}}$$

for all $u\in D^-$ and $m=1,2,\ldots$. Taking into account that, by (19.54), $|\phi_-(w,u)|\ge c|u-w|^2$ for all $u,w\in D^-$, this implies that

$$I_m^1(u)+I_m^2(u)+I_m^3(u) \le C\|d\varrho_j(u)\| \int\limits_{w\in S_m} \frac{dS_m}{|\phi_-(w,u)|\,|u-w|^{2n-2}} + C \int\limits_{w\in S_m} \frac{dS_m}{|\phi_-(w,u)|\,|u-w|^{2n-3}}$$

$$+ C\|d\varrho_j(u)\|^2 \int\limits_{w\in S_m} \frac{dS_m}{|\phi_-(w,u)|^2|u-w|^{2n-3}} \qquad (19.67)$$

for all $u\in D^-$ and $m=1,2,\ldots$.

Now let some $\delta>0$ be given. We want to prove (19.47) for this δ. Without loss of generality we may assume that $\delta<1/2$. In view of (19.54), we have

$$|\phi_-(w,u)|\,|u-w|^{2n-3} \ge c|\varrho_j(u)|^{\delta/2}|u-w|^{2n-1-\delta} ,$$

$$|\phi_-(w,u)|\,|u-w|^{2n-2} \ge c|\varrho_j(u)|^{1/2+\delta/2}|u-w|^{2n-1-\delta} ,$$

$$|\phi_-(w,u)|^2|u-w|^{2n-3} \ge c|\varrho_j(u)|^{1+\delta/2}|u-w|^{2n-1-\delta} ,$$

for all $u\in D^-$, $w\in S_m$ ($m=1,2,\ldots$). Further, by Proposition 3 (i) in Appendix B, we have

$$\sup_m \int_{S_m} \frac{dS_m}{|u-w|^{2n-1-\delta}} < \infty.$$

Therefore it follows from (19.67) that

$$I_m^1(u)+I_m^2(u)+I_m^3(u) \leq C\left(\frac{1}{|\varrho_j(u)|^{\delta/2}}+\frac{\|d\varrho_j(u)\|}{|\varrho_j(u)|^{1/2+\delta/2}}+\frac{\|d\varrho_j(u)\|^2}{|\varrho_j(u)|^{1+\delta/2}}\right) \tag{19.68}$$

for all $u\epsilon D^-$ and $m=1,2,\ldots$.

Since $\varrho_j(z)>0$ and $\varrho_j(u)<0$ if $z\epsilon D^+$ and $u\epsilon D^-$, there is a constant $K'<\infty$ such that

$$\{u\epsilon D^-: |u-z|\leq K\varepsilon\} \subseteq \{u\epsilon D^-: \varrho_j(u)\leq K'\varepsilon\} \tag{19.69}$$

for all $z\epsilon D^+$. Recall that, by (19.44),

$$|A(z,u)| \geq c\left(|\varrho_j(u)| + |\mathrm{Im}\, A(z,u)| + |u-z|^2\right) \tag{19.70}$$

for all $z\epsilon D^+$ and $u\epsilon D^-$. For $z\epsilon D^+$ and $\varepsilon>0$, we introduce the following abbreviations

$$J_1(\varepsilon,z) = \int_{\substack{u\epsilon D^- \\ |u-z|\leq K\varepsilon}} \frac{|\varrho_j(u)|^{-\delta/2}}{|u-z|^{2n-1}}\, d\sigma,$$

$$J_2(\varepsilon,z) = \int_{\substack{u\epsilon D^- \\ |\varrho_j(u)|\leq K'\varepsilon}} \frac{\|d\varrho_j(u)\|\, |\varrho_j(u)|^{-1/2-\delta/2}}{(|\varrho_j(u)|+|u-z|^2)^2\, |u-z|^{2n-5}}\, d\sigma,$$

$$J_3(\varepsilon,z) = \int_{\substack{u\epsilon D^- \\ |\varrho_j(u)|\leq K'\varepsilon}} \frac{\|d\varrho_j(u)\|^2\, |\varrho_j(u)|^{-1-\delta/2}}{(|\varrho_j(u)|+|\mathrm{Im}\, A(z,u)|+|u-z|^2)^2\, |u-z|^{2n-5}}\, d\sigma,$$

$$J_4(\varepsilon,z) = \int_{\substack{u\epsilon D^- \\ |\varrho_j(u)|\leq K'\varepsilon}} \frac{\|d\varrho_j(u)\|\left(1 + \left|\ln|\varrho_j(u)|\right|\right)}{\left(|\varrho_j(u)|+|u-z|^2\right)^2\, |u-z|^{2n-4}}\, d\sigma,$$

$$J_5(\varepsilon,z) = \int\limits_{\substack{u\in D^- \\ |\varrho_j(u)|\le K'\varepsilon}} \frac{\|d\varrho_j(u)\|^2 |\varrho_j(u)|^{-1/2}}{(|\varrho_j(u)|+|\mathrm{Im}\ A(z,u)|+|u-z|^2)^2 |u-z|^{2n-4}}\, d\sigma.$$

Taking into account (19.69) and (19.70), then it follows from (19.65), (19.66), and (19.68) that

$$I(\tilde{R}^l,\varepsilon,z) \le C\sum_{i=1}^{5} J_i(\varepsilon,z)$$

for all $z\in D^+$, all sufficiently small $\varepsilon>0$, and $l=0,1,2$. So it is sufficient to prove that, for $i=1,\dots,5$,

$$J_i(\varepsilon,z) \le C\varepsilon^{1/2-\delta} \tag{19.71}$$

for all $z\in D^+$ and all $\varepsilon>0$.

Since ϱ_j does not have degenerate critical points, it is easy to show (for instance, by means of Proposition 0.7 in Appendix B) that $\mathrm{dist}(u,S) \le C|\varrho_j(u)|^{1/2}$ and hence

$$J_1(\varepsilon,z) \le C \int\limits_{\substack{u\in D^- \\ |u-z|\le K\varepsilon}} \frac{[\mathrm{dist}(u,S)]^{-\delta}}{|u-z|^{2n-1}}\, d\sigma$$

for all $z\in D^+$ and $\varepsilon>0$. Therefore, if $i=1$, then (19.71) follows from Proposition 5 (i) in Appendix B (with $\alpha=\delta$ and $\beta=1/2-\delta$). If $i=2$, then (19.71) follows from Proposition 5 (ii) in Appendix B (with $\alpha=1/2+\delta/2$, $\beta=1/2-\delta$, and $d=2$), and if $i=4$, then, in view of the estimate

$$1 + \big|\ln|\varrho_j(u)|\big| \le C|\varrho_j(u)|^{-\delta/2} \qquad (u\in D^-),$$

(19.71) follows also from Proposition 5 (ii) (with $\alpha=\delta/2$, $\beta=1/2-\delta$, and $d=1$).

It remains to consider the cases $i=3$ and $i=5$. Since

$$A(z,u) = \langle w_1(\beta(z),\beta(u)),\beta(u)-\beta(z)\rangle = -\langle w_1^*(\beta(u),\beta(z)),\beta(u)-\beta(z)\rangle$$

for some $w_1^*\in \mathrm{Div}(-\sigma_k)$, where $\sigma_k=-\varrho_j\circ\beta^{-1}$ and β is biholomorphic, it follows from the lemma stated in the proof of Theorem 14.1 that the following statements (i)-(iv) hold true:

(i)

$$c\|d\varrho_j(z)\| \leq \| d_u \operatorname{Im} A(z,u)|_{u=z}\| \leq C\|d\varrho_j(z)\|$$

for all $z,u \in Y^2$.

(ii)

$$\| d\varrho_j(z) \wedge d_u \operatorname{Im} A(z,u)|_{u=z} \| \geq c\|d\varrho_j(z)\|^2$$

for all $z,u \in Y^2$.

(iii)

$$\left| \frac{\partial \operatorname{Im} A(z,\cdot)}{\partial t_m}(u) - \frac{\partial \operatorname{Im} A(z,\cdot)}{\partial t_m}(z) \right| \leq C|u-z|$$

for all $z,u \in Y^2$ and $m=1,\dots,2n$, where $t_1,\dots,t_{2n}$ are the real coordinates in $\mathbb{C}^n$.

(iv) If $d\varrho_j(u_0)=0$ for some $u_0 \in S$ (this is possible only when $[U_j, \varrho_j, \varphi_j, D_j]$ is of type II), then

$$|\operatorname{Im} A(z,u)| \leq C(|u-u_0||u-z| + |u-z|^2)$$

for all $z,u \in Y^2$.

Hence, for i=3, estimate (19.71) follows from Proposition 5 (iii) in Appendix B (with $\alpha=1+\delta/2$, $\beta=1/2-\delta$, and d=2), and, for i=5, (19.71) follows from the same proposition (with $\alpha=1/2$, $\beta=1/2-\delta$, and d=1). Thus, estimate (19.47) and hence Lemma 19.9 is proved.

19.13. Proof of Theorem 19.2'. Since, by Lemma 19.6, $Z^0_{n,n-q-1}(\bar{G},E^*)$ is a finitely codimensional subspace of Im A, $(\operatorname{Im} A)^{\perp}$ is a finitely codimensional subspace of $\hat{Z}^0_{0,q}(\bar{D},E)$.

Let $\tilde{T},\tilde{K},L$ be the operators from construction 19.8. Since $\tilde{K}$ is bounded as an operator from $Z^0_{0,q}(\bar{D},E)$ into $C^{1/2}_{0,q}(\bar{D},E)$ (cf. assertion (19.22)), then it follows from Lemma 19.9 that the restriction of $\tilde{K}$+L to $\hat{Z}^0_{0,q}(\bar{D},E)$ is compact as an operator with values in $C^0_{0,q}(\bar{D},E)$. We denote this restriction by K. Further, let T be the restriction of $\tilde{T}$ to $\hat{Z}^0_{0,q}(\bar{D},E)$. Then (19.23) implies (19.1) and, except for relation (19.2), assertion

(ii) of Theorem 19.2' is proved. Since, by Stokes' theorem $E_{0,q}^{\alpha\longrightarrow 0}(\bar{D},E)$ is contained in $\hat{Z}_{0,q}^{0}(\bar{D},E)$ for all $0\leq\alpha\leq 1/2$, this implies assertion (i) of Theorem 19.2' (cf. Remark III following Def. 11.1).

It remains to prove (19.2). Let $f\in Z_{0,q}^{0}(\bar{D},E)\cap E_{0,q}^{0}(D,E)$ and $g\in Z_{n,n-q-1}^{0}(\bar{G},E^{*})$. We have to show that

$$\int_{\partial G} f\wedge g = 0. \tag{19.72}$$

Let $\varrho:\Theta \longrightarrow \mathbb{R}$ be as at the beginnig of construction 19.3. Take $\varepsilon>0$ so small that

$$G_{\varepsilon} := G\cup\{z\in\Theta:\ \varrho(z)<\varepsilon\} \subset\subset G'$$

is a non-degenerate strictly q-convex extension of G. By Proposition 12.10 (iii), we can find a sequence $g_k\in Z_{n,n-q-1}^{0}(\bar{G}_{\varepsilon},E)$ which converges to g uniformly on $\bar{G}$. Then

$$\int_{\partial G} f\wedge g = \lim_{k} \int_{\partial G} f\wedge g_k .$$

Since $f\in Z_{0,q}^{0}(\bar{D},E)$ and, therefore, the forms $f\wedge g_k$ are closed on $G_{\varepsilon}\setminus\bar{G}$, this implies, by Stokes' formula, that

$$\int_{\partial G} f\wedge g = \lim_{k} \int_{\partial G_{\varepsilon}} f\wedge g_k .$$

Since f is $\bar{\partial}$-exact on D and hence the forms $f\wedge g_k$ are exact on $\partial G_{\varepsilon} \subseteq D$, so we see (again by Stokes' theorem) that the integrals on the right hand side of the last relation vanish for all k. Hence (19.72) holds true.

<u>19.14. Proof of Theorem 19.1'.</u> Since G is relatively compact in Y, and Y is (n-q)-convex, we can find a relatively compact domain $G' \subset\subset Y$ such that $G \subset\subset G'$, Y is an (n-q)-convex extension of G' and hence $Y\setminus\bar{G}$ is an (n-q)-convex extension of $G'\setminus\bar{G}$. Set $D=G'\setminus\bar{G}$. Then, by Theorem 12.14, the restriction map

$$H^{0,q}(Y\setminus\bar{G},E) \longrightarrow H^{0,q}(D,E)$$

is an isomorphism. On the other hand, X is a q-concave extension of $Y\setminus\bar{G}$

and thus, by Theorem 16.1, the map

$$H^{0,q}(X,E) \longrightarrow H^{0,q}(Y\setminus\bar{G},E)$$

is injective. Together this implies that the restriction map

$$H^{0,q}(X,E) \longrightarrow H^{0,q}(D,E) \qquad (19.73)$$

is injective. Further, by Proposition 0.5 in Appendix B and Observation 4.15, we may assume that ∂G and $\partial G'$ are smooth. Then, by Theorem 19.2', the space $E_{0,q}^{1/2\to 0}(\bar{D},E)$ is closed with respect to the norm $\|\cdot\|_{0,D}$. Since (19.73) is injective, this implies that $E_{0,q}^{0}(X,E)$ is closed with respect to uniform convergence on the compact subsets of X. Theorem 19.1' is proved.

CHAPTER V. SOME APPLICATIONS

20. Solvability criterions for $\bar{\partial}u=f$ and duality between the Dolbeault cohomology with compact support and the usual Dolbeault cohomology

Let X be an n-dimensional complex manifold, and $1\le q\le n$ an integer.

If f is a $\bar{\partial}$-closed continuous (0,q)-form on X and the equation $\bar{\partial}u = f$ has a solution, then, by Stokes' theorem,

$$\int_X f\wedge g = 0 \qquad \text{for all } g\in[Z^{\infty}_{n,n-q}(X)]_0. \tag{20.1}$$

It arises the question whether, conversely, (20.1) implies solvability of $\bar{\partial}u=f$? Of course, a necessary condition for an affirmative answer is that $E^0_{0,q}(X)$ is closed with respect to uniform convergence on the compact subsets of X. We shall see that this is also sufficient (Proposition 20.1).

Perhaps it should be recalled here that we know already the following three conditions on X, each of which guarantees that $E^0_{0,q}(X)$ is closed:

1) X is (n-q)-convex (cf. the Andreotti-Grauert Theorem 12.16).
2) X is (q+1)-concave (cf. the Andreotti-Grauert Theorem 15.13).
3) X is a q-concave domain in a compact complex manifold (cf. the Andreotti-Vesentini Theorem 19.1) or, more generally, X is as in Theorem 19.1'.

In particular, if X is compact, then $E^0_{0,q}(X)$ is closed.

Now let f be a $\bar{\partial}$-closed continuous (0,q)-form with compact support in X. Then, again by Stokes' formula, for the solvability of $\bar{\partial}u=f$ with $u\in[C^0_{0,q-1}(X)]_0$ it is necessary that

$$\int_X f\wedge g = 0 \qquad \text{for all } g\in Z^{\infty}_{n,n-q}(X). \tag{20.2}$$

We shall see that condition (20.2) is also sufficient for the solvability of $\bar\partial u=f$ with $u\in[C^0_{0,q-1}(X)]_0$ if and only if the space $\bar\partial[C^\infty_{0,q-1}(X)]_0$ is closed with respect to the topology of $[C^\infty_{0,q}(X)]_0$ (cf. (0.8)). Further, we shall see that this is the case if and only if $E^0_{n,n-q+1}(X)$ is closed with respect to uniform convergence on the compact subsets of X (Propositions 20.2 and 20.3).

It is the aim of the present section to prove these solvability criterions, and then, by means of them, to establish duality relations between the Dolbeault cohomology with compact support and the usual Dolbeault cohomology. As an application, we prove the Hartogs' extension theorem of Kohn-Rossi.

20.1. Proposition. Let E be a holomorphic vector bundle over an n-dimensional complex manifold, and $1\le q\le n$ an integer. Suppose the space $E^0_{0,q}(X,E)$ is closed with respect to uniform convergence on the compact subsets of X. Let $f\in Z^0_{0,q}(X,E)$. Then the equation $\bar\partial u=f$ has a solution $u\in C^0_{0,q-1}(X,E)$ if and only if

$$\int_X g\wedge f = 0 \qquad \text{for all } g\in[Z^\infty_{n,n-q}(X,E^*)]_0. \tag{20.3}$$

Proof. That condition (20.3) is necessary follows from Stokes' theorem. To prove the converse, we assume that f does not belong to $E^0_{0,q}(X,E)$. Since $E^0_{0,q}(X,E)$ is closed, then, by the Hahn-Banach theorem, we can find a continuous linear functional F on $C^0_{0,q}(X,E)$, i.e. an E^*-valued (n,n-q)-current with compact support (cf. Sect. 0.14), such that $F(f)\neq 0$ and $F(E^0_{0,q}(X,E)) = \{0\}$, i.e. $\bar\partial F=0$. By Corollary 2.15 (i), we can find a form $g\in[Z^\infty_{n,n-q}(X,E^*)]_0$ and an E^*-valued (n,n-q-1)-current S with compact support such that $F = \langle g\rangle + \bar\partial S$. If we interpret $\bar\partial S$ as a continuous linear functional on $C^0_{0,q}(X,E)$ (cf. Sect. 0.14), then $\bar\partial S$ vanishes on $(Z^0_{0,q}(X,E)) = \{0\}$. Hence $\langle g\rangle(f) = F(f) \neq 0$, i.e. condition (20.3) is violated. □

20.2. Propositon. Let E be a holomorphic vector bundle over the n-dimensional complex manifold X, and $1\le q\le n$ an integer. Suppose $E^0_{n,n-q+1}(X,E^*)$ is closed with respect to uniform convergence on the compact subsets of X. Let $f\in[Z^0_{0,q}(X,E)]_0$. Then the equation $\bar\partial u=f$ can be solved with $u\in[C^0_{0,q-1}(X,E)]_0$ if and only if

$$\int_X g\wedge f = 0 \qquad \text{for all } g\in Z^\infty_{n,n-q}(X,E^*). \tag{20.4}$$

Proof. That condition (20.4) is necessary follows from Stokes' theorem. To prove the converse, we assume that condition (20.4) is fulfilled.

Then we define a linear functional F on $E^0_{n,n-q+1}(X,E^*)$ as follows: If $h\in E^0_{n,n-q+1}(X,E^*)$, then we take $v\in C^0_{n,n-q}(X,E^*)$ with $\bar\partial v=h$, and set

$$F(h) = \int_X f\wedge v.$$

This definition is correct. In fact, let $v,w\in C^0_{n,n-q}(X,E^*)$ with $\bar\partial v=\bar\partial w=h$. Then by Theorem 2.14 (smoothing of the Dolbeault cohomology), we can find $g\in Z^\infty_{n,n-q}(X,E^*)$ such that $v-w-g = \bar\partial p$ for some $p\in C^0_{n,n-q-1}(X,E^*)$. Hence, by Stokes' formula,

$$\int_X f\wedge(v-w-g) = 0,$$

i.e., by condition (20.4),

$$\int_X f\wedge v = \int_X f\wedge w.$$

Since, by hypothesis, the space $E^0_{n,n-q+1}(X,E^*)$ is closed, it follows from Banach's open mapping theorem that F is continuous. By the Hahn-Banach theorem, there exists a continuous linear extension $\tilde F$ of F onto $C^0_{n,n-q+1}(X,E^*)$. If we interpret this extension as a current with compact support (cf. Sect. 0.14), then , for all $h\in[C^\infty_{n,n-q}(X,E^*)]_0$, we have

$$(-1)^{q-1}(\bar\partial\tilde F)(h) = F(\bar\partial h) = \int_X f\wedge h = \langle f\rangle(h),$$

i.e. $\langle f\rangle = \bar\partial[(-1)^{q-1}\tilde F]$. By Corollary 2.15 (ii), this implies that $f=\bar\partial u$ for some $u\in[C^0_{0,q-1}(X,E)]_0$. □

If E is a holomorphic vector bundle over an n-dimensional complex manifold X and $E^0_{n,n-q+1}(X,E^*)$ is closed, then it follows from Proposition 20.2 that the space $\bar\partial[C^\infty_{0,q-1}(X,E)]_0$ is closed with respect to the topology of $[C^\infty_{0,q}(X,E)]_0$ (cf. (0.8)). The converse is also true:

<u>20.3. Proposition.</u> Let E be a holomorphic vector bundle over an n-dimensional complex manifold X, and let $1\le q\le n$ be an integer. Then

$\bar{\partial}[C^{\infty}_{0,q-1}(X,E)]_0$ is closed with respect to the topology of $[C^{\infty}_{0,q}(X,E)]_0$ if and only if $E^{0}_{n,n-q+1}(X,E^*)$ is closed with respect to uniform convergence on the compact subsets of X.

<u>Proof.</u> As already observed, it follows from Proposition 20.2 that this condition is sufficient. We prove the converse. Assume that $\bar{\partial}[C^{\infty}_{0,q-1}(X,E)]_0$ is closed. By Stokes' theorem, $E^{0}_{n,n-q+1}(X,E^*)$ is contained in the closed subspace of all $f \in Z^{0}_{n,n-q+1}(X,E^*)$ satisfying the condition

$$\int_X f \wedge g = 0 \qquad \text{for all } g \in [Z^{\infty}_{0,q-1}(X,E)]_0. \tag{20.5}$$

We shall prove that this subspace is even equal to $E^{0}_{n,n-q+1}(X,E^*)$.

Fix $f \in Z^{0}_{n,n-q+1}(X,E^*)$ such that (20.5) is fulfilled. Setting

$$F(h) = (-1)^q \int_X f \wedge h,$$

we define a continuous linear functional on $[C^{\infty}_{0,q-1}(X,E)]_0$. By (20.5) there exists a uniquely determined linear functional H on $\bar{\partial}[C^{\infty}_{0,q-1}(X,E)]_0$ such that the diagram

$$\begin{array}{ccc} [C^{\infty}_{0,q-1}(X,E)]_0 & \xrightarrow{\quad F \quad} & \mathbb{C} \\ \Big\downarrow \bar{\partial} & \overset{H}{\nearrow} & \\ \bar{\partial}[C^{\infty}_{0,q-1}(X,E)]_0 & & \end{array} \tag{20.6}$$

is commutative. Since, by hypothesis, the space $\bar{\partial}[C^{\infty}_{0,q-1}(X,E)]_0$ is closed with respect to the topology of $[C^{\infty}_{0,q}(X,E)]_0$ and since the operator $\bar{\partial}$ is continuous with respect to this topology, it follows from Banach's open mapping theorem that H is continuous. By the Hahn-Banach theorem, H admits a continuous extension to some E^*-valued (n,n-q)-current $\tilde{H}$ on X. Since (20.6) is commutative, we have

$$\bar{\partial}\tilde{H}(h) = (-1)^q \tilde{H}(\bar{\partial}h) = (-1)^q F(h) = \int_X f \wedge h = \langle f \rangle (h),$$

for all $h \in [C^0_{0,q-1}(X,E)]_0$, i.e. $\bar{\partial}\tilde{H} = \langle f \rangle$. By Theorem 2.14 (smoothing of the Dolbeault cohomology), this implies that $f \in E^0_{n,n-q+1}(X,E)$. □

Let E be a holomorphic vector bundle over an n-dimensional complex manifold X, and let $0 \leq q \leq n$ be an integer.

Recall the smoothing isomorphism (Theorem 2.14)

$$H^{0,q}(X,E) \cong H^{0,q}_0(X,E) = Z^0_{0,q}(X,E)/E^0_{0,q}(X,E).$$

Further on, we assume that $H^{0,q}(X,E)$ is endowed with the factor topology of the Fréchet space $Z^0_{0,q}(X,E)$ with respect to the subspace $E^0_{0,q}(X,E)$. If $E^0_{0,q}(X,E)$ is closed in $Z^0_{0,q}(X,E)$, then (and only then) $H^{0,q}(X,E)$ is also a Fréchet space. By $(H^{0,q}(X,E))^*$ we denote the space of continuous linear functionals on $H^{0,q}(X,E)$.

For any fixed $h \in [Z^{\infty}_{n,n-q}(X,E^*)]_0$, the map

$$Z^0_{0,q}(X,E) \ni u \longrightarrow \int_X u \wedge h$$

is continuous and linear on $Z^0_{0,q}(X,E)$. By Stokes' theorem, this map vanishes on $E^0_{0,q}(X,E)$ and hence it defines a continuous linear functional on $H^{0,q}(X,E)$. Thus, a homomorphism

$$[Z^{\infty}_{n,n-q}(X,E^*)]_0 \longrightarrow (H^{0,q}(X,E))^*$$

is defined. Again by Stokes' theorem, this hompmorphism vanishes on $\bar{\partial}[C^{\infty}_{n,n-q-1}(X,E^*)]_0$ (if $q \leq n-1$). Hence, a homomorphism

$$[H^{n,n-q}(X,E^*)]_0 \longrightarrow (H^{0,q}(X,E))^* \tag{20.7}$$

is induced, where

$$[H^{n,n-q}(X,E^*)]_0 := [Z^{\infty}_{n,n-q}(X,E^*)]_0/\bar{\partial}[C^{\infty}_{n,n-q-1}(X,E^*)]_0. \tag{20.8}$$

The homomorphism (20.7) will be called the <u>canonical homomorphism</u> from $[H^{n,n-q}(X,E^*)]_0$ into $(H^{0,q}(X,E))^*$.

20.4. Theorem. Let E be a holomorphic vector bundle over an n-dimensional complex manifold X. Then, for all $0 \le q \le n$, the canonical homomorphism (20.7) is surjective. Moreover, the following assertions hold true:

(i) The canonical homomorphism

$$[(H^{n,0}(X,E^*)]_0 \longrightarrow (H^{0,n}(X,E))^* \tag{20.9}$$

is an isomorphism. If $E^0_{0,n}(X,E)$ is closed with respect to uniform convergence on the compact subsets of X, then, moreover,

$$\dim\,[H^{n,0}(X,E^*)]_0 = \dim\,H^{0,n}(X,E). \tag{20.10}$$

(ii) The canonical homomorphism

$$[(H^{n,n}(X,E^*)]_0 \longrightarrow (\mathcal{O}(X,E))^* \tag{20.11}$$

is an isomorphism if and only if $E^0_{0,1}(X,E)$ is closed with respect to uniform convergence on the compact subsets of X. In this case

$$\dim\,[H^{n,n}(X,E^*)]_0 = \dim \mathcal{O}(X,E). \tag{20.12}$$

(iii) Let $1 \le q \le n-1$. Then the canonical homomorphism (20.7) is an isomorphism if and only if $E^0_{0,q+1}(X,E)$ is closed with respect to the uniform convergence on the compact subsets of X. If both $E^0_{0,q+1}(X,E)$ and $E^0_{0,q}(X,E)$ are closed with respect to this topology, then, moreover,

$$\dim\,[H^{n,n-q}(X,E^*)]_0 = \dim\,H^{0,q}(X,E). \tag{20.13}$$

<u>Proof of: (20.7) is always surjective.</u> Let $F \in (H^{0,q}(X,E))^*$ be given. Denote by p the canonical projection from $Z^0_{0,q}(X,E)$ onto $H^{0,q}(X,E)$. Set $H = F \circ p$. By definition of the topology of $H^{0,q}(X,E)$ given above, p is continuous. Hence H is continuous. By the Hahn-Banach theorem, H admits continuous linear extension onto $C^0_{0,q}(X,E)$, i.e. H admits an extension to some E^*-valued (n,n-q)-current $\tilde{H}$ with compact support on X (cf. Sect. 0.14). Since $\tilde{H}$ vanishes on $E^0_{0,q}(X,E)$, we have $\bar{\partial}\tilde{H}=0$. By Corollary 2.15 (i), we can find $g \in [Z^\infty_{n,n-q}(X,E^*)]_0$ with

$$\tilde{H} - \langle g \rangle = \bar{\partial} S \tag{20.14}$$

for some E^*-valued (n,n-q-1)-current S with compact support on X. In order to prove that (20.7) is surjective, now it is sufficient to show that

$$\tilde{H}(h) = \int_X g \wedge h$$

for all $h \in Z^0_{0,q}(X,E)$. Let a form h $Z^0_{0,q}(X,E)$ be given. By Theorem 2.14 (smoothing of the Dolbeault cohomology), we can find $h_{\infty} \in Z^{\infty}_{0,q}(X,E)$ with $h - h_{\infty} \in E^0_{0,q}(X,E)$. Since $\tilde{H}$ vanishes on $E^0_{0,q}(X,E)$, then $\tilde{H}(h) = \tilde{H}(h_{\infty})$, and, by Stokes' theorem,

$$\int_X g \wedge h = \int_X g \wedge h_{\infty}.$$

Thus, it remains to prove that

$$\tilde{H}(h_{\infty}) = \int_X g \wedge h_{\infty}. \tag{20.15}$$

Take a C^{∞} function χ on X such that supp $\chi \subset\subset X$ and $\chi = 1$ in some neighborhood of supp $\tilde{H}$ ∪ supp g ∪ supp S. Then

$$\tilde{H}(\chi h_{\infty}) = \tilde{H}(h_{\infty}), \qquad \int_X g \wedge \chi h_{\infty} = \int_X g \wedge h_{\infty}, \qquad \text{and } \bar{\partial} S(\chi h_{\infty}) = 0.$$

Hence (20.15) follows from (20.14).

Proof of assertion (i). Since $Z^0_{0,n}(X,E) = C^0_{0,n}(X,E)$, it is trivial that (20.9) is injective. Since (20.7) is always surjective, so it is proved that (20.9) is an isomorphism. If $E^0_{0,n}(X,E)$ is closed, then $H^{0,n}(X,E)$ is a Fréchet space, and, by the Hahn-Banach theorem, $\dim H^{0,n}(X,E) = \dim (H^{0,n}(X,E))^*$. Since (20.9) is an isomorphism, this implies (20.10).

Proof of assertions (ii) and (iii). First we prove that, for all $0 \le q \le n-1$, the following two conditions are equivalent:

$(A)_q$ The homomorphism (20.7) is an isomorphism.

$(B)_q$ The space $E^0_{0,q+1}(X,E)$ is closed.

Assume $(A)_q$ is true. Then, in particular, (20.7) is injective, i.e.

$$\bar\partial[C^{\infty}_{n,n-q-1}(X,E^*)]_0$$

$$= \{g\in[Z^{\infty}_{n,n-q}(X,E^*)]_0 : \int_X g\wedge h = 0 \quad \text{for all } h\in Z^0_{0,q}(X,E) \}. \qquad (20.16)$$

Therefore $(A)_q$ implies that $\bar\partial[C^{\infty}_{n,n-q-1}(X,E^*)]_0$ is closed with respect to the topology of $[C^{\infty}_{n,n-q}(X,E^*)]_0$. In view of Proposition 20.3 (applied to the bundle $E^*\otimes K$, where K is the bundle of holomorphic (n,0)-forms on X), this means that $(A)_q$ implies $(B)_q$.

Assume that $(B)_q$ is true. Then, by Proposition 20.2 (again applied to $E^*\otimes K$), the space

$$M := [Z^{\infty}_{n,n-q}(X,E^*)]_0 \cap \bar\partial[C^0_{n,n-q-1}(X,E^*)]_0$$

is equal to the right hand side of (20.16). Since by Corollary 2.15 (ii), M is always equal to the left hand side of (20.16), this means that (20.16) holds true, i.e. (20.7) is injective. Since (20.7) is always surjective, so it is proved that $(B)_q \Longrightarrow (A)_q$.

Now we prove assertion (ii). The equivalence of $(A)_0$ and $(B)_0$ means that (20.11) is an isomorphism if and only if $E^0_{0,1}(X,E)$ is closed. This implies (20.12), because $\dim \mathcal{O}(X,E) = \dim(\mathcal{O}(X,E))^*$ by the Hahn-Banach theorem.

Finally, we prove assertion (ii). The first part of this assertion means that $(A)_q \Longleftrightarrow (B)_q$. Assume $E^0_{0,q+1}(X,E)$ is closed, i.e. $(B)_q$ holds true. Then also $(A)_q$ is fulfilled, i.e. (20.7) is an isomorphism. If, moreover, $E^0_{0,q}(X,E)$ is closed, then $H^{0,q}(X,E)$ is a Fréchet space, and (20.13) follows from the Hahn-Banach theorem and the fact that (20.7) is an isomorphism. □

<u>20.5. Corollary.</u> If X is an n-dimensional complex manifold without compact connected components, then, for any holomorphic vector bundle E over X,

$$(H^{0,n}(X,E))^* = 0,$$

i.e. $E^0_{0,n}(X,E)$ is dense in $Z^0_{0,n}(X,E)$.

<u>Proof.</u> This follows immediately from part (i) of Theorem 20.4, because, by uniqueness of holomorphic functions,

$$[H^{n,0}(X,E^*)]_0 = 0. \qquad \square$$

<u>20.6. Theorem.</u> If X is a 0-convex n-dimensional complex manifold without compact connected components, then, for any holomorphic vector bundle E

over X,

$$H^{0,n}(X,E) = 0.$$

Proof. By the Andreotti-Grauert Theorem 12.16, $E^0_{0,n}(X,E)$ is closed. Thus, by Corollary 20.5, $E^0_{0,n}(X,E) = Z^0_{0,n}(X,E)$. □

Remark to Theorem 20.6. By Corollary 5.10, each domain $X \subset\subset Y$ with C^2 boundary in an arbitrary complex manifold Y is 0-convex. It is easy to show that, moreover, a complex manifold X is 0-convex if it admits an exhausting function which has only a finite number of critical points.

Question. Does there exist complex manifolds which are not 0-convex?

20.7. Theorem. If E is a holomorphic vector bundle over a q-convex n-dimensional complex manifold X ($0 \le q \le n-1$), then

$$\dim\,[H^{0,r}(X,E)]_0 = \dim H^{0,n-r}(X,E^*\otimes K) < \infty \tag{20.17}$$

for all $0 \le r \le q$, where K denotes the bundle of holomorphic (n,0)-forms on X. If $q=n-1$, then, moreover,

$$\dim\,[H^{0,n}(X,E)]_0 = \dim \mathcal{O}(X,E^*\otimes K). \tag{20.18}$$

Proof. Let $0 \le r \le q$. Then $n-q \le n-r \le n$ and, by the Andreotti-Grauert Theorem 2.16,

$$\dim H^{0,n-r}(X,E^*\otimes K) < \infty \qquad \text{and} \qquad \dim H^{0,n-r+1}(X,E^*\otimes K) < \infty.$$

Hence, by Theorem 20.4 (iii),

$$\dim\,[H^{n,r}(X,E\otimes K^*)]_0 = \dim H^{0,n-r}(X,E^*\otimes K) < \infty.$$

Since $E\otimes K^*$-valued (n,r)-forms may be viewed as (0,r)-forms with values in $(E\otimes K^*)\otimes K = E$, this coincides with (20.17).

Let $q=n-1$. Then, by the Andreotti-Grauert Theorem 12.16, $\dim H^{0,1}(X,E^*\otimes K) < \infty$ and it follows from Theorem 20.4 (ii) that

$$\dim\,[H^{n,n}(X,E\otimes K^*)]_0 = \dim \mathcal{O}(X,E^*\otimes K).$$

This coincides with (20.18), identifying E K^*-valued (n,n)-forms and E-valued (0,n)-forms. □

20.8 Theorem. If E is a holomorphic vector bundle over a q-concave n-dimensional complex manifold X ($2 \le q \le n-1$), then

$$\dim [H^{0,r}(X,E)]_0 = \dim H^{0,n-r}(X,E^* K) < \infty \tag{20.19}$$

for all $n-q+2 \le r \le n$, where K is the bundle of holomorphic (n,0)-forms on X.

Proof. Let $n-q+2 \le r \le n$. Then $0 \le n-r \le q-2$ and, by the Andreotti-Grauert Theorem 15.13,

$$\dim H^{0,n-r}(X,E^*\otimes K) < \infty \qquad \text{and} \qquad \dim H^{0,n-r+1}(X,E^*\otimes K) < \infty.$$

Hence, by Theorem 20.4 (iii) and (ii),

$$\dim [H^{n,r}(X,E\otimes K^*)]_0 = \dim H^{0,n-r}(X,E^*\otimes K) < \infty.$$

Identifying E K^*-valued (n,r)-forms and E-valued (0,r)-forms, this implies (20.19). □

If the manifold X in Theorem 20.8 is a domain in some compact complex manifold Y and E is a holomorphic vector bundle over Y, then the conclusion of Theorem 20.8 can be strengthened:

20.9. Theorem. Let $X \subseteq Y$ be a q-concave domain in an n-dimensional compact complex manifold Y ($1 \le q \le n-1$). Then, for any holomorphic vector bundle E over Y, relation (20.19) holds true for all $n-q+1 \le r \le n$.

Proof. In view of Theorem 20.8 we may assume that $r=n-q+1$. Then, by the Andreotti-Grauert Theorem 15.13,

$$\dim H^{0,n-r}(X,E^*\otimes K) < \infty,$$

and, by the Andreotti-Vesentini theorem 19.1, the space $E^0_{0,n-r+1}(X,E^*\otimes K)$ is closed. If $q=1$ and hence $n-r=0$, then this implies, by Theorem 20.4 (ii), that

$$\dim [H^{n,n}(X,E\otimes K^*)]_0 = \dim H^{0,0}(X,E^*\otimes K) < \infty. \tag{20.20}$$

If $2 \le q \le n-1$, then we can apply part (iii) of Theorem 20.4, and we obtain

$$\dim [H^{n,r}(X,E\otimes K^*)]_0 = \dim H^{0,n-r}(X,E^*\otimes K) < \infty. \tag{20.21}$$

Identifying $E\otimes K^*$-valued (n,r)-forms and E-valued $(0,r)$-forms, (20.20) and (20.21) imply that (20.19) holds true also for $r=n-q+1$. □

<u>20.10. Theorem.</u> Let E be a holomorphic vector bundle over an n-dimensional <u>compact</u> complex manifold Y, and let $X \subset\subset Y$ be a domain which is both q-concave and $(n-q-1)$-convex ($1\le q\le n-1$). Then

$$\dim\, [H^{0,r}(X,E)]_0 = \dim H^{0,n-r}(X,E^*\otimes K) \qquad \text{for all } 0\le r\le n. \tag{20.22}$$

If $r\neq n-q$, then, moreover,

$$\dim\, [H^{0,r}(X,E)]_0 < \infty. \tag{20.23}$$

If X is a q-concave extension of some strictly $(n-q-1)$-convex domain $D \subset\subset X$, then

$$\dim\, [H^{0,n-q}(X,E)]_0 = \infty. \tag{20.24}$$

<u>Proof.</u> If $0\le r\le n-q-1$, then (20.22) and (20.23) hold true by (20.17) in Theorem 20.7. If $n-q+1\le r\le n$, then (20.22) and (20.23) follow from Theorem 20.9. Let $r=n-q$. By the Andreotti-Grauert Theorem 12.6, then

$$\dim H^{0,q+1}(X,E^*\ K) < \infty,$$

and, by the Andreotti-Vesentini Theorem 19.1, $E^0_{0,q}(X,E^*\otimes K)$ is closed. Thus, we can apply part (iii) of Theorem 20.4 and obtain that

$$\dim\, [H^{n,n-q}(X,E\ \ K^*)]_0 = \dim H^{n,q}(X,E^*\otimes K).$$

Since $E\otimes K^*$-valued $(n,n-q)$-forms may be interpreted as E-valued $(0,n-q)$-forms, this implies (20.22) for $r=n-q$. If X is a q-concave extension of a strictly $(n-q-1)$-convex domain, then (20.24) follows from (20.22) and the Andreotti-Norguet Theorem 18.4. □

As an interesting application of Theorem 20.7, now we obtain the following version of the Hartogs' extension phenomenon:

<u>20.11. Theorem</u> (Kohn-Rossi theorem). Let X be a connected non-compact 1-convex complex manifold such that there exists at least one non-constant holomorphic function on X.[1)] Then, for any compact set $M \subset\subset X$ such

---

1) There exist connected non-compact pseudoconvex manifolds without non-constant global holomorphic functions (see [Grauert 1963])

that $X\setminus M$ is connected, the restriction map

$$\mathcal{O}(X) \longrightarrow \mathcal{O}(X\setminus M)$$

is an isomorphism.

Proof. The injectivity is clear, because X is connected. We prove the surjectivity. Let $f\in \mathcal{O}(X\setminus M)$ be given. Since X is 1-convex, by Theorem 20.7,

$$N:= \dim\,[H^{0,1}(X)]_0 < \infty. \tag{20.25}$$

Choose a C^{∞} function χ on X such that $\operatorname{supp}\chi \subset\subset X$ and $\chi = 1$ in a neighborhood of M. Without loss of generality we may assume that f is not constant. Then the forms $f^k\bar\partial\chi\in[Z^{\infty}_{0,1}(X)]_0$ $(k=0,1,\dots)$ are linearly independent. Therefore, by (20.25), we can find a vector $0 \neq (a_0,\dots,a_N)$ in $\mathbb{C}^{N+1}$ such that

$$-\sum_{k=0}^{N} a_k f^k\bar\partial\chi = \bar\partial u$$

for some $u\in[C^{\infty}(X)]_0$. Setting

$$F = (1-\chi)\sum_{k=0}^{N} a_k f^k - u,$$

we obtain a holomorphic function F on X with

$$F = \sum_{k=0}^{N} a_k f^k \qquad \text{on } X\setminus[(\operatorname{supp}\chi)\cup(\operatorname{supp} u)].$$

By hypothesis, there is a non-constant holomorphic function h on X. Then also the forms $h^k f\bar\partial\chi\in[Z^{\infty}_{0,1}(X)]_0$ $(k=0,1,\dots)$ are linearly independent. Again by (20.25), we can find a vector $0 = (b_0,\dots,b_N)\in\mathbb{C}^{N+1}$ with

$$-\sum_{k=0}^{N} b_k h^k f\bar\partial \quad = \bar\partial v$$

for some $v\in[C^{\infty}(X)]_0$. Setting

$$G = \sum_{k=0}^{N} b_k h^k \qquad \text{and} \qquad H = (1-\chi)Gf - v$$

we obtain holomorphic functions $G \not\equiv 0$ and $H \not\equiv 0$ on X such that

$$H = Gf \qquad \text{on } X\setminus[(\operatorname{supp}\chi)\cup(\operatorname{supp} v)], \tag{20.26}$$

and

$$\sum_{k=0}^{N} a_k H^k G^{N-k} = G^N \sum_{k=0}^{N} a_k f^k = G^N F \qquad \text{on } X\setminus[(\operatorname{supp}\chi)\cup(\operatorname{supp} u)\cup(\operatorname{supp} v)].$$

Since X is connected, it follows that

$$\sum_{k=0}^{N} a_k H^k G^{N-k} = G^N F \qquad \text{on } X.$$

Therefore

$$\sum_{k=0}^{N} a_k \left(\frac{H}{G}\right)^k = F \qquad \text{on } X\setminus\{z\in X:\ G(z)=0\}.$$

Since $G \not\equiv 0$ and $(a_0,\dots,a_N) \neq 0$, this implies that H/G is a locally bounded meromorphic function on X, i.e. H/G is holomorphic on X. Since, by (20.26), H/G=f in some open part of X\M, and X\M is connected, it follows that H/G=f on X\M. □

20.12. Remark to Theorem 20.11. If $X \subset\subset Y$ is a non-degenerate strictly pseudoconvex domain with $\partial X \neq \emptyset$ in some complex manifold Y, then there exist non-constant holomorphic functions on X. In this case, even $\dim \mathcal{O}(X) = \infty$. This well-known fact can be proved, for instance, in the same way as statement (A) in the proof of Theorem 23.3 below. On the other hand, this is a special case of Theorem 18.4.

20.13. Theorem. Let E be a holomorphic vector bundle over an n-dimensional complex manifold Y, and $X \subset\subset Y$ a domain with C^1 boundary. Suppose, for some $0 \le q \le n$, at least one of the following conditions is fulfilled:

(i) X is q-convex;
(ii) X is (n-q+1)-concave;
(iii) Y is compact and X is (n-q)-concave.

Let U be a neighborhood of ∂X, and $f \in Z^0_{0,q}(U,E)$. Then there exists a form $F \in Z^0_{0,q}(X,E)$ with F=f in some neighborhood of ∂X if and only if

$$\int_{\partial X} f \wedge g = 0 \qquad \text{for all } g \in Z^0_{n,n-q-1}(X,E^*). \tag{20.27}$$

Proof. By Stokes' theorem it is clear that condition (20.27) is necessary. Assume (20.27) is fulfilled. Take a C^{∞} function χ on Y such that $\chi = 1$ in a neighborhood of ∂X and supp $\chi \subset\subset U$. Then $\bar\partial\chi \wedge f \in [Z^{0}_{0,q+1}(X,E)]_{0}$ and, by (20.27),

$$\int_X g\wedge(\bar\partial\chi\wedge f) = \int_{\partial X} g\wedge f = 0 \tag{20.28}$$

for all $g \in Z^{0}_{0,n-q-1}(X,E^*)$. Since at least one of the conditions (i), (ii), (iii) is fulfilled, it follows from at least of one of the Theorems 12.16, 15.13, and 19.1 that $E^{0}_{n,n-q}(X,E^*)$ is closed with respect to uniform convergence on the compact subsets of X. By this and by (20.28), it follows from Proposition 20.2 that the equation $\bar\partial u = \bar\partial\chi\wedge f$ can be solved with $u \in [C^{0}_{0,q}(X,E)]_{0}$. Set $F = \chi f - u$. □

21. The domain $|z^0|^2+\ldots+|z^q|^2 < |z^{q+1}|^2+\ldots+|z^n|^2$

Let $\mathbb{P}^n$ be the n-dimensional complex projective space, and let E be a holomorphic vector bundle over $\mathbb{P}^n$. For $z \in \mathbb{P}^n$, by $[z^0:\ldots:z^n]$ we denote the homogeneous coordinates of z. Let $0 \le q \le n-1$ be an integer. Set

$$D := \{z \in \mathbb{P}^n : |z^0|^2+\ldots+|z^q|^2 < |z^{q+1}|^2+\ldots+|z^n|^2\}.$$

As observed in Sects. 5.16 and 6.12, D is completely strictly q-convex and strictly (n-q-1)-concave. Hence the theory developed in the preceding sections admits several applications to this domain. Let us notice some of them.

21.1. Theorem. (i) For all integers $1 \le r \le n$, the space $E^{0}_{0,r}(D,E)$ is closed with respect to uniform convergence on the compact subsets of D.

(ii) $\dim H^{0,r}(D,E) < \infty$ for all $0 \le r \le n$ with $r \ne n-q-1$.

(iii) $H^{0,r}(D,E) = 0$ if $n \ge r \ge n-q$.

(iv) $\dim H^{0,n-q-1}(D,E) = \infty$.

Proof. (i) follows from the Andreotti-Grauert Theorems 12.16 and 15.13, and the Andreotti-Vesentini Theorem 19.1.

(ii) follows from the Andreotti-Grauert Theorems 12.16 and 15.13.
(iii) follows from the Andreotti-Grauert Theorem 12.16.
(iv) follows from the Andreotti-Norguet Theorem 18.4. □

21.2. Theorem. (i) For all integers $1 \leq r \leq n$, the operator $\bar\partial$ is 1/2-regular at $Z^0_{0,r}(\bar D, E)$.

(ii) $\dim H^{0,r}_{1/2 \longrightarrow 0}(\bar D, E) < \infty$ if $0 \leq r \leq n$ with $r \neq n-q-1$.[1]

(iii) $H^{0,r}_{1/2 \longrightarrow 0}(\bar D, E) = 0$ if $n \geq r \geq n-q$.

(iv) $\dim H^{0,n-q-1}_{1/2 \longrightarrow 0}(\bar D, E) = \infty$.

Proof. (i) follows from Theorems 11.2, 14.5 and 19.2.
(ii) follows from Theorems 11.2 and 14.5.
(iii) follows from Theorem 12.7.
(iv) follows from Theorem 18.6. □

21.3. Theorem. For all integers $0 \leq r \leq n$, the canonical homomorphism

$$[H^{n,r}(D,E)]_0 \longrightarrow \left(H^{0,n-r}(D,E^*)\right)^*$$

is an isomorphism and

$$\dim\, [H^{n,r}(D,E)]_0 = \dim\, (H^{0,n-r}(D,E^*))^*. \tag{21.1}$$

Hence, by assertions (ii)-(iv) in Theorem 21.1,

$$\dim\, [H^{0,r}(D,E)]_0 < \infty \quad \text{if } 0 \leq r \leq n \text{ with } r \neq q+1,$$

$$\dim\, [H^{0,r}(D,E)]_0 = 0 \quad \text{if } 0 \leq r \leq q,$$

$$\dim\, [H^{0,q+1}(D,E)]_0 = \infty.$$

Proof. This follows from Theorem 21.1 (i) and Theorem 20.4. □

22. The condition Z(r)

It is clear that the results on q-convex and q-concave manifolds presented above admit several generalizations to manifolds which satisfy appropriate combinations of convexity and concavity conditions. An example we met already in Sect. 19: The boundary of the domain D in Theorem 19.2' consists of two pieces, $\partial G'$ and $-\partial G$, where $\partial G'$ is strictly (n-q)-convex and $-\partial G$ is strictly q-concave.

---

[1] $H^{0,r}_{1/2 \longrightarrow 0}(\bar D, E) := Z^0_{0,r}(\bar D, E) / E^{1/2 \longrightarrow 0}_{0,r}(\bar D, E)$

More generally, one can give the following

22.1. Definition. Let $D \subset\subset X$ be a domain in an n-dimensional complex manifold X. We say D is a non-degenerate strictly Z(r) domain ($1 \le r \le n$) if, for each $x \in \partial D$, there exists a C^2 function $\varrho : U \longrightarrow \mathbb{R}$ without degenerate critical points in some neighborhood U of x such that $D \cap U = \{z \in U: \varrho(z) < 0\}$ and at least one of the following two conditions is fulfilled:

(i) ϱ is strictly (n-r+1)-convex;

(ii) ϱ is strictly (r+2)-concave.

Example. Let $\varrho : X \longrightarrow \mathbb{R}$ be a strictly plurisubharmonic function without degenerate critical points on an n-dimensional complex manifold X such that the surface $\{z \in X: \varrho(z)=0\}$ is compact (for instance, $X=\mathbb{C}^n$ and $\varrho = |z|^2 - 1$). Then, for any $1 \le r \le n-2$ and all sufficiently small $\varepsilon > 0$, the set $\{z \in X: -\varepsilon < \varrho(z) < \varepsilon\}$ is a non-degenerate strictly Z(r) domain.

22.2. Theorem. Let $D \subset\subset X$ be a non-degenerate strictly Z(r) domain in an n-dimensional complex manifold X ($1 \le r \le n$). Then, for any holomorphic vector bundle E over X, the operator $\bar\partial$ is 1/2-regular at $Z^0_{0,r}(\bar D, E)$, and

$$\dim H^{0,r}_{1/2 \longrightarrow 0}(\bar D, E) < \infty. \text{ [1]}$$

Proof. This is a straightforward combination of the proofs of Theorems 11.2 and 14.5. □

For references, we notice also the following obvious

22.3. Corollary to Theorem 22.2 and Lemma 12.8 (iii) and (iv). Let $D \subset\subset X$ be a non-degenerate strictly Z(r) domain in an n-dimensional complex manifold X ($1 \le r \le n$), and let G be a non-degenerate strictly (n-r)-convex extension of D in X. Then, for any holomorphic vector bundle E over X,

$$\text{(i)} \quad E^{1/2 \longrightarrow 0}_{0,r}(\bar G, E) = Z^0_{0,r}(\bar G, E) \cap E^{1/2 \longrightarrow 0}_{0,r}(\bar D, E).$$

In particular, for any $f \in Z^0_{0,r}(\bar G, E)$ with $f=0$ on D, the equation $\bar\partial u = f$ can be solved with $u \in C^{1/2}_{0,r-1}(\bar G, E)$.

(ii) The image of the restriction map

$$Z^0_{0,r-1}(\bar G, E) \longrightarrow Z^0_{0,r-1}(\bar D, E)$$

is dense in $Z^0_{0,r-1}(\bar D, E)$ (with respect to uniform convergence on $\bar D$).

---

[1] $H^{0,r}_{1/2 \longrightarrow 0}(\bar D, E) := Z^0_{0,r}(\bar D, E) / E^{1/2 \longrightarrow 0}_{0,r}(\bar D, E)$

23. The Rossi theorem on attaching complex manifolds to a complex manifold along a strictly pseudoconcave boundary of real dimension ≥ 5

Throughout this section, X is an n-dimensional complex manifold, and $\varrho: X \longrightarrow \mathbb{R}$ is a strictly plurisubharmonic function without degenerate critical points such that the set $D := \{z \in X: 0 < \varrho(z) < 1\}$ is relatively compact in X.

23.1. Theorem (see [Rossi 1965]). If $n \geq 3$, then D is biholomorphically equivalent to a neighborhood of the boundary of some relatively compact domain $Y \subset\subset Z$ in a complex manifold Z (i.e. one can "fill in" the pseudoconcave "holes" of D).

In the present section we prove the following result which is the first step in the proof of Theorem 23.1:

23.2. Theorem. If $n \geq 3$, then we can find an integer N such that D is biholomorphically equivalent to some complex submanifold of $\mathbb{C}^N$ (of course, this submanifold is not closed in $\mathbb{C}^N$).

We do not present here the arguments which lead from Theorem 23.2 to Theorem 23.1. Instead, following [Rossi 1965], we refer to Theorem XII D 8 in [Gunning/Rossi 1965], which gives a version of Theorem 23.1 with isolated singularities in Z. We remark also that, modulo isolated singularities, the implication "Theorem 23.2 $\Longrightarrow$ Theorem 23.1" is a special case of the more recent (and more powerful) Harvey-Lawson theorem (see [Harvey/Lawson 1975] and [Harvey 1977]).

In [Rossi 1965] a counterexample is given (which will be reproduced in the next section) which shows that the conclusion of Theorem 23.1 need not be true if n=2. Since the arguments which lead from Theorem 23.2 to Theorem 23.1 work also for n=2, so also the conclusion of Theorem 23.2 need not be true if n=2. We shall prove (Theorem 23.3 below) that, for n=2, the conclusion of Theorem 23.2 (and therefore the conclusion of Theorem 23.1) holds true if and only if the operator $\bar\partial$ is 0-regular at $Z^0_{0,1}(\bar D_t)$ for all $0<t<1$, where $D_t := \{z \in X: 0 < \varrho(z) < t\}$.

Thus, Rossi's counterexample shows also that the separation Theorem 19.2' is not true without the hypothesis that the q-concave "holes" in D can be "filled in" by the relatively compact complex manifold G.

23.3. Theorem. Let n=2. Then there exists an integer N such that D is biholomorphically equivalent to a complex submanifold of $\mathbb{C}^N$ if and only if, for any $0<t<1$, the $\bar\partial$-operator is 0-regular at $Z^0_{0,1}(\bar D_t)$, where $D_t := \{z \in X: 0 < \varrho(z) < t\}$.

Proof of Theorems 23.2 and 23.3. If n=2 and D is biholomorphically equivalent to a complex submanifold of $\mathbb{C}^N$, then, by the arguments mentioned above (for instance by the Harvey-Lawson theorem), it follows that D is contained as an open subset in some larger complex manifold Y, such that, for any 0<t<1, the set $\{z\in D: \varrho(z)=t\}$ is the boundary of a non-degenerate strictly pseudoconvex domain $G_t \subset\subset Y$. In view of the separation Theorem 19.2', this implies that if n=2 then the $\bar\partial$-operator is 0-regular at $Z^0_{0,1}(\bar D_t)$ for all 0<t<1. If $n\geq 3$ then, obviously, the domains D_t are non-degenerate strictly Z(1) domains. Hence it follows from Theorem 22.2 that the operator $\bar\partial$ is 0-regular at $Z_{0,1}(\bar D_t)$.

Therefore, it remains to prove that, for any $n\geq 2$, the following statement holds true: If the operator $\bar\partial$ is 0-regular at $Z^0_{0,1}(\bar D_t)$ for all 0<t<1, then there exists an integer N such that D is biholomorphically equivalent to a submanifold of $\mathbb{C}^N$.

Assume the $\bar\partial$-operator is 0-regular at $Z^0_{0,1}(\bar D_t)$ for all 0<t<1. In order to prove that D is biholomorphically equivalent to a submanifold of some $\mathbb{C}^N$, obviously, it is sufficient to prove the following two statements:

(A) If x,y are different points in D, then there exists a sequence f_k (k=1,2,...) of holomorphic functions on D such that $\lim f_k(y) = 0$ and $\lim f_k(x) = 1$.

(B) If $x\in D$ and $v\in T'_x(D)$ is a holomorphic tangent vector (cf. Sect. 4.1), then there exists a sequence F_k (k=1,2,...) of holomorphic functions on D such that $\lim \partial F_k(x) = v$.

We shall prove (A) and (B) simultaneously.

Let $x,y\in D$ such that $x\neq y$ and, without loss of generality, $\varrho(y)\leq\varrho(x)$. Further, let $v\in T'_x(D)$.

Take a neighborhood $U \subset\subset D$ of x such that there exist holomorphic coordinates in a neighborhood U of x. Then we can find a holomorphic function on U with h(x)=0 and $\partial h(x)=v$, and, using the Levi polynomial (cf. Sect. 7.1), we can find a holomorphic function g on U such that g(x)=0 and

$$\operatorname{Re} g(z) \geq \varrho(x) - \varrho(z) + |x-z|^2 \tag{23.1}$$

for all $z\in U$. Take a C^∞ function χ on X with supp $\chi \subset\subset U$ and $\chi = 1$ in a neighborhood of x. Setting $s_k = e^{-kg}\bar\partial\chi$ and $S_k=he^{-kg}\bar\partial\chi$, we obtain two sequences $s_k, S_k \in Z^0_{0,1}(\bar D)$ (k=1,2,...). Since $\bar\partial\chi = 0$ in a neighborhood of x and outside supp χ, it follows from (23.1) that there is a number $\varrho(x)<b<1$ with

$$\lim_k \| s_k \|_{0,D_b} = \lim_k \| S_k \|_{0,D_b} = 0. \tag{23.2}$$

Since s_k and S_k vanish outside supp χ and since supp $\chi \subset\subset D$, we can find $0<\varepsilon<b$ such that $s_k=S_k=0$ on D_ε. In particular, s_k and S_k belong to $E^0_{0,1}(\bar{D}_\varepsilon)$. Since the operator $\bar\partial$ is 0-regular at $Z^0_{0,1}(\bar{D}_\varepsilon)$, and D_b is a non-degenerate strictly (n-1)-convex extension of D_ε, now it follows from Lemma 12.8 (ii) and (iii) that $\bar\partial$ is 0-regular also at $Z^0_{0,1}(\bar{D}_b)$, and $s_k, S_k \in E^0_{0,1}(\bar{D}_b)$. Taking into accont (23.2), so we obtain sequences w_k, W_k of continuous functions on $\bar{D}_b$ with $\bar\partial w_k = s_k$, $\bar\partial W_k = S_k$ and

$$\lim_k \| w_k \|_{0,D_b} = \lim \| W_k \|_{0,D_b} = 0.$$

Setting $\tilde{f}_k = \chi e^{-kg} - w_k$ and $\tilde{F}_k = \chi h e^{-kg} - W_k$, we obtain sequences of continuous functions on $\bar{D}_b$ which are holomorphic in D_b and such that

$$\lim_k \| \tilde{f}_k - \chi e^{-kg} \|_{0,D_b} = \lim_k \| \tilde{F}_k - \chi h e^{-kg} \|_{0,D_b} = 0. \tag{23.3}$$

Since D is a non-degenerate strictly pseudoconvex extension of D_b and $\bar\partial$ is 0-regular at $Z^0_{0,1}(\bar{D}_b)$, from part (iv) of Lemma 12.8 we obtain sequences f_k and F_k of continuous functions on $\bar{D}$ which are holomorphic in D and such that

$$\| f_k - \tilde{f}_k \|_{0,D_b} \le \frac{1}{k} \qquad \text{and} \qquad \| F_k - \tilde{F}_k \|_{0,D_b} \le \frac{1}{k}.$$

By (23.3), then

$$\lim_k \| f_k - \chi e^{-kg} \|_{0,D_b} = \lim_k \| F_k - \chi h e^{-kg} \|_{0,D_b} = 0. \tag{23.4}$$

Since $\varrho(y) \le \varrho(x)$ and $|x-y|>0$, it follows from (23.1) that $\lim \chi(y) e^{-kg(y)} = 0$ for $k \longrightarrow \infty$. Moreover, $\chi(x) e^{-kg(x)} = 1$ for all k. Therefore, it follows from (23.4) that the sequence f_k has the properties as in statement (A).

Since $h(x)=g(x)=0$ and $\chi = 1$ in a neighborhood of x, it follows that $\partial(\chi h e^{-kg})(x) = \partial h(x) = v$ for all k. Therefore, it follows from (23.4) that the sequence F_k has the properties as in statement (B). □

24. Rossi's example of a real 3-dimensional strictly pseudoconcave boundary which cannot be embedded into $\mathbb{C}^N$

Set

$$\Theta = \{z\in\mathbb{C}^2: \ |z|<3 \text{ and } |z_1| > \frac{1}{128}\}.$$

Choose $\varepsilon>0$ so small that $\varepsilon<1/128$ and the functions $\varrho_+,\varrho_-:\mathbb{C}^2 \longrightarrow \mathbb{R}$ defined by

$$\varrho_{\pm}(z) = \left|2z_1 \pm \frac{\bar{z}_2}{1+|z_2|^2}\right|(1 + |z_2|^2), \qquad z\in\mathbb{C}^2,$$

are strictly plurisubharmonic on Θ. This is possible by Observation 4.15 and since $2|z_1|(1+|z_2|^2)$ is strictly plurisubharmonic for all $z\in\mathbb{C}^2$ with $z_1\neq 0$.

Set

$$U_+ = \{z\in\mathbb{C}^2: \ z_1 \neq 0\} \qquad \text{and} \qquad U_- = \{z\in\mathbb{C}^2: \ z_2 \neq 0\}.$$

We introduce a new complex structure on $\mathbb{C}^2\setminus\{0\} = U_+\cup U_-$ by saying that the functions

$$u_+ := \frac{z_2}{z_1} \qquad \text{and} \qquad v_+ := \frac{z_1^2}{2} - \frac{\varepsilon\bar{u}_+}{1+|u_+|^2}$$

are holomorphic coordinates on U_+, and the functions

$$u_- := \frac{z_1}{z_2} \qquad \text{and} \qquad v_- := \frac{z_2^2}{2} - \frac{\varepsilon\bar{u}_-}{1+|u_-|^2}$$

are holomorphic coordinates on U_-. This is correct, because

$$u_{\pm} = \frac{1}{u_{\mp}} \qquad \text{and} \qquad v_{\pm} = v_{\mp}u_{\mp}^2 \mp \varepsilon u_{\mp} \quad \text{on } U_+\cap U_-. \tag{24.1}$$

The complex manifold formed by $\mathbb{C}^2\setminus\{0\}$ with this new complex sructure will be denoted by M.

24.1. Lemma. On $G := \{z \in \mathbb{C}^2 : 1/2 < |z| < 2\}$, the function $|z|^2$ is strictly plurisubharmonic also with respect ot the complex structure of M.

Proof. Set

$$\tilde{U}_+ = \{z \in U_+ : 2|z_1| > |z_2|\} \quad \text{and} \quad \tilde{U}_- = \{z \in U_- : 2|z_2| > |z_1|\}.$$

Then $|u_\pm| < 2$ and $1/128 < |v_\pm| < 2$ on $\tilde{U}_\pm \cap G$. Therefore, by setting $h_\pm(z) = (u_\pm(z), v_\pm(z))$, we obtain some maps $h_\pm$ defined on $\tilde{U}_\pm \cap G$ and with values in Θ. Since $u_\pm, v_\pm$ are holomorphic coordinates on $U_\pm$ with respect to the complex structure of M, the following assertion holds true: If the domains $\tilde{U}_\pm \cap G$ are endowed with the complex structure of M, and Θ carries the standard structure of $\mathbb{C}^n$, then the maps $h_\pm$ are biholomorphic from $\tilde{U}_\pm \cap G$ onto some open subsets of Θ. Since

$$|z|^2 = 2\left|v_\pm + \frac{\bar{u}_\pm}{1+|u_\pm|^2}\right|(1+|u_\pm|^2)$$

on $U_\pm$, i.e. $|z|^2 = \rho_\pm \circ h_\pm$ on $U_\pm$, and since the functions $\rho_\pm$ are strictly plurisubharmonic with respect to the standard structure of $\mathbb{C}^n$, this implies that $|z|^2$ is strictly plurisubharmmonic with respect to the complex structure of M. $\square$

24.2. Theorem. There does not exist a complex manifold $\tilde{M}$ with the following properties:

(i) M is an open subset of $\tilde{M}$;

(ii) there exists a relatively compact domain $\tilde{D} \subset\subset \tilde{M}$ such that $\partial\tilde{D} = \{z \in \mathbb{C}^n : |z| = 1\}$.

Proof. By (2.41), $v_+u_+ = v_-u_- - \varepsilon$ on $U_+ \cap U_-$. Therefore, the two sides of this equation together define a holomorphic function on M, which will be denoted by v_0. Moreover it follows from (24.1) that the functions v_+ and v_- are holomorphic on M. Consider the holomorphic map $\phi: M \longrightarrow \mathbb{C}^3$ which is defined by

$$\phi(z) = (v_0(z), v_+(z), v_-(z)).$$

First we prove that rank $d\phi = 2$ everywhere on M.

Set $N_\pm = \{z \in M : v_\pm(z) = 0\}$ and notice that $N_\pm \subseteq U_\mp$. Since $\varepsilon u_+ \neq 0$ on $U_+ \cap U_-$ ($\supseteq N_+ \cap N_-$), it follows from (24.1) that $N_+ \cap N_- = \emptyset$. Since v_+ and u_+

are holomorphic coordinates on U_+, and v_- and u_- are holomorphic coordinates on U_-, we see that v_+ and $v_0=v_+u_+$ form holomorphic coordinates on $U_+\setminus N_+$, and v_- and $v_0= v_-u_- -\varepsilon$ form holomorphic coordinates on $U_-\setminus N_-$. Since $N_\pm \subseteq U_\mp$ and $N_+\cap N_- = \emptyset$, this implies that, everywhere on M, at least one of the pairs (v_+,v_0) or (v_-,v_0) is a system of local holomorphic coordinates. Hence rank $d\phi = 2$ everywhere on M.

Now we assume that there exists a complex manifold $\tilde{M}$ with the properties (i) and (ii). Without loss of generality, we can assume that $\tilde{M}=M\cup\tilde{D}$. Then, by Lemma 24.1, $\tilde{D}$ is strictly pseudoconvex, and, by Theorem 20.11 (see also Remark 20.12), the restriction map $\mathcal{O}(\tilde{M}) \longrightarrow \mathcal{O}(M)$ is surjective. Therefore, ϕ admits a holomorphic extension $\tilde{\phi}$ onto $\tilde{M}$. Denote by Crit $\tilde{\phi}$ the set of all $z\in\tilde{M}$ with rank $d\tilde{\phi}(z) < 2$. Since Crit $\tilde{\phi} \subseteq \tilde{M}\setminus M$, and $\tilde{M}\setminus M$ is compact, we see that Crit $\tilde{\phi}$ (being an analytic set) consists of a finite number of connected compact sets. Hence v_0(Crit $\tilde{\phi}$) is finite.

Take an arbitrary complex number c, and set $L_c := \{z\in\tilde{M}: v_0(z)=cv_+(z)\}$. L_c is an 1-dimensional closed analytic subset of $\tilde{M}$. If $z\in M\setminus U_+$, i.e. $z_1=0$ and $z_2\neq 0$, then $u_-(z) = 0$ and therefore it follows from (24.1) that $cv_+(z) = 0$. On the other hand, $v_0(z) = v_-(z)u_-(z) - \varepsilon \neq 0$ for $z\in M\setminus U_+$. Hence $L_c\cap M \subseteq U_+$. Since $v_0=v_+z_2/z_1$ and $v_+=0$ on U_+, we have

$$L_c\cap M = \{z\in\mathbb{C}^2\setminus\{0\}: z_2=cz_1\}. \tag{24.2}$$

In particular, $L_c\cap M$ has the topological type of a punctured plane. Since $v_0=u_+v_+$ on U_+, we have

$$v_0 = \frac{z_1z_2}{2} - \frac{z_2}{z_1}\,\frac{\bar{z}_2/\bar{z}_1}{1+|z_2/z_1|^2} \qquad \text{on } U_+.$$

Therefore, v_0 is not constant on L_c, and, for any point x which belongs to the boundary of $L_c\cap(\tilde{M}\setminus M)$ in L_c, we obtain

$$v_0(x) = \lim_{L_c\ni z\longrightarrow x} v_0(z) = -\frac{\varepsilon|c|^2}{1+|c|^2}. \tag{24.3}$$

Hence v_0 is a non-constant holomorphic function on L_c which is constant on the boundary of the compact set $L_c\cap(\tilde{M}\setminus M)$. Since L_c is of complex dimension 1, this is possible only if $L_c\cap(\tilde{M}\setminus M)$ is finite. Since, topologically, $L_c\cap M$ is a punctured plane, and since L_c is closed in $\tilde{M}$, it follows that $L_c\cap(\tilde{M}\setminus M)$ consists of precisely one point x_c. Since $\tilde{\phi}(z)=\tilde{\phi}(-z)$ for all $z\in M$, and since, by (24.2), $-z\in L_c\setminus\{x_c\}$ for all $z\in L_c\setminus\{x_c\}$, we see that $x_c\in$ Crit $\tilde{\phi}$. By (24.3), this implies

$$- \frac{|c|^2}{1+|c|^2} \in v_0(\mathrm{Crit}\ \tilde{\Phi}).$$

This is impossible, because c is an arbitrary complex number, whereas $v_0(\mathrm{Crit}\ \tilde{\Phi})$ is finite. □

NOTES

<u>To Sections 1 and 2.</u> For some historical remarks concerning the Martinelli-Bochner-Koppelman and the Cauchy-Fantappie formulas we refer to the Notes at the end of Chapter 1 in [H/L] (in [H/L] these formulas are called the Leray resp. the Koppelman-Leray formula). The Kodaira theorem on finiteness of the cohomology of compact complex manifolds with coefficients in holomorphic vector bundles (which is equivalent to Theorem 1.15, via Dolbeault isomorphism) was obtained in [Kodaira 1953] by means of the theory of harmonic integrals. The elementary proof by means of the Martinelli-Bochner formula given here is due to [Toledo/Tong 1976] (see also [Henkin 1977]). In [Cartan/Serre 1953] this theorem was generalized to cohomology with coefficients in coherent analytic sheaves.

The Poincare $\bar{\partial}$-lemma for smooth forms (Theorems 2.12) is due to Grothendieck and Dolbeault (see [Cartan 1953/54, Dolbeault 1953,1956]). The regularity of the $\bar{\partial}$-operator (Theorem 1.13), the Poincaré $\bar{\partial}$-lemma for currents (Theorem 2.13), and the Dolbeault isomorphism

$$H_{\infty}^{s,r}(X) \cong H^{r}(\text{sheaf of holomorphic } (s,0)\text{-forms on } X) \cong H_{cur}^{s,r}(X)$$

(Theorem 2.14) were obtained in [Dolbeault 1953,1956]. With respect to these results we refer also to [de Rham 1955, Schwartz 1955, Griffiths/Harris 1978, Čirka 1979].

<u>To Section 3.</u> For historical remarks concerning the piecewise Cauchy--Fantappie formula (Theroem 3.12 and Corollary 3.13) corresponding to the canonical combination of a Leray <u>data</u> (cf. Sect. 3.4), we refer to the Notes at the end of Chapter 4 in [H/L] (in [H/L] these formulas are called the Leray-Norguet resp. the Koppelman-Leray-Norguet formula). For the case of a general Leray <u>map</u> (in the sense of Sect. 3.4), the piecewise Cauchy-Fantappie formula was established and applied to the tangent Cauchy-Riemann equation in [Ajrapetjan/Henkin 1984]. We remark that, in the present monograph, we use only piecewise Cauchy-Fantappie formulas corresponding to the canonical combination of a Leray data. The general formula will be used only in the subsequent parts of the pending book mentioned in the Preface.

<u>To Sections 4-6.</u> The concepts of q-convexity and q-concavity were developed in the works [Rothstein 1955] and [Andreotti/Grauert 1962]. The

fact that q-convex domains in Stein manifolds are <u>completely</u> q-convex (Theorem 5.3) was observed in [Coen 1969]. For more recent developments of these concepts we refer to [Grauert 1981, Diederich/Fornaess 1985, 1986, Peternell 1985].

<u>To Sections 7 and 9.</u> Local solutions with uniform estimates of the equation $\bar{\partial}u=f_{0,r}$ on strictly q-convex domains with $r \geq n-q$ (cf. Theorems 7.8 and 9.1), for the first time, were obtained in the case of C^2 domains in [Fischer/Lieb 1974] as a generalization of the corresponding results proved in [Grauert/Lieb 1970, Henkin 1970, Lieb 1970, Henkin/Romanov 1971] for strictly pseudoconvex domains (for such estimates on pseudoconvex domains, cf. also the Notes at the ends of Chapters 2, 3, and 4 in [H/L]).

<u>To Sections 11 and 12</u>. Theorem 11.2, for the first time, was proved for domains with C^2 boundary in [Fischer/Lieb 1974]. The Andreotti-Grauert Theorem 12.16 was otained in [Andreotti/Grauert 1962]. Moreover, in that work, a more general result for cohomology with coefficients in arbitrary coherent analytic sheaves is proved. In the more special situation of (s,r)-forms (with values in the trivial bundle), another proof of this theorem was given in [Hörmander 1965], which leads to a sharpened result, with weighted L^2 estimates.

Notice that the proof of Theorem 11.2 obtained in [Fischer/Lieb 1974] makes use of the Andreotti-Grauert Theorem 12.16, whereas the proof given here is independent of Theorem 12.16; moreover, we prove Theorem 12.16 by means of Theorem 11.3. Our proof of Theorem 12.16 consists in an inductive procedure with respcet to the levels of an exhausting function which is strictly (q+1)-convex at infinity. Theorem 12.14 on the extension of the Dolbeault cohomology classes of order $\geq n-q$ along q-convex levels, obtained by this procedure, may be considered as a natural supplement to the classical Hartogs extension theorem for holomorphic functions. Observe that this inductive procedure was also used in Sect. 2.12 of [H/L] in solving the $\bar{\partial}$-equation on completely pseudoconvex manifolds. Sections 11 and 12 of the present work contain all results of Sections 2.11 and 2.12 in [H/L] (as the special case q=n-1).

Notice also that, in distinction to [Fischer/Lieb 1974], we do not assume that the boundary of the domain D in Theorem 11.2 is smooth. However, in order to limit the technical difficulties, we restrict ourselves to the non-degenerate case. The extra work which we have to do (in order to obtain all necessary estimates also in the case of non-degenerate critical points on the boundary) pays off. For instance, in the inductive procedure mentioned above, we need not worry about the levels with critical points. In [H/L] extra arguments were necessary in order to "jump over" such critical levels.

In our opinion, there is no doubt that Theorem 11.2' holds true also for arbitrary strictly q-convex domains. For q=n-1, i.e. for strictly pseudoconvex domains, this is proved (see Theorem 3.2.1 in [H/L] if $\alpha=0$, and [Bruna/Burgues 1986] if $\alpha=1/2$).

Also, the authors are sure that the following theorem holds true:

If, under the hypotheses of Theorem 12.7, the boundary of D is of class C^{∞}, then, for all $n-q\leq r\leq n$,

$$T_r(Z^k_{0,r}(\bar{D},E) \subseteq C^{k+1/2}_{0,r-1}(\bar{D},E) \qquad \text{for all } k=1,2,\ldots ,$$

where T_r is the operator from Theorem 12.7.

For q=n-1, ,i.e. for strictly pseudoconvx domains D, this is proved (see [Siu 1974, Greiner/Stein 1977, Lieb/Range 1986]). Notice also the following fundamental consequence of this theorem, which was obained, for the first time, for q=n-1, in [Kohn 1965]:

If, under the hypotheses of Theorem 12.7, the boundary of D is of class C^{∞}, then, for any $n-q\leq r\leq n$, there exist a linear operator $T_r:Z^{\infty}_{0,r}(\bar{D},E) \longrightarrow C^{\infty}_{0,r-1}(\bar{D},E)$ and a sequence of constants $C_k< \infty$ such that, for all $f\in Z^{\infty}_{0,r}(\bar{D},E)$, $\bar{\partial}T_rf = f$ and

$$T_r(Z^k_{0,r}(\bar{D},E) \subseteq C^{k+1/2}_{0,r-1}(\bar{D},E) \qquad \text{for all } k=1,2,\ldots .$$

To Sections 13-15. Local solutions with uniform estimates of the equation $\bar{\partial}u=f_{0,r}$ on strictly q-concave domains with $1\leq r\leq q-1$, which are close to the estimates given in Theorem 14.1, for the first time, were announced for domains with C^2 boundary in [Ovrelid 1976]. Also the global Theorem 14.5 was announced in [Ovrelid 1976]. The first complete proof of these results was given in [Lieb 1979], also for C^2 domains.

The Andreotti-Grauert Theorem 15.13 was otained in [Andreotti/Grauert 1962]. Moreover, in that work, a more general result for cohomology with coefficients in arbitrary coherent analytic sheaves is proved. In the more special situation of (s,r)-forms (with values in the trivial bundle), another proof of this theorem was given in [Hörmander 1965], which leads to a sharpened result, with weighted L^2 estimates.

Notice that the proof of Theorem 14.5 obtained in [Lieb 1979] makes use of the Andreotti-Grauert Theorem 15.13, whereas the proof given here is independent of Theorem 15.13; moreover, we prove Theorem 15.13 by means of Theorem 14.5. Our proof of Theorem 15.13 consists in an inductive procedure with respcet to the levels of an exhausting function which is strictly (q+1)-concave at infinity, similar as in the proof of Theorem 12.16.

Notice also that, in distinction to [Lieb 1979], we do not assume that the boundary of the domain D in Theorem 14.5 is smooth. As in the case of Theorem 11.2 we admit non-degenerate critical points. In our opinion, there is no doubt that Theorem 14.5 holds true also for arbitary strictly q-concave domains.

Finally, let us notice that, in distinction to [Lieb 1979], we construct a formula which solves the $\bar\partial$-equation locally "without shrinking of the neighborhood" (Theorem 13.10 (i)). Moreover, the following theorem holds true:

If $[U,\varrho,\varphi,H,D]$ is a q-concave configuration in $\mathbb{C}^n$ ($1\le q\le n-1$), then for any $1\le r\le q-1$ and for each continuous $\bar\partial$-closed $(0,r)$-form f on $\bar D$, there exists a solution u of $\bar\partial u=f$ which is also continuous on $\bar D$. Moreover, this solution can be given by a bounded (with respect to the supnorm) linear operator the norm of which tends to zero if the diameter of D tends to zero.

To prove this theorem it is necessary to strengthen the estimates given in Theorem 14.1. This can be done by means of a construction from [Henkin 1977].

Corollary 13.9 and Theorem 15.2 on the extension of holomorphic functions along 1-concave levels generalize the classical Hartogs extension theorem. A further natural generalization of this fact is given by Theorems 13.10 and 15.11 on the extension of the Dolbeault cohomology classes of order r along q-concave levels with $0\le r\le q-1$ (the corresponding interpretation of Theorem 12.14 was already mentioned above).

To Section 17. The Martineau Theorem 17.9 was obtained in [Martineau 1962,1966]. By means of this theorem, Martineau proved also the following important result: If $K \subset\subset \mathbb{C}^n$ is a convex compact set, then the space $\mathcal{O}((\mathbb{C}^n\setminus K)^*)$ is isomorphic to the dual of the space $\mathcal{O}(K)$ of all holomorphic functions in some neighborhoods of K. A proof of the latter result was obtained independently also in [Ajzenberg 1966].

To Section 18. The Andreotti-Norguet theorem (Theorems 18.2 and 18.3) was proved in [Andreotti/Norguet 1966]. The proof given here and Lemma 18.1 are taken from [Grauert 1981]. The corresponding fact with uniform estimates (Theorems 18.5 and 18.6), for the first time, was obtained in [Andreotti/Hill 1972] in the case when D is of class C^∞ (then the form f_x in Theorems 18.5 and 18.6 can be chosen to be also of class C^∞).

To Section 19. The Andreotti-Vesentini separation theorem (Theorem 19.1') was obtained in [Andreotti/Vesentini 1965]. The corresponding result with uniform estimates (Theorem 19.2'), in the case when D is strictly pseudoconcave (i.e. q=n-1) and of class C^2, was proved in [Henkin 1977]. In its general form, Theorem 19.2' seems to be new.

We remark that, in our opinion, there is no doubt that Theorem 19.2' holds true also for arbitrary strictly q-concave domains D (i.e. the assumption that the domain D in Theorem 19.2' is non-degenerate can be omitted).

Moreover, the authors are sure that also the following theorem holds true:

If, under the hypotheses of Theorem 19.2', D is of class C^{∞}, then, moreover

$$T\left(\hat{Z}^{0}_{0,q}(\bar{D},E)\cap Z^{k}_{0,q}(\bar{D},E)\right) \subseteq C^{k+1/2}_{0,q-1}(\bar{D},E) \qquad \text{for all } k=1,2,\ldots ,$$

where $\hat{Z}^{0}_{0,q}(\bar{D},E)$ and T are as in part (ii) of Theorem 19.2'.

<u>To Section 20.</u> The general facts concerning the concept of duality were obtained in [Serre 1955]. Theorems 20.6-20.10 follow from these general facts combined with the Andreotti-Grauert Theorems 12.16, 15.13, the Andreotti-Vesentini Theorem 19.1', and the Andreotti-Norguet Theorem 18.4.

The Hartogs extension Theorem 20.11 was proved in [Kohn/Rossi 1965].

Theorem 20.13 on the extension of $\bar{\partial}$-closed forms, in the case of C^{∞} forms, follows from the results of [Andreotti/Grauert 1962] and [Andreotti/Vesentini 1965]. Let us mention also the following two statements A and B, which complete Theorem 20.13:

<u>A.</u> Let $D \subset\subset X$ be a completely strictly q-convex domain with C^{∞} boundary in an n-dimensional complex manifold X $(0\le q\le n-1)$, E a holomorphic vector bundle over X, U a neighborhood of ∂D, and $U_+:=(X\setminus D)\cap U$. Then, for any $0\le r\le q-1$, there exists a continuous linear operator S acting from $Z^{1/2}_{0,r}(U_+,E)$ into $Z^{0}_{0,r}(D\cup U,E)$ such that $Sf|_{U_+} = f$ for all $f\in Z^{1/2}_{0,r}(U_+,E)$, and, moreover,

$$S\left(Z^{\alpha}_{0,r}(U_+,E)\right) \subseteq Z^{\alpha-1/2}_{0,r}(D\cup U,E) \qquad \text{for all } \tfrac{1}{2} \le \alpha \le \infty.$$

This statement follows from the results of [Kohn/Rossi 1965] (if $\alpha=\infty$), [Folland/Stein 1974], and [Rothschild/Stein 1976].

<u>B.</u> Let $D \subset\subset X$ be a C^{∞} domain in an n-dimensional compact complex manifold X. Suppose D is both strictly q-convex and strictly (n-q-1)--concave $(0\le q\le n-1)$. Let E be a holomorphic vector bundle over X, U a neighborhood of ∂D, $U_+:=(X\setminus D)\cap U$, and f an E-valued $\bar{\partial}$-closed (0,r)-form of class C^{α} over U_+ $(0\le r\le q,\ 1/2 \le \alpha\le \infty)$. Then the following two conditions are equivalent:

(i) There exists an E-valued $\bar{\partial}$-closed (0,r)-form F of class $C^{\alpha-1/2}$ on $D\cup U$ with $F|_{U_+} = f$.

(ii) $$\int_{\partial D} f \wedge g = 0 \qquad \text{for all } g \in Z^{\infty}_{n,n-r-1}(\bar{D}, E^*).$$

For $q=n-1$ and $D \subset\subset \mathbb{C}^n$, this result was obtained in [Dautov 1972] (see also [Ajzenberg/Dautov 1983]). Then, in [Henkin 1977], it was proved for the case when (also) $q=n-1$ but D is not necessarily contained in $\mathbb{C}^n$.

PROBLEMS

1. Linearly q-concave domains. A domain $D \subseteq \mathbb{C}^n$ (resp. a domain D contained in the n-dimensional complex projective space $\mathbb{P}^n$) is said to be linearly q-convex if, for any $z \in \mathbb{C}^n \setminus D$ (resp. $z \in \mathbb{P}^n \setminus D$), there exists a complex q-dimensional plane $\mathbb{C}^q_z \subseteq \mathbb{C}^n$ (resp. a q-dimensional complex projective plane $\mathbb{P}^q_z \subseteq \mathbb{P}^n$) such that $z \in \mathbb{C}^q_z \subseteq \mathbb{C}^n \setminus D$ (resp. $z\ P^q_z \subseteq P^n \setminus D$). If, moreover, this plane can be chosen continuously depending on z, then we say D is continuously linearly q-convex.

Example. The domains $|z^0|^2+\ldots+|z^q|^2 < |z^{q+1}|^2+\ldots+|z^n|^2$, considered in Sect. 21, are continuously linearly q-convex.

Problems. (i) Is any linearly q-convex domain q-convex? (ii) Is any continuously linearly q-convex domain completely q-convex?

The authors believe that, at least under the hypothesis of the second question, the answer to the first one is affirmative, i.e.: any continuously linearly q-convex domain is q-convex. It seems that in this case an exhausting function which is (q+1)-convex at infinity can be constructed patching together, by means of procedure $\max_\beta$ (cf. Def. 4.12), appropriate local (q+1)-convex functions with joint (q+1)-dimensional subspaces of $T'_z(D)$ on which their Levi forms are positive definite.

2. Embedding of q-convex manifolds into linearly (q+k)-convex domains. Let $D \subset\subset X$ be a strictly q-convex domain in an n-dimensional complex manifold X.

Problem. Does there exist an integer k and a linearly (q+k)-convex $\tilde{D} \subsetneq \mathbb{C}^{n+k}$ such that D is biholomorphically equivalent to a closed submanifold of $\tilde{D}$?

Remark. If q=n-1, the answer is affirmative. This was proved by Henkin and Fornaess (see [Čirka/Henkin 1975, Fornaess 1976]).

3. Extension of $\bar{\partial}$-closed forms from complex submanifolds into q-convex domains. Let E be a holomorphic vector bundle over an n-dimensional complex manifold X, $D \subset\subset X$ a completely strictly q-convex domain ($0 \le q \le n-1$), and M a closed complex submanifold of codimension k in X.

Problem. Prove that, for all $n-q-1 \le r \le n$, the restriction map

$$Z^0_{0,r}(\bar{D},E) \longrightarrow Z^0_{0,r}(\overline{D \cap M},E)$$

is surjective.

Remarks. For r=0 and q=n-1 this is proved (cf. Sect. 4.11 in [H/L]).

If M is a smooth complete intersection in X, ∂D is of class C^2, and the intersection $M \cap \partial D$ is transversal, then the problem is also solved (together with P.L.Poljakov, unpublished). Let us outline the idea in the following more special situation: Assume X is a domain in $\mathbb{C}^n$, E is the trivial line bundle, k=1, and $D \subset\subset X$ is a strictly q-convex domain with C^2 boundary which has, moreover, the following property:

There exist C^1 functions $F_1(x,z),\dots,F_{n-q}(x,z)$ defined for $x \in \partial D$ and $z \in C^n$ which are complex linear with respect to z such that, for any point $x \in \partial D$, the complex plane

$$P_x := \{z \in X: F_1(x,x-z)=\dots=F_{n-q}(x,x-z)=0\}$$

satisfies the following three conditions:

(i) P_x is complex q-dimensional and contained in the complex tangent plane of ∂D at x;

(ii) $P_x \cap D = \emptyset$;

(iii) $|z-x| = O([\mathrm{dist}(z,\partial D)]^{1/2})$ for $P_x \ni z \longrightarrow x$.

Set $A = |F_1|^2+\dots+|F_{n-q}|^2$. Let $w(z,x) = ((w_1(z,x),\dots,w_n(z,x))$ be the vector of functions such that

$$A(z,x) = \sum_{j=1}^{n} w_j(z,x)(x_j-z_j).$$

Then w(z,x) is a Leray data for D. Set

$$v^0(z,x,t) = t\frac{\bar{x}-\bar{z}}{|x-z|^2}+(1-t)\frac{w(z,x)}{A(z,x)} .$$

Now let M be a smooth complete intersection of codimension 1 in X such that the intersection $M \cap \partial D$ is transversal, i.e. $M = \{z \in X: g(z)=0\}$, where g is a holomorphic function on X with $dg(z) \neq 0$ for all $z \in M$, and $dg|_{\partial D}(z) \neq 0$ for $z \in \partial D \cap M$. Further, let $f \in Z^0_{0,r}(\overline{D \cap M})$. Set

$$H_f(z) = c_1 \int\limits_{x\in M\cap D} f(x)\wedge \frac{\det(\bar{x}-\bar{z}, \overbrace{d\bar{x}}^{n-r-1}, \overbrace{d\bar{z}}^{r})}{|x-z|^{2n}} \wedge \frac{\omega(x)}{dg(x)}$$

$$+ c_2 \int\limits_{\substack{x\in M\cap\partial D \\ 0\le t\le 1}} f(x)\wedge\det(v^0, d_t v^0, \overbrace{\bar{\partial}_z v^0}^{r}, \overbrace{\bar{\partial}_x v^0}^{n-r-2}) \wedge \frac{\omega(x)}{dg(x)}, \quad z\in D\setminus M,$$

where c_1, c_2 are appropriate constants. In this way we obtain a continuous $(0,r)$-form H_f on $D\setminus M$, and, estimating the integrals in the definition of H_f, one can prove that gH_f admits a continuous extension onto $\bar{D}$ such that, after the correct choice of the constants c_1 and c_2,

$$gH_f\big|_M = f.$$

4. Uniform approximation. Let $K \subset\subset X$ be a compact set in the n-dimensional complex manifold X such that, for some (q+1)-convex function $\varrho: X \longrightarrow \mathbb{R}$ $(0\le q\le n-1)$,

$$K = \{z\in X:\ \varrho(z)\le 0\}.$$

(Here it is permitted that ϱ has degenerate critical points.)

Problem. Prove that any $(0,n-q-1)$-form f of class C^k in a neighborhood of K which is $\bar{\partial}$-closed in the interior of K can be approximated uniformly on K, together with all derivatives of order $\le k$, by $\bar{\partial}$-closed C^∞ forms in a neighborhood of K $(k=0,1,2,\dots)$.

For q=n-1, this is proved (see Chapter 3 in [H/L] for k=0, and [Bruna/Burgues 1986] for the general case).

5. The $\bar{\partial}$-equation on analytic sets. Let X be a k-dimensional closed analytic subset of the unit ball B in $\mathbb{C}^n$ which is a local complete intersection, i.e. for any $z\in X$ there are holomorphic functions $f_1,\dots,f_{n-k}$ in a neighborhood $U \subseteq B$ of z such that

$$X\cap U = \{x\in U:\ f_1(x)=\dots=f_{n-k}(x)=0\}.$$

Let S(X) be the singular locus of X and

$$d := \dim X - \dim S(X).$$

It is well-known that then any $\bar{\partial}$-closed $(0,r)$-form on $X\setminus S(X)$ with $r\leq d-2$ is $\bar{\partial}$-exact. This is a special case of the following cohomological generalization of the second Riemann extension theorem [Scheja 1961]:

If Y is a pure dimensional closed analytic subset of an open set in $\mathbb{C}^n$ which is a local complete intersection, and A is a closed analytic subset of Y, then, for all

$$0 \leq r \leq \dim Y - \dim A - 2,$$

the restriction map

$$H^{0,r}(Y) \longrightarrow H^{0,r}(Y\setminus A)$$

is bijective.

The following problem is related to the first Riemann extension theorem:

Problem. Suppose f is a $\bar{\partial}$-closed $(0,r)$-form on $X\setminus S(X)$ with $r=d-1$ which admits a C^∞ extension into B. Does it follow that f is $\bar{\partial}$-exact on $X\setminus S(X)$?

If X is of codimension 1, i.e.

$$X = \{z\in B:\ g(z)=0\} \quad \text{and} \quad S(X) = \{z\in X:\ dg(z)=0\}$$

for some holomorphic function g on B, then the answer is affirmative. Let us outline the proof of this fact (obtained together with P.L. Poljakov): In view of the cohomological generalization of the second Riemann extension theorem mentioned above it is sufficient to prove that, for any $\varepsilon>0$, f admits a $\bar{\partial}$-closed C^∞ extension into $B_\varepsilon\setminus S(X)$, where $B_\varepsilon:=\{z\in\mathbb{C}^n:\ |z|<1-\varepsilon\}$. To obtain this extension, first we define a C^∞ form H_f on $B_\varepsilon\setminus X$ as follows:

$$H_f(z) = \lim_{\delta\to 0} c_1 \int\limits_{\substack{x\in X\cap B_\varepsilon \\ \|dg(x)\|>\delta}} f(x)\wedge \frac{\det(\bar{x}-\bar{z}, \overbrace{d\bar{z}}^{r}, \overbrace{d\bar{x}}^{n-r-1})}{|x-z|^{2n}} \wedge \frac{\omega(x)}{dg(x)}$$

$$+ \lim_{\delta\to 0} c_2 \int\limits_{\substack{x\in X\cap\partial B \\ \|dg(x)\|>\delta \\ 0\leq t\leq 1}} f(x)\wedge \det(v^0, d_t v^0, \overbrace{\bar{\partial}_z v^0}^{r}, \overbrace{\bar{\partial}_x v^0}^{n-r-2}) \wedge \frac{\omega(x)}{dg(x)},$$

for $z \in B_\varepsilon \setminus M$, where

$$v^0 = v^0(z,x,t) = t\frac{\bar{x}-\bar{z}}{|x-z|^2} + (1-t)\frac{\bar{x}}{\langle \bar{x}, x-z\rangle},$$

and c_1, c_2 are appropriate constants. The limits for $\delta \longrightarrow 0$ exist in view of a theorem of Coleff and Herrera (see [Coleff/Herrera 1978]). By this theorem, moreover:

$$\lim_{\delta \longrightarrow 0} \int_{\substack{x \in X \cap B_\varepsilon \\ \|dg(x)\| > \delta}} f(x) \wedge \frac{h(x)}{dg(x)} = \lim_{\lambda \longrightarrow 0} \int_{\substack{x \in B_\varepsilon \\ |g(x)| = \lambda}} f(x) \wedge \frac{h(x)}{g(x)}$$

for any smooth form h with compact support in B. Using the latter relation and the fact that f is $\bar{\partial}$-closed on $X \setminus S(X)$ one can prove that H_f is $\bar{\partial}$-closed in $B_\varepsilon \setminus X$. Further, estimating the integrals in the definition of H_f, it can be shown that the form $F := gH_f$ admits a continuous extension into $B_\varepsilon \setminus S(X)$ such that, after the correct choice of c_1 and c_2,

$$F\big|_{X \setminus S(X)} = f.$$

6. Let $D \subset\subset X$ be a non-degenerate strictly q-concave domain in a compact complex manifold X. Then we know from the Andreotti-Vesentini Theorem 19.2 that $E^0_{0,q}(\bar{D})$ is a closed subspace of the Banach space $Z^0_{0,q}(\bar{D})$.

Problem. Prove that $E^0_{0,q}(\bar{D})$ is not complemented in $Z^0_{0,q}(\bar{D})$, i.e. prove that there does not exist a closed linear subspace F of $Z^0_{0,q}(\bar{D})$ such that $F \cap E^0_{0,q}(\bar{D}) = \{0\}$ and $F + E^0_{0,q}(\bar{D}) = Z^0_{0,q}(\bar{D})$.

7. Let $D \subset\subset X$ be a domain in an n-dimensional complex manifold X of the form

$$D = \{z \in X: \varrho_1(z) < 0,\ \varrho_2(z) > 0\}$$

where ϱ_1 and ϱ_2 are strictly plurisubharmonic functions with

$$d\varrho_1 \wedge d\varrho_2 \neq 0 \qquad \text{if } \varrho_1 = \varrho_2 = 0.$$

Problems. (i) Prove that, for $n \geq 4$, the space $E^0_{0,1}(D)$ is a closed subspace of the Fréchet space $Z^0_{0,1}(D)$, and $E^0_{0,1}(\bar{D})$ is a closed subspace of the Banach space $Z^0_{0,1}(\bar{D})$.

(ii) Find an example which shows that this is not true for n=3.

<u>Remark.</u> If $X = \mathbb{C}^3$ and if there exists a pluriharmonic function ϱ_0 such that

$$\{\varrho_0 = 0\} \cap \bar{D} = \{\varrho_1 = \varrho_2 = 0\} \cap \bar{D},$$

then the spaces $E^0_{0,1}(D)$ and $E^0_{0,1}(\bar{D})$ are closed (cf. [Henkin 1977] and Sect. 13 of the present work). It seems that the solution of the first problem can be reduced to this case by means of the Rossi theorem (cf. Sect. 23). An example which solves the second problem, probably, can be found by means of Rossi's example (cf. Sect 24).

A positive solution of the first problem would give a new approach to the deep theorem of M. Kuranishi (see [Kuranishi 1982]) and T. Akahori (see [Akahori 1987]) on local embedding into $\mathbb{C}^N$ of strongly pseudoconvex CR-structures M with $\dim_{\mathbb{R}} M \geq 7$; a counterexample for $\dim_{\mathbb{R}} M = 5$ could be obtained, probably, by means of an example as in the second problem.

APPENDIX A. ESTIMATION OF SOME INTEGRALS (THE SMOOTH CASE)

In this Appendix A and the following Appendix B we prove estimates for some integrals over hypersurfaces resp. domains in $\mathbb{R}^n$ which are used in the basic text of the present monograph. Appendix A is devoted to the case of smooth hypersurfaces resp. domains with smooth boundary. In Appendix B we consider the more general case of surfaces with non--degenerate singularities. In both appendixes the notations introduced in Sects. 0.1-0.4 of the basic text will be used.

<u>0. Preliminaries.</u> We use the following notations:

$m\geq 2$ is an integer, and $x_1,\ldots,x_m$ are the canonical coordinates in $\mathbb{R}^m$.

R is a positive number and B_R is the open ball with radius R centered at the origin in $\mathbb{R}^m$.

ϱ is a real-valued C^2 function in some neighborhood of $\bar{B}_R$ such that $d\varrho(x)\neq 0$ for all $x\in\bar{B}_R$ with $\varrho(x)=0$.

$D:=\{x\in B_R\colon\ \varrho(x)<0\}$.

$S:=\{x\in B_R\colon\ \varrho(x)=0\}$, and we assume that $S\neq\emptyset$.

dS is the Euclidean volume form of S, i.e. dS is the C^1 form of degree $m-1$ on S which is defined by the following condition:

$$\left.\begin{array}{l}\text{If } v \text{ is a differential form of degree } m-1 \text{ in a neighbor-}\\ \text{hood of } S \text{ such that } d\varrho(x)\wedge v(x) = d\varrho(x)\wedge dx_1\wedge\ldots\wedge dx_m\\ \text{for all } x\in S, \text{ then } v|S=dS.\end{array}\right\} \qquad (1)$$

We orient S by dS. □

In this Appendix A we prove the following three propositions:

<u>1. Proposition.</u> (i) There exists $C<\infty$ such that, for all $y\in D$ and $\varepsilon>0$,

$$\int\limits_{x\in S}\frac{dS}{(\varepsilon+|x-y|^2)|x-y|^{m-1}} \leq \begin{cases} C+C|\ln\varepsilon| & \text{if } l=3\\ C\varepsilon^{-1/2} & \text{if } l=2.\end{cases} \qquad (2)$$

(ii) Suppose $h(y,x)$ is a C^1-function which is defined for x,y in some neighborhood of $\bar{B}_R$ such that the following two conditions are fulfilled:

$$h(y,y) = 0 \qquad \text{for all } y\in\bar{B}_R$$

and

$$d\varrho(y)\wedge d_x h(y,x)\Big|_{x=y} \neq 0 \qquad \text{if } y\in\bar{S}. \tag{3}$$

Then there exists a constant $C<\infty$ such that, for all $y\in D$ and $\varepsilon>0$,

$$\int\limits_{x\in S} \frac{dS}{(\varepsilon+|h(y,x)|+|x-y|^2)^l\,|x-y|^{m-3}} \leq \begin{cases} C & \text{if } l=1 \\ C\varepsilon^{-1/2} & \text{if } l=2. \end{cases} \tag{4}$$

2. Proposition. There exists a constant $C<\infty$ such that if f is a C^1 function on D with

$$\|df(x)\| \leq |\varrho(x)|^{-1/2} \qquad \text{for all } x\in D,$$

then

$$|f(x)-f(y)| \leq C|x-y|^{1/2} \qquad \text{for all } x,y\in D.$$

3. Proposition. (i) There exists a constant $C<\infty$ such that, for all $y\in\bar{B}_R$,

$$\int\limits_{x\in B_R} \frac{dx_1\wedge\ldots\wedge dx_m}{(|\varrho(x)|+|x-y|^2)^2\,|x-y|^{m-4}} \leq C. \tag{5}$$

(ii) Let $h(y,x)$ be as in Proposition 1 (ii), and let $d>0$ and $\alpha,\beta \geq 0$ be numbers with

$$\alpha+\beta < \frac{d+1}{2}.$$

Then there exists a constant $C<\infty$ such that, for all $y\in\bar{B}_R$ and $r\geq 0$,

$$\int\limits_{\substack{x\in B_R \\ |\varrho(x)|<r}} \frac{|\varrho(x)|^{-\alpha} dx_1\wedge\ldots\wedge dx_m}{(|\varrho(x)|+|h(y,x)|+|x-y|^2)^2 |x-y|^{m-3-d}} \leq Cr^{\beta}. \tag{6}$$

<u>4. Proof of Proposition 1 (i).</u> It is sufficient to prove that for each $z\in\bar{S}$ there exist a neighborhood U of z and a constant $C<\infty$ such that, for all $y\in D$ and $\varepsilon>0$, the integral

$$\int\limits_{x\in S\cap U} \frac{dS}{(\varepsilon+|x-y|^2)|x-y|^{m-1}} \tag{7}$$

can be estimated by the right hand side of (2).

Fix $z\in\bar{S}$. Since $d\varrho(z)\neq 0$, we can find a neighborhood U of z and an index $1\leq j\leq m$ such that

$$dS = F dx_1\wedge\ldots\wedge\widehat{dx_j}\wedge\ldots\wedge dx_m\Big|_S \qquad \text{on } S\cap U,$$

where F is a real-valued function which is bounded by some constant $K<\infty$ on $S\cap U$. Then the integral (7) is bounded by

$$K\int\limits_{\substack{t\in\mathbb{R}^{m-1} \\ |t|<2R}} \frac{dt_1\wedge\ldots\wedge dt_{m-1}}{(\varepsilon+|t|^2)|t|^{m-1}}$$

Therefore it is sufficient to prove the following statement: For any number $b<\infty$, there exists a constant $L<\infty$ such that, for all $\varepsilon>0$,

$$\int\limits_{\substack{t\in\mathbb{R}^{m-1} \\ |t|<b}} \frac{dt_1\wedge\ldots\wedge dt_{m-1}}{(\varepsilon+|t|^2)|t|^{m-1}} < \begin{cases} L + L|\ln\varepsilon| & \text{if } l=3 \\ L\varepsilon^{-1/2} & \text{if } l=2. \end{cases} \tag{8}$$

If s_{m-2} is the volume of the (m-2)-dimensional unit sphere, then the integral on the left hand side of (8) is equal to

$$s_{m-2}\int_0^d \frac{x^{l-2}}{\varepsilon+x^2}\,dx.$$

Since

$$\int_0^b \frac{x}{\varepsilon+x^2}\,dx = \frac{1}{2}\ln(1+\frac{d^2}{\varepsilon}),$$

and

$$\int_0^b \frac{dx}{\varepsilon+x^2} \le \int_0^{\varepsilon^{1/2}} \frac{dx}{\varepsilon} + \int_{\varepsilon^{1/2}}^{\infty} \frac{dx}{x^2} = 2\varepsilon^{-1/2},$$

this implies (8). □

5. Proof of Proposition 1 (ii). It is sufficient to prove that for each $z\in\bar{S}$ there exist a neighborhood U of z and a constant $C<\infty$ such that, for all $y\in D\cap U$ and $\varepsilon>0$, the integral

$$\int_{x\in S\cap U} \frac{dS}{(\varepsilon+|h(y,x)|+|x-y|^2)^1|x-y|^{m-3}} \tag{9}$$

can be estimated by the right hand side of (4). Fix $z\in\bar{S}$. Since h and ϱ are C^1 functions and in view of (3), we can find a neighborhood U of z and integers $1\le j<k\le m$ such that for all $y\in U$

$$dS \le K\left|d_x h(y,x)\wedge dx_1\wedge\ldots\widehat{j,k}\ldots\wedge dx_m\right|_S$$

on $U\cap S$, where $K<\infty$ is a constant which is independent of $y\in U$. Since h is a C^1 function with $h(y,y)=0$, we can find a constant $L<\infty$ such that, for all $x,y\in U$,

$$|h(y,x)|^2 + \sum_{s\ne j,k}|x_s-y_s|^2 \le L|x-y|^2.$$

Therefore, the integral (9) can be estimated by

$$K(1+L)^1L \int_{\substack{t\in\mathbb{R}^{m-1}\\ |t|<b}} \frac{dt_1\wedge\ldots\wedge dt_{m-1}}{(\varepsilon+|t_1|+|t|^2)^1|t|^{m-3}},$$

where

$$b = \max_{x,y\in\overline{U\cap B_R}} \left[|h(y,x)|^2 + \sum_{s=j,k} |x_s - y_s|^2\right]^{1/2}.$$

So it remains to find a constant $M<\infty$ (which depends only on b and m) such that, for all $\varepsilon>0$,

$$\int_{\substack{t\in\mathbb{R}^{m-1}\\ |t|<b}} \frac{dt_1\wedge\ldots\wedge dt_{m-1}}{(\varepsilon+|t_1|+|t|^2)^l |t|^{m-3}} < \begin{cases} M & \text{if } l=1 \\ M\varepsilon^{-1/2} & \text{if } l=2. \end{cases} \tag{10}$$

If m=2 and l=1, then (10) holds true with M=b.
If m=2 and l=2, then the integral in (10) can be estimated by

$$2\int_0^b \frac{xdx}{\varepsilon^2+x^2} = \ln(1+\frac{d}{\varepsilon^2}) \le M\varepsilon^{-1/2},$$

where

$$M = \max_{0<\varepsilon<\infty} \varepsilon^{1/2} \ln(1+\frac{b^2}{\varepsilon^2}) < \infty.$$

Now we consider the case when $m\geq 3$ and l=2. Then the integral in (10) can be estimated by

$$2\int_{\substack{s\in\mathbb{R}^{m-2}\\ |s|<b}} ds_1\wedge\ldots\wedge ds_{m-2} \int_0^b \frac{dx}{(\varepsilon+x+|s|^2)^2|s|^{m-3}} \le 4\int_{\substack{s\in\mathbb{R}^{m-2}\\ |s|<b}} \frac{ds_1\wedge\ldots\wedge ds_{m-2}}{(\varepsilon+|s|^2)|s|^{m-3}}.$$

Therefore, in this case, (10) follows from (8).

Finally, let $m\geq 3$ and l=1. Then (10) follows from the more precise estimate

$$\int_{\substack{t\in\mathbb{R}^{m-1}\\ |t_1|<r}} \frac{dt_1\wedge\ldots\wedge dt_{m-1}}{(|t_1|+|t|^2)|t|^{m-3}} \le 8s_{m-3}r^{1/2} \qquad \text{for all } r>0, \tag{11}$$

where s_{m-3} is the Euclidean volume of the (m-3)-dimensional unit sphere in $\mathbb{R}^{m-2}$. To prove the latter estimate we observe that the integral on the left hand side of (11) is bounded by

$$2\int_0^r dx \int_{s\in\mathbb{R}^{m-2}} \frac{ds_1\wedge\ldots\wedge ds_{m-2}}{(x+|s|^2)|s|^{m-3}} \le 2s_{m-3}\int_0^r dx \int_0^\infty \frac{du}{x+u^2},$$

where

$$\int_0^\infty \frac{du}{x+u^2} \le \int_0^{x^{1/2}} \frac{du}{x} + \int_{x^{1/2}}^\infty \frac{du}{u^2} = 2x^{-1/2}. \quad \square$$

6. Proof of Proposition 2. After passing to appropriate local coordinates, we may assume the $\varrho=x_1$. Then $D=\{x\in B_R: x_1<0\}$ and

$$\|df(x)\| \le |x_1|^{-1/2} \qquad \text{for all } x\in D. \tag{12}$$

If $x,y\in D$ and $\min(|x_1|,|y_1|) \ge |x-y|$, then this implies

$$|f(x)-f(y)| = \left|\int_0^1 \frac{d}{dt}f(tx+(1-t)y)dt\right| \le \int_0^1 \frac{|x-y|dt}{|tx_1+(1-t)y_1|^{1/2}} \le |x-y|^{1/2}.$$

If $x,y\in D$ with $x_j=y_j$ for all $2\le j\le m$, then from (12) we obtain

$$|f(x)-f(y)| = \left|\int_0^1 \frac{d}{dt}f(tx_1+(1-t)y_1,x_2,\ldots,x_m)dt\right|$$

$$\le \int_0^1 \frac{|x_1-y_1|dt}{|tx_1+(1-t)y_1|^{1/2}} = 2\left||x_1|^{1/2}-|y_1|^{1/2}\right| \le 2|x-y|^{1/2}.$$

Together this implies that, for all x,y D,

$$\begin{aligned}|f(x)-f(y)| &\le |f(x)-f(-|x-y|,x_2,\ldots,x_m)| \\ &\quad +|f(-|x-y|,x_2,\ldots,x_m)-f(-|x-y|,y_2,\ldots,y_m)| \\ &\quad +|f(-|x-y|,y_2,\ldots,y_m) - f(y)| \\ &\le 5|x-y|^{1/2}. \quad \square\end{aligned}$$

7. Proof of Proposition 3 (i). Since $d\varrho(x) \neq 0$ if $\varrho(x)=0$, we can find $\delta>0$ such that $\|d\varrho(x)\| \geq \delta$ if $|\varrho(x)| \leq \delta$ $(x \in \bar{B}_R)$. It is enough to prove that

$$\sup_{y \in B_R} \int_{\substack{x \in B_R \\ |\varrho(x)|<\delta}} \frac{dx_1 \wedge \ldots \wedge dx_m}{(|\varrho(x)|+|x-y|^2)^2 |x-y|^{m-4}} < \infty. \tag{13}$$

Since $\|d\varrho(x)\| \geq \delta$ if $|\varrho(x)| \leq \delta$, we can find a constant $K<\infty$ and a finite number of open sets $U_1, \ldots, U_N \subseteq B_R$ such that

$$\{x \in B_R : |\varrho(x)|<\delta\} \subseteq U_1 \cup \ldots \cup U_N$$

and, for each $1 \leq l \leq N$, there exists an index $1 \leq j(l) \leq m$ such that the functions

$$\varrho, x_1, \ldots \widehat{j(l)} \ldots, x_m$$

form a system of C^1 coordinates on U_l with

$$dx_1 \wedge \ldots \wedge dx_m = h_l |d\varrho \wedge dx_1 \wedge \ldots \widehat{j(l)} \ldots \wedge dx_m| \qquad \text{on } U_l,$$

where h_l is a positive continuous function $\leq K$ on U_l. Therefore, the integral in (13) is bounded by

$$\sup_{y \in B_R} \sum_{l=1}^{N} K \int_{x \in U_l} \frac{|d\varrho \wedge dx_1 \wedge \ldots \widehat{j(l)} \ldots \wedge dx_m|}{(|\varrho|+|x-y|^2)^2 |x-y|^{m-4}}$$

$$\leq NK \int_{\substack{s \in \mathbb{R}^{m-1} \\ |s|<2R}} ds_1 \wedge \ldots \wedge ds_{m-1} \int_{t \in \mathbb{R}^1} \frac{dt}{(|t|+|s|^2)^2 |s|^{m-4}}$$

$$\leq 2NK \int_{\substack{s \in \mathbb{R}^{m-1} \\ |s|<2R}} \frac{ds_1 \wedge \ldots \wedge ds_{m-1}}{|s|^{m-2}}. \quad \square$$

8. Proof of Proposition 3 (ii). For any $M<\infty$, we can find $C_0<\infty$ such that, for all $r>0$,

$$\int_{\substack{s\in\mathbb{R}^{m-2}\\|s|<M}} ds_1\wedge\ldots\wedge ds_{m-2}\int_{-r}^{r} dx\int_{-\infty}^{\infty}\frac{|x|^{-\alpha}dt}{(|x|+|t|+|s|^2)^2|s|^{m-3-d}}$$

$$= 2\int_{\substack{s\in\mathbb{R}^{m-2}\\|s|<M}} ds_1\wedge\ldots\wedge ds_{m-2}\int_{-r}^{r}\frac{|x|^{-\alpha}dx}{(|x|+|s|^2)|s|^{m-3-d}}$$

$$\leq 2\int_{\substack{s\in\mathbb{R}^{m-2}\\|s|<M}}\frac{ds_1\wedge\ldots\wedge ds_{m-2}}{|s|^{m-2-(d+1-2\alpha-2\beta)}}\int_{-r}^{r}|x|^{\beta-1}dx \leq C_0 r^{\beta}. \tag{14}$$

For $a>0$ and $y\in\bar{B}_R$ we set

$$F_a(y) = \{x\in B_R:\ \max(\varrho(x),|x-y|)<a\}.$$

Then, by (3), we can find constants $0<\delta<1$ and $C_1<\infty$ such that: for each $y\in\bar{B}_R$, there exist two indices $1\leq k<l\leq m$ such that the functions

$$\varrho,\ h(y,\cdot),\ x_1,\ \ldots\widehat{k,l}\ldots,\ x_m$$

form a system of C^1 coordinates on $F_\delta(y)$ with

$$dx_1\wedge\ldots\wedge dx_m \leq C_1|d\varrho(x)\wedge d_x h(y,x)\wedge dx_1\wedge\ldots\widehat{k,l}\ldots\wedge dx_m| \qquad \text{on } F_o(y).$$

Then, by (14), for some $C_2<\infty$,

$$\int_{\substack{x\in F_\delta(y)\\|\varrho(x)|<r}}\frac{|\varrho(x)|^{-\alpha}dx_1\wedge\ldots\wedge dx_m}{(|\varrho(x)|+|h(y,x)|+|x-y|^2)^2|x-y|^{m-3-d}} \leq C_2 r^{\beta}. \tag{15}$$

Since $d\varrho(x)=0$ for all $x\in\bar{B}_R$ with $\varrho(x)=0$, it is clear that there is a constant $C_3<\infty$ such that, for all $r>0$,

$$\int_{\substack{x\in B_R \\ |\varrho(x)|<r}} |\varrho(x)|^{-\alpha} dx_1\wedge\ldots\wedge dx_m \le C_3 r^{1-\alpha} \le C_3 r^{\beta}.$$

Hence, for all $0<r<\delta$ and $y\in\bar{B}_R$, one obtains

$$\int_{\substack{x\in B_R\setminus F_\delta(y) \\ |\varrho(x)|<r}} \frac{|\varrho(x)|^{-\alpha} dx_1\wedge\ldots\wedge dx_m}{(|\varrho(x)|+|h(y,x)|+|x-y|^2)^2|x-y|^{m-4}} \le \frac{C_3}{\delta^m} r^{\beta}.$$

Together with (15) this implies (6). □

APPENDIX B. ESTIMATION OF SOME INTEGRALS (THE NON-SMOOTH CASE)

Here we generalize the estimates from Appendix A to the case of a hypersurface defined by a function which may have non-degenerate critical points. We use again the notations introduced in Sects. 0.1–0.4 of the basic text of the present monograph.

First we recall some facts about non-degenerate critical points (for more information about such points, see, for instance, [Milnor 1963, 1965a, 1965b, Hirsch 1976, Wallace 1968]).

0. The notion of a non-degenerate critical point. If $A: X \longrightarrow Y$ is a C^1 map between two C^1 manifolds X and Y, then $z \in X$ is called a *critical point* of A if the rank of the differential of A at z is not maximal.

Now let $f: X \longrightarrow \mathbb{R}^1$ be a C^2 function on a C^2 manifold X, and let $z \in X$ be a critical point of f, i.e. $df(z)=0$. Then z is called a *non-degenerate* critical point of f if, for any system of C^2 coordinates $x_1, \ldots, x_n$ in a neighborhood of z,

$$\det \left[\frac{\partial^2 f(z)}{\partial x_i \partial x_j} \right]_{i,j=1}^{n} = 0. \tag{1}$$

Otherwise, z is called a *degenerate* critical point of f.

Remark. A direct computation shows that, under the hypothethis $df(z)=0$, condition (1) is independent of the choice of the coordinates $x_1, \ldots, x_n$.

0.1. Proposition. Non-degenerate critical points are isolated, i.e. if $f: X \longrightarrow \mathbb{R}^1$ is a C^2 function on a C^2 manifold X and $z \in X$ is a non--degenerate critical point of f, then there exists a neighborhood U of z such that $df(x) \neq 0$ for all $x \in U \setminus \{z\}$.

Proof. Condition (1) means that the differential of the map

$$X \ni p \longrightarrow \left(\frac{\partial f}{\partial x_1}(p), \ldots, \frac{\partial f}{\partial x_n}(p) \right) \in \mathbb{R}^n$$

is of maximal rank at z. Hence, by the implicit function theorem, this map is 1-1 in some neigborhood U of z, and, in particular, z is the only point in U which is mapped onto zero. □

From Proposition 0.1 one obtains immediately the following

<u>0.2. Corollary.</u> Let $f:X \longrightarrow \mathbb{R}$ be a C^2 function without degenerate critical points on a C^2 manifold X, and $K \subset\subset X$ a compact set. If $g:X \longrightarrow \mathbb{R}$ is a C^2 function such that the first- and second-order derivatives of g are sufficiently small, uniformly on K, then f+g does not have degenerate critical points in K. □

The following important lemma is due to M. Morse (cf. Lemma A in §2 of [Milnor 1965a]):

<u>0.3. Lemma.</u> If $D \subseteq \mathbb{R}^k$ is a domain and $f:D \longrightarrow \mathbb{R}$ is a C^2 function, then, for almost all real-linear maps $L: \mathbb{R}^k \longrightarrow \mathbb{R}$, the function f+L does not have degenerate critical points in D. (Here "almost all" means that the Lebesgue measure of the set of all real-linear maps $L: \mathbb{R}^k \longrightarrow \mathbb{R}$ such that f+L <u>has</u> degenerate critical points is zero.)

<u>Proof.</u> Let $A:D \longrightarrow \mathbb{R}^k$ be the C^1 map defined by

$$A(x) = -\left(\frac{\partial f(x)}{\partial x_1},\ldots,\frac{\partial f(x)}{\partial x_k}\right).$$

Denote by Crit(A) the set of all critical points of A in D, and by L_y, $y \in \mathbb{R}^k$, the linear map $L_y: \mathbb{R}^k \longrightarrow \mathbb{R}$ defined by $L_y(x) = x_1y_1 + \ldots + x_ny_n$.

Since , by Sard's theorem (cf., for instance, Proposition 1.4.7 in [Narasimhan 1968]), the Lebesgue measure of A(Crit(A)) is zero, it is sufficient to prove that, for all y $\mathbb{R}^k \setminus A(\mathrm{Crit}(A))$, $f+L_y$ has only non--degenerate critical points. To do this, we consider a point y $\mathbb{R}^k$ such that $f+L_y$ has at least one degenerate critical point in D. If x is such a point, then $y = A(x)$ and $dA(x)=0$, i.e. y A(Crit(A)). □

We need also the following generalization of this lemma:

<u>0.4. Proposition.</u> Let $X \subseteq \mathbb{R}^n$ be a C^2 submanifold of $\mathbb{R}^n$ and $f:X \longrightarrow \mathbb{R}$ a C^2 function. Then, for almost all real-linear maps $L: \mathbb{R}^n \longrightarrow \mathbb{R}$, the function $f+L|X$ does not have degenerate critical points in X.

<u>Proof.</u> Fix $z \in X$ and choose linear coordinates $x_1,\ldots,x_n$ in $\mathbb{R}^n$ such that, for some neighborhood U of z, $x_1,\ldots,x_k$ are C^2 coordinates in $U \cap X$. It is sufficient to prove the following

<u>Statement.</u> For almost all linear maps $L: \mathbb{R}^n \longrightarrow \mathbb{R}$, $(f+L)|U \cap X$ does not have degenerate critical points in $U \cap X$.

<u>Proof of the statement.</u> Let $x_1^*,\ldots,x_n^*: \mathbb{R}^n \longrightarrow \mathbb{R}^1$ be the linear maps with $x_i^*(x)=x_i$, E' the space of all linear maps on $\mathbb{R}^n$ spanned by $x_1^*,\ldots,x_k^*$, and E" the space of all linear maps on $\mathbb{R}^n$ spanned by

$x^*_{k+1},\dots,x^*_n$. Since $h:=(x_1,\dots,x_k)$ is a C^2 diffeomorphism from on $U\cap X$ onto $D:=h(U\cap X)\subseteq \mathbb{R}^k$, then for any fixed $L''\in E''$, it follows from Lemma 0.3 applied to $(f+L'')\circ h^{-1}$ that, for almost all linear maps $\tilde{L}: \mathbb{R}^k \longrightarrow \mathbb{R}$, the function $(f+L'')+\tilde{L}\circ h$ does not have degenerate critical points in $U\cap X$. Since E' coincides with the space of all maps of the form $\tilde{L}\circ h$, so the following statement is obtained: If, for $L''\in E''$, $M(L'')$ denotes the set of all $L'\in E'$ such that $(f+L''+L')|U\cap X$ has at least one degenerate critical point in $U\cap X$, then the Lebesgue measure of $M(L'')$ is zero for all $L''\in E''$. In view of Fubini's theorem, this completes the proof. ☐

0.5. Proposition. Let X be a C^∞ manifold which is countable at infinity, and $f:X \longrightarrow \mathbb{R}$ a C^∞ function. Then there exists a sequence of C^∞ functions $f_j:X\longrightarrow \mathbb{R}$ $(j=1,2,\dots)$ without degenerate critical points which converges to f together with the first- and second-order derivatives, uniformly on the compact subsets of X.

Moreover, if there is a compact set $K \subset\subset X$ such that f does not have degenerate critical points on K, then this sequence f_j can be chosen so that $f_j = f$ on K.

Proof. Since X is countable at infinity, we may assume that X is a C^∞ submanifold of some $\mathbb{R}^n$. By Proposition 0.1 we can find a neighborhood $U \subset\subset X$ of K such that f does not have degenerate critical points in U. Choose a C^∞ function χ on X such that $\chi = 0$ on K and $\chi = 1$ in some neighboorhood of $X\setminus U$. By Popbsition 0.4 we can find a sequence of linear maps $L_i: \mathbb{R}^n\longrightarrow \mathbb{R}$ which tends to zero and such that all $f+L_i|X$ do not have degenerate critical points in X. By Proposition 0.2, for each $j\geq 1$, we can choose an index i_j so large that the function

$$f_j := f + \chi L_{i_j}\big|_X$$

does not have degenerate critical points in X too. ☐

0.6. Proposition. Let f be a real-valued C^2 function in a neighborhood U of the origin in $\mathbb{R}^n$ such that 0 is a non-degenerate critical point of f which is neither the point of a local minimum nor the point of a local maximum of f. Then there exists $\varepsilon>0$ such that

$$df(x)\wedge d|x| \neq 0 \qquad \text{for all } x \text{ with } 0<|x|<\varepsilon \text{ and } f(x)=f(0).$$

Proof. Let q be the number of positive eigenvalues of the matrix of second-order derivatives of f at 0. Since 0 is a non-degenerate critical point of f, then this matrix has $n-q$ negative eigenvalues and, after an appropriate change of the linear coordinates in $\mathbb{R}^n$, Taylor's formula gives

$$f = f(0) + \sum_{j=1}^{q} x_j^2 - \sum_{j=q+1}^{n} x_j^2 + o(|x|^2) \qquad \text{for } x \longrightarrow 0. \qquad (2)$$

Hence

$$df = 2\sum_{j=1}^{q} x_j dx_j - 2\sum_{j=q+1}^{n} x_j dx_j + o(|x|) \qquad \text{for } x \longrightarrow 0, \qquad (3)$$

and so

$$df(x) \wedge d|x| = 8\sum_{j=1}^{q} \sum_{k=q+1}^{n} \frac{x_j x_k}{|x|} dx_j \wedge dx_k + o(|x|) \qquad \text{for } x \longrightarrow 0. \qquad (4)$$

Since the origin is neither the point of a local minimum nor the point of a local maximum of f and hence $1 \leq q \leq n-1$, this completes the proof. □

It is more difficult to prove the following result:

0.7. Proposition. Let X be an n-dimensional real C^2 manifold, $f: X \longrightarrow \mathbb{R}$ a C^2 function, $z \in X$ a non-degenerate critical point of f, and q the number of positive eigenvalues of the matrix of second-order derivatives of f at z. Then there exist C^1 coordinates $x=(x_1, \ldots, x_n): U \longrightarrow \mathbb{R}^n$ in some neighborhood U of z such that $x(z)=0$ and

$$f = \sum_{j=1}^{q} x_j^2 - \sum_{j=q+1}^{n} x_j^2 \qquad \text{on } U.$$

References for the proof. For the case of a C^∞ function, this result is due to M. Morse and the proof can be found, for instance, in [Milnor 1963]. For the general case, see Lemma 1.1 and Exercise **1 in Section 6.1 of [Hirsch 1976]. □

Remark. All estimates given below can be proved also without Proposition 0.7 (using only relation (2)). However, we shall use Proposition 0.7, in view of the technical simplifications obtained in this way.

1. Notations. Further on, in this appendix, we use the following notations:

$m \geq 2$ is an integer.

R is a positive integer and B_R is the open ball with radius R centered at the origin in $\mathbb{R}^m$.

ϱ is a real-valued C^2 function in some neighborhood of $\bar{B}_R$ such that the following conditions are fulfilled:

(i) $d\varrho(x) \neq 0$ for all $x \in \bar{B}_R$ with $x \neq 0$;

(ii) $\varrho(0)=0$, $d\varrho(0)=0$ and the Hessian matrix

$$\left[\frac{\partial^2 \varrho(0)}{\partial x_j \partial x_k} \right]_{j,k=1}^{m}$$

has q positive and m-q negative eigenvalues where $1 \le q \le m-1$.

$D := \{x \in B_R : \varrho(x) < 0\}$.

$S_b := \{x \in B_R : \varrho(x) = b\}$ for any real number $b \neq 0$.

$S := S_0 := \{x \in B_R \setminus \{0\} : \varrho(x) = 0\}$.

dS_b is the Euclidean volume form of S_b, i.e. the form of degree m on S_b which is defined by the following condition:

> If v is a differential form of degree m-1 in a neighbor-borhood of S_b such that $d\varrho(x) \wedge v(x) = d\varrho(x) \wedge dx_1 \wedge \ldots \wedge dx_m$ for all $x \in S_b$, then $v|S_b = dS_b$. (5)

We orient the surfaces S_b by dS_b. □

In this appendix we prove the following four propositions.

2. Proposition. (i) There exists a constant $C < \infty$ such that, for all $b \in \mathbb{R}$, $y \in B_R$ and $\varepsilon > 0$,

$$\int_{x \in S_b} \frac{dS_b}{(\varepsilon + |x-y|^2)\,|x-y|^{m-3}} \le C(1 + |\ln \varepsilon|) \tag{6}$$

and

$$\int_{x \in S_b} \frac{\|d\varrho(x)\|\, dS_b}{(\varepsilon + |x-y|^2)\,|x-y|^{m-2}} \le C[1 + |\ln \varepsilon| + \|d\varrho(y)\|\, \varepsilon^{-1/2}]. \tag{7}$$

(ii) Suppose h(x,y) is a C^1 function defined for x,y in a neighborhood of $\bar{B}_R$ such that, for some constant $K < \infty$, the following estimates are fulfilled:

$$\frac{1}{K} \| d\varrho(y) \| \le \left\| d_x h(y,x) \Big|_{x=y} \right\| \le K \| d\varrho(y) \| \qquad (y \in \bar{B}_R), \tag{8}$$

$$\frac{1}{K} \| d\varrho(y) \|^2 \le \left| d\varrho(y) \wedge d_x h(y,x) \Big|_{x=y} \right| \qquad (y \in \bar{B}_R), \tag{9}$$

$$\left| \left.\frac{\partial h(y,x)}{\partial x_j}\right|_{x=y} - \frac{\partial h(y,x)}{\partial x_j} \right| \leq K|x-y| \qquad (x,y\in\bar{B}_R,\ 1\leq j\leq m), \tag{10}$$

$$|h(y,x)| \leq K(|x||x-y| + |x-y|^2) \qquad (x,y\in\bar{B}_R). \tag{11}$$

Then there exists $C<\infty$ such that, for all $b\in\mathbb{R}$, $y\in B_R$, and $\varepsilon>0$,

$$\int_{x\in S_b} \frac{\|d\varrho\|^2 dS_b}{(\varepsilon+|h(y,x)|+|x-y|^2)^2|x-y|^{m-3}} \leq C[1+|\ln\varepsilon|+\|d\varrho(y)\|\varepsilon^{-1/2}] \tag{12}$$

3. Proposition. (i) For each $\alpha>0$, there exists a constant $C_\alpha<\infty$ such that, for all $b\in\mathbb{R}$, $0<r<R$ and $y,z\in B_R$,

$$\int_{\substack{x\in S_b\\ |x-z|<r}} \frac{dS_b}{|x-y|^{m-1-\alpha}} \leq C_\alpha r^\alpha. \tag{13}$$

(ii) If $h(y,x)$ is as in Proposition 2 (i), then there exists a constant $C<\infty$ such that, for all $0<r<R$ and $y\in D$,

$$\int_{\substack{x\in S\\ |x|<r}} \frac{\|d\varrho\| dS}{(|h(y,x)|+|x-y|^2)|x-y|^{m-3}} \leq Cr. \tag{14}$$

4. Proposition. There exists a constant $C<\infty$ such that: If f is a C^1 function on D with the property that

$$\|df(y)\| \leq \big|\ln|\varrho(y)|\big| + \frac{\|d\varrho(y)\|}{|\varrho(y)|^{1/2}} \qquad (y\in D) \tag{15}$$

and

$$f(0) := \lim_{D\ni y\to 0} f(y) \tag{16}$$

exists (for $2\leq q\leq m-2$ the existence of this limes follows from (15)), then, for all $x,y\in D$,

$$|f(x)-f(y)| \leq C(1+|\ln r|)r^{1/2}|x-y|^{1/2}, \qquad \text{where } r=\max(|x|,|y|). \tag{17}$$

5. Proposition. (i) For all $\alpha,\beta\geq 0$ with $\alpha+\beta<1$, there exists a constant $C<\infty$ such that, for all $0<r<\infty$ and $y\in B_R$,

$$\int\limits_{\substack{x\in B_R\\ |x-y|<r}} \frac{dx_1\wedge\ldots\wedge dx_m}{[\mathrm{dist}(x,S)]^{\alpha}\,|x-y|^{m-1}} \leq Cr^{\beta}, \tag{18}$$

where $\mathrm{dist}(x,S) = \inf\{|x-s|:\ s\in S\}$.

(ii) Let $h(y,x)$ be as in Proposition 2 (ii), and let $d>0$ and $\alpha,\beta \geq 0$ be numbers with

$$\alpha + \beta < \frac{d}{2}.$$

Then there exists a constant $C<\infty$ such that, for all $y\in\bar{B}_R$ and $r>0$,

$$\int\limits_{\substack{x\in B_R\\ |\varrho(x)|<r}} \frac{\|d\varrho(x)\|\,|\varrho(x)|^{-\alpha}dx_1\wedge\ldots\wedge dx_m}{(|\varrho(x)|+|x-y|^2)^2\,|x-y|^{m-3-d}} \leq Cr^{\beta}. \tag{19}$$

(iii) Let $h(y,x)$ be as in Proposition 2 (ii), and let $d>0$ and $\alpha,\beta \geq 0$ be numbers with

$$\alpha + \beta < \frac{d+1}{2}.$$

Then there exists a constant $C<\infty$ such that, for all $y\in\bar{B}_R$ and $r>0$,

$$\int\limits_{\substack{x\in B_R\\ |\varrho(x)|<r}} \frac{\|d\varrho(x)\|^2\,|\varrho(x)|^{-\alpha}\,dx_1\wedge\ldots\wedge dx_m}{(|\varrho(x)|+|h(y,x)|+|x-y|^2)^2\,|x-y|^{m-3-d}} \leq Cr^{\beta}. \tag{20}$$

6. Preparations for the proofs of Propositions 2 - 5. By Proposition 0.7 and by Propositions 1 and 2 in Appendix A, in the proofs of Propositions 2 - 5, we may assume that

$$\varrho = \sum_{j=1}^{q} x_j^2 - \sum_{j=q+1}^{n} x_j^2, \tag{21}$$

and therefore

$$d\varrho = 2\sum_{j=1}^{q} x_j dx_j - 2\sum_{j=q+1}^{n} x_j dx_j. \tag{22}$$

For $j=1,\ldots,m$, we set

$$U_j^+ = \{x\in B_R:\ x_j > \tfrac{|x|}{m}\} \qquad \text{and} \qquad U_j^- = \{x\in B_R:\ x_j < -\tfrac{|x|}{m}\}.$$

Then, by (22), for all $1\le j\le m$ and $b\in\mathbb{R}$, the functions $x_1,\ldots,\hat{x}_j,\ldots,x_m$ form a system of coordinates on $U_j^+\cap S_b$ as well as on $U_j^-\cap S_b$. From (5) and (22) it follows that

$$dS_b = \left|\ \frac{|x|}{x_j} dx_1\wedge\ldots\hat{j}\ldots\wedge dx_m\Big|_{S_b}\ \right| \qquad \text{on } (U_j^+\cup U_j^-)\cap S_b \text{ for all } b\in\mathbb{R}. \tag{23}$$

Further, notice that

$$S_b \subseteq U_1^+\cup\ldots\cup U_m^+\cup U_1^-\cup\ldots\cup U_m^- \qquad \text{for all } b\in\mathbb{R}. \tag{24}$$

<u>7. Proof of Proposition 2 (i).</u> Since, by (22),

$$\|d\varrho(x)\| \le \|d\varrho(y)\| + 2|x-y|,$$

estimate (7) is a consequence of (6) and the following estimate:

$$\int_{x\in S_b} \frac{dS_b}{(\varepsilon+|x-y|^2)\,|x-y|^{m-2}} \le C\varepsilon^{-1/2} \tag{25}$$

In view of (23) and (24), the integrals in (6) and (25) are bounded by

$$\sum_{j=1}^{m}\ \int_{(U_j^+\cup U_j^-)\cap S_b} \frac{|x|}{|x_j|}\ \frac{\left|dx_1\wedge\ldots\hat{j}\ldots\wedge dx_m\right|}{(\varepsilon+|x-y|^2)\,|x-y|^{m-l}}$$

with $l=3$ resp. $l=2$. Since $x_1,\ldots,\hat{x}_j,\ldots,x_m$ are coordinates on $U_j^+\cap S_b$ as well as on $U_j^-\cap S_b$ and since $|x|/|x_j|\le m$ on $U_j^+\cup U_j^-$, so it follows that the integrals in (6) and (25) are bounded by

$$2m^2 \int\limits_{\substack{t\in \mathbb{R}^{m-1} \\ |t|<2R}} \frac{dt_1 \dots dt_{m-1}}{(\varepsilon+|t|^2)|t|^{m-1}}$$

with l=3 resp. l=2. Therefore (6) and (25) follow from estimate (8) in Appendix A. □

8. Proof of Proposition 3 (i). By (23) and (24), the integral in (13) is bounded by

$$\sum_{j=1}^{m} \int\limits_{\substack{(U_j^+\cup U_j^-)\cap S_b \\ |x-z|<r}} \frac{|x|}{|x_j|} \frac{\left| dx_1\wedge\dots\wedge\widehat{\ }_j\dots\wedge dx_m \right|}{|x-y|^{m-1-\alpha}}.$$

Since $x_1,\dots,\widehat{\ }_j,\dots,x_m$ are coordinates on $U_j^+\cap S_b$ as well as on $U_j^-\cap S_b$ and since $|x|/|x_j|\leq m$ on $U_j^+\cup U_j^-$, this implies that the integral in (13) can be estimated by

$$2m^2 \int\limits_{\substack{t\in \mathbb{R}^{m-1} \\ |t|<r}} \frac{dt_1\wedge\dots\wedge dt_{m-1}}{|t|^{m-1-\alpha}} = 2m^2 s_{m-2} r^{\alpha},$$

where s_{m-2} is the Euclidean volume of the (m-2)-dimensional unit sphere. □

9. Proof of Proposition 4. For $x\in D$ and $0\leq t\leq 1$, we set

$$v(x,t) = \big((1-t)x_1,\dots,(1-t)x_q,(1+t)x_{q+1},\dots,(1+t)x_m\big).$$

Then

$$\varrho(v(x,t)) = (1+t^2)\varrho(x) - 2t|x|^2$$

and hence

$$\varrho(v(x,t)) \leq -2t|x|^2 \qquad (x\in D,\ 0\leq t\leq 1). \tag{26}$$

Further,

$$|v(x,t)| \leq 2|x| \qquad (x\in D,\ 0\leq t\leq 1) \qquad (27)$$

and

$$|v(x,t)-x| \leq t|x| \qquad (x\in D,\ 0\leq t\leq 1). \qquad (28)$$

By (26) and (27),

$$\ln|\varrho(v(x,t))| \leq K(1+|\ln|x||+|\ln t|)$$

for some constant $K<\infty$. Since $\|d\varrho(y)\| \leq 2|y|$, this implies

$$\frac{\|d\varrho(v(x,t))\|}{|\varrho(v(x,t))|^{1/2}} \leq Kt^{-1/2}.$$

Therefore it follows from (15) that, for some constant $C_1<\infty$,

$$\| df(v(x,t))\| \leq C_1(1+|\ln|x||+t^{-1/2}) \qquad (x\in D,\ 0\leq t\leq 1). \qquad (29)$$

Since $|\partial v(x,t)/\partial t|\leq|x|$, this implies that

$$|f(v(x,t))-f(x)| = \left|\int_0^t \frac{d}{ds}f(v(x,s))ds\right| \leq 4C_1|x|\left(1+|\ln|x||\right)t^{1/2} \qquad (30)$$

for all $x\in D$ and $0\leq t\leq 1$. From (27) and (29) it follows that

$$|f(v(x,1))-f(0)| = \left|\int_0^1 \frac{d}{dt}f(tv(x,1))dt\right| \leq 4C_1|x|(1+|\ln|x||) \qquad (31)$$

for all $x\in D$.

Now let $x,y\in D$ be fixed, and $r=\max(|x|,|y|)$. In order to prove (17), we distinguish the cases $|x|\leq 4|x-y|$ and $|x|\geq 4|x-y|$.

Let $|x|\leq 4|x-y|$. Then $|y|\leq 5|x-y|$ and it follows from (30) and (31) that

$$\begin{aligned}|f(x)-f(y)| &\leq |f(x)-f(v(x,1))| + |f(v(x,1))-f(0)| \\ &\quad + |f(0)-f(v(y,1))| + |f(v(y,1))-f(y)| \\ &\leq 20\ C_1 \left(|x|^{1/2}(1+|\ln|x||) + |y|^{1/2}(1+|\ln|y||)]|x-y|^{1/2}\right).\end{aligned}$$

This implies (17), because $|x|,|y|\le r$ and the function $t^{1/2}(1+|\ln t|)$ is increasing for $0<t<e^{-1/2}$.

Now let $|x|>4|x-y|$. Then we can find a C^1 map $w:[0,1]\longrightarrow D$ such that $w(0)=x$, $w(1)=y$, $|w(t)|\le r$ for all $0\le t\le 1$, and, for some constant $C_2<\infty$,

$$\frac{|x|}{C_2} \le |w(t)| \le C_2|x| \qquad (0\le t\le 1) \qquad (32)$$

and

$$\left|\frac{dw(t)}{dt}\right| \le C_2|x-y| \qquad (0\le t\le 1). \qquad (33)$$

We define

$$g(t) = v\left(w(t), \frac{|x-y|}{|x|}\right) \qquad (0\le t\le 1).$$

Then it follows from (26) and (32) that

$$\varrho(g(t)) \le - 2C_2^2|x-y||x| \qquad (0\le t\le 1). \qquad (34)$$

Since $d\varrho(y) \le 2|y|$, (27) and (32) imply that $d\varrho(g(t)) \le 4C_2|x|$. Hence

$$\frac{\| d\varrho(g(t))\|}{|\varrho(g(t))|^{1/2}} \le 4C_2^2|x|^{1/2}|x-y|^{-1/2},$$

and it follows from (15) and (34) that, for some constant $C_3<\infty$,

$$\|df(g(t))\| \le C_3(1 + |\ln|x-y||x|| + |x|^{1/2}|x-y|^{-1/2}).$$

Since, by (33), $|dg(t)/dt| \le 2C_2|x-y|$ and since $|x|,|x-y|\le r$, the last inequality implies that, for some constant $C_4<\infty$,

$$|f(g(1)) - f(g(0))| = \left|\int_0^1 \frac{d}{dt}f(g(t))dt\right| \le C_4(1+|\ln r|)r^{1/2}|x-y|^{1/2}.$$

By (30), we have

$$|f(x) - f(g(0))| \le 4C_1(1+|\ln|x||)|x|^{1/2}|x-y|^{1/2}$$

and (since $|y| \leq 2|x|$)

$$|f(g(1)) - f(y)| \leq 8C_1(1+|\ln|y||)|x|^{1/2}|x-y|^{1/2}.$$

Taking into account that $|x|, |y|, |x-y| \leq r$, we see that the last three estimates imply (17). □

<u>10. Further preparations for the proofs of Propositions 2 (ii), 3 (ii), and 5.</u> Set

$$\lambda = \frac{1}{8K^2m^2} \qquad \text{and} \qquad \delta = \lambda^5,$$

where K is as in (8) - (11). We may assume that $\delta < 1$.
For each $y \in B_R$, we define

$$F(y) = \{x \in B_R : |x-y| < \lambda|y|\} \qquad \text{and} \qquad E(y) = \{x \in B_R : |x-y| < \delta|y|\}.$$

<u>10.1. Lemma.</u> For any $y \in B_R \setminus \{0\}$, we can find indices $1 \leq k < l \leq m$ such that the following two assertions hold true:

(i) For s=k, l, the functions $\varrho, x_1, \ldots, \hat{s}, \ldots, x_m$ form a system of C^1 coordinates on F(y), and

$$dx_1 \wedge \ldots \wedge dx_m \leq \frac{1}{\lambda|y|}\left|d\varrho \wedge dx_1 \wedge \ldots \hat{s} \ldots \wedge dx_m\right| \qquad \text{for } x \in F(y). \tag{35}$$

(ii) The functions

$$\varrho, h(y,\cdot), x_1, \ldots \widehat{k,l} \ldots, x_m$$

form a system of C^1coordinates on E(y), and hence, for all $b \in \mathbb{R}$, the functions

$$h(y,\cdot), x_1, \ldots \widehat{k,l} \ldots, x_m$$

form a system of C^1 coordinates on $S_b \cap E(y)$. Moreover,

$$dx_1 \wedge \ldots \wedge dx_m \leq \frac{1}{\delta|y|^2}\left|d\varrho \wedge d_x h(y,x) \wedge dx_1 \wedge \ldots \widehat{k,l} \ldots \wedge dx_m\right| \qquad \text{on } E(y) \tag{36}$$

and, for all $b \in \mathbb{R}$,

$$dS_b \le \frac{1}{\delta|y|}|d_x h(y,x)\wedge dx_1\wedge\ldots\widehat{k,l}\ldots\wedge dx_m| \qquad \text{on } S_b\ E(y). \tag{37}$$

<u>Proof.</u> Fix $y\epsilon B_R\setminus\{0\}$ and set

$$v_{ij}(x) = 2x_i\frac{\partial h(y,x)}{\partial x_j} - 2x_j\frac{\partial h(y,x)}{\partial x_i}.$$

Since $\|d\varrho(y)\|=2|y|$ and

$$d\varrho(y)\wedge d_x h(y,x)\Big|_{x=y} = \sum_{1\le i<j\le m} v_{ij}(y)dx_i(y)\wedge dx_j(y),$$

it follows from (9) that, for some $1\le k<l\le m$,

$$|v_{kl}(y)| \ge \frac{4}{Km^2}|y|^2. \tag{38}$$

Choose $s\epsilon\{k,l\}$ such that

$$|y_s| \ge \frac{1}{4}|v_{kl}(y)|\left|\frac{\partial h(y,x)}{\partial x_s}\Big|_{x=y}\right|^{-1}.$$

Then, by (38) and (8), and taking again into account that $\|d\varrho(y)\|$ $= 2|y|$, we obtain that

$$|y_s| \ge \frac{|y|}{2K^2m^2} = 4\lambda|y|$$

and hence

$$|x_s| \ge 3\lambda|y| \neq 0 \qquad \text{if } x\epsilon F(y). \tag{39}$$

This implies (35). In particular, $d\varrho\wedge dx_1\wedge\ldots\underset{s}{\wedge}\ldots\wedge dx_m \neq 0$ on $F(y)$, i.e. ($F(y)$ is convex!) the functions $\varrho, x_1,\ldots\underset{s}{\wedge}\ldots x_m$ form a system of C^1 coordinates on $F(y)$.

In view of (8) and (10), it follows from (38) that

$$|v_{k,l}(x)| \ge \frac{3}{Km^2}|y|^2 \qquad \text{if } x\epsilon F(y). \tag{40}$$

Since $d\varrho \wedge d_x h(y,x) \wedge dx_1 \wedge \ldots \widehat{k,l} \ldots \wedge dx_m = v_{k,l} dx_1 \wedge \ldots \wedge dx_m$ on $F(y)$, this implies (36). In particular,

$$d\varrho \wedge d_x h(y,x) \wedge dx_1 \wedge \ldots \widehat{k,l} \ldots \wedge dx_m \neq 0 \qquad \text{on } F(y). \tag{40a}$$

Set

$$\Theta = \left\{ x \in B_R : \frac{|\varrho(x)-\varrho(y)|^2}{\lambda^2 |y|^2} + \sum_{i \neq s} |x_i - y_i|^2 < \lambda^6 |y|^2 \right\}.$$

Since $\| d\varrho \| < 3|y|$ on $F(y)$ and by (35), then it is easy to show that

$$E(y) \subset\subset \Theta \subset\subset F(y).$$

In view of (40a) and since Θ is convex with respect to the coordinates $\varrho, x_1, \ldots \widehat{s} \ldots, x_m$, it follows that the functions $\varrho, h(y,\cdot), x_1, \ldots \widehat{k,l} \ldots, x_m$ form a system of C^1 coordinates on $E(y)$.

It remains to prove (37). By (5), we have

$$dS_b = \frac{\| d\varrho \|}{|v_{k,l}|} \left| d_x h(y,x) \wedge dx_1 \wedge \ldots \widehat{k,l} \ldots \wedge dx_m |_{S_b} \right| \qquad \text{on } S_b \cap E(y).$$

Taking into account that $\| d\varrho \| \leq 3|y|$ on $E(y)$, together with (40) this implies (37). □

<u>11. Proof of Proposition 2 (ii).</u> Since

$$\| d\varrho(x) \|^2 = 4|x|^4 \leq 4|y|^2 + 8|y||x-y| + 4|x-y|^2$$

and since (6) and (25) are already proved, it remains to find a constant $C < \infty$ such that

$$\int_{x \in S_b} \frac{|y| dS_b}{(\varepsilon + |h(y,x)| + |x-y|^2)^2 |x-y|^{m-3}} \leq C \varepsilon^{-1/2}$$

for all $b \in \mathbb{R}$, $\varepsilon > 0$ and $y \in D$. Taking into account again (25) (and the definition of $E(y)$), we see that there is a constant $C < \infty$ such that

$$\int_{x\in S_b\setminus E(y)} \frac{|y|dS_b}{(\varepsilon + |h(y,x)| + |x-y|^2)^2|x-y|^{m-3}}$$

$$\leq \frac{1}{\delta}\int_{x\in S_b} \frac{dS_b}{(\varepsilon + |x-y|^2)^2|x-y|^{m-2}} \leq C\varepsilon^{-1/2}$$

for all $b\in \mathbb{R}$, $\varepsilon>0$ and $y\in D$. Therefore it is sufficient to find a constant $C<\infty$ with

$$\int_{x\in S_b\cap E(y)} \frac{|y|dS_b}{(\varepsilon + |h(y,x)| + |x-y|^2)^2|x-y|^{m-3}} \leq C\varepsilon^{-1/2}. \tag{41}$$

Since h is a C^1 function in a neighborhood of $\bar{B}_R\times\bar{B}_R$ and $h(z,z)=0$ for all $z\in\bar{B}_R$, there is a constant $c>0$ such that

$$|z-v| \geq c(|z-v| + |h(z,v)|) \tag{42}$$

for all $z,v\in\bar{B}_R$. Therefore and by Lemma 10.1, the left hand side of (41) is bounded by

$$\frac{1}{\delta c^{m+1}}\int_{\substack{t\in \mathbb{R}^{m-1}\\ |t|<b}} \frac{dt_1\wedge\ldots\wedge dt_{m-1}}{(\varepsilon + |t_1| + |t|^2)^2|t|^{m-3}},$$

where

$$b:= \max_{z,v\in\bar{B}_R} [|h(z,v)|^2 + |z-v|^2]^{1/2}.$$

Now (42) follows from estimate (10) in Appendix A.

12. Proof of Proposition 3 (ii). Since $d\varrho(x) = 2|x| \leq 2|y| + 2|x-y|$ and since (13) is already proved, it is sufficient to find a constant $C<\infty$ with

$$\int_{\substack{x\in S\\ |x|<r}} \frac{|y|dS}{(|h(y,x)|+|x-y|^2)|x-y|^{m-3}} \leq Cr.$$

Notice that, for all $0<r<R$ and $y\in D$, by the definition of $E(y)$,

$$\int_{\substack{x\in S\setminus E(y)\\ |x|<r}} \frac{|y|dS}{(|h(y,x)|+|x-y|^2)|x-y|^{m-3}} \leq \frac{1}{\delta} \int_{\substack{x\in S\\ |x|<r}} \frac{dS}{|x-y|^{m-2}}.$$

By (13) it is therefore sufficient to find a constant $C<\infty$ such that, for all $0<r<R$ and $y\in D$,

$$\int_{\substack{x\in S\cap E(y)\\ |x|<r}} \frac{|y|dS}{(|h(y,x)|+|x-y|^2)|x-y|^{m-3}} \leq Cr. \tag{43}$$

If c is as in (42), then it follows from Lemma 10.1 that the left hand side of (43) is bounded by

$$\frac{1}{\delta c^{m-1}} \int_{\substack{t\in \mathbb{R}^{m-1}\\ |t_1|<k(r)}} \frac{dt_1\wedge\ldots\wedge dt_{m-1}}{(|t_1|+|t|^2)|t|^{m-3}},$$

where

$$k(r):=\sup_{x\in E(y),\ |x|<r} |h(y,x)|.$$

Since $|x-y|<\delta|y|<2\delta r$ if $x\in E(y)$ and $|x|<r$, it follows from (11) and the definition of δ that $k(r)<r^2$. Therefore (43) follows from estimate (11) in Appendix A. □

<u>13. Proof of Proposition 5 (i).</u> Clearly,

$$\int_{\substack{x\in B_R\\ |x-y|<r}} \frac{dx_1\wedge\ldots\wedge dx_m}{[\mathrm{dist}(x,S)]^{\alpha}|x-y|^{m-1}}$$

$$\leq r^{\beta} \int_{\substack{x\in B_R\\ |x-y|<r}} \frac{dx_1\wedge\ldots\wedge dx_m}{[\mathrm{dist}(x,S)]^{\alpha}|x-y|^{m-1+\beta}}.$$

Therefore, it is sufficient to prove that

$$\sup_{y\in B_R} \int_{x\in B_R} \frac{{}_1\wedge\ldots\wedge dx_m}{[\mathrm{dist}(x,S)]^{\alpha}\,|x-y|^{m-1+\beta}} < \infty .$$

To do this, we fix some $\varepsilon>0$ such that $\alpha+\beta+\varepsilon<1$. Since

$$\sup_{y\in B_R} \int_{\substack{x\in B_R\\ \mathrm{dist}(x,S)\geq|x-y|}} \frac{dx_1\wedge\ldots\wedge dx_m}{[\mathrm{dist}(x,S)]^{\alpha}\,|x-y|^{m-1+\beta}}$$

$$\leq \sup_{y\in B_R} \int_{x\in B_R} \frac{dx_1\wedge\ldots\wedge dx_m}{|x-y|^{m-1+\alpha+\beta}} < \infty ,$$

then it is enough to prove that

$$\sup_{y\in B_R} \int_{x\in B_R} \frac{dx_1\wedge\ldots\wedge dx_m}{[\mathrm{dist}(x,S)]^{\alpha+\beta+\varepsilon}\,|x-y|^{m-1-\varepsilon}} < \infty .$$

Let U_j^+ and U_j^- be the sets introduced in point 6 of this appendix. Fix $1\leq j\leq m$, $\sigma\in\{+,-\}$, and set

$$W_k = \{x\in U^{\sigma}:\ 2^{-k-1}\leq|x|\leq 2^{-k}\}$$

for $k=0,1,\ldots$. Then, by (22) and the definition of U_j^{σ},

$$dx_1\wedge\ldots\wedge dx_m \leq m2^k\left|d\varrho\wedge dx_1\wedge\ldots\wedge\widehat{\ }_j\ldots\wedge dx_m\right| \qquad \text{on } W_k .$$

Since $\varrho=0$ on S, $0\in S$ and $\|d\varrho(x)\|=2|x|$, we have

$$|\varrho(x)| \leq 2^{1-k}|\mathrm{dist}(x,S)| \qquad \text{for } x\in W_k .$$

The last two estimates imply that, for all $y\in B_R$,

$$\int_{x\in W_k} \frac{dx_1\wedge\ldots\wedge dx_m}{[\mathrm{dist}(x,S)]^{\alpha+\beta+\varepsilon}\,|x-y|^{m-1-\varepsilon}}$$

$$\leq m2^{k+(1-k)(\alpha+\beta+\varepsilon)} \int_{x\in W_k} \frac{|d\varrho(x)\wedge dx_1\wedge\ldots\hat{j}\ldots\wedge dx_m|}{|\varrho(x)|^{\alpha+\beta+\varepsilon}|x-y|^{m-1-\varepsilon}}$$

$$\leq 2m2^{k(1-\alpha-\beta-\varepsilon)} \int_{-d_k}^{d_k} \frac{ds}{|s|^{\alpha+\varepsilon+\beta}} \int_{\substack{t\in \mathbb{R}^{m-1}\\ |t|<2R}} \frac{dt_1\wedge\ldots\wedge dt_{m-1}}{|t|^{m-1-\varepsilon}},$$

where

$$d_k := \max_{x\in W_k} |\varrho(x)| \leq \max_{x\in W_k} |x|^2 \leq 2^{-2k}.$$

Since

$$\int_{\substack{t\in \mathbb{R}^{m-1}\\ |t|<2R}} \frac{dt_1\wedge\ldots\wedge dt_{m-1}}{|t|^{m-1-\varepsilon}} < \infty$$

and

$$\int_{-2^{-2k}}^{2^{-2k}} \frac{ds}{|s|^{\alpha+\varepsilon+\beta}} = 2(2^{-2k})^{1-\alpha-\varepsilon-\beta},$$

it follows that, for some constant $C<\infty$,

$$\sup_{y\in B_R} \int_{x\in W_k} \frac{dx_1\wedge\ldots\wedge dx_m}{[\mathrm{dist}(x,S)]^{\alpha+\beta+\varepsilon}|x-y|^{m-1-\varepsilon}} \leq C(2^{\alpha+\beta+\varepsilon-1})^{2k}.$$

Since $\alpha+\beta+\varepsilon<1$ and hence $2^{\alpha+\beta+\varepsilon-1}<1$, this implies that, for some $C<\infty$,

$$\sup_{y\in B_R} \int_{x\in B_R} \frac{dx_1\wedge\ldots\wedge dx_m}{[\mathrm{dist}(x,S)]^{\alpha+\beta+\varepsilon}|x-y|^{m-1-\varepsilon}} \leq C\sum_{k=0}^{\infty} (2^{\alpha+\beta+\varepsilon-1})^{2k} < \infty. \quad \square$$

13. Proof of Proposition 5 (ii) and (iii). Let U_j^+ and U_j^- be the sets introduced in point 6 of this appendix. Then, by (21) and (22), the functions $\varrho, x_1, \ldots, \hat{\ }_j, \ldots, x_m$ form a system of coordinates on U_j^+ as well as on U_j^-, and

$$dx_1\wedge\ldots\wedge dx_m \;\le\; \frac{m}{\|d\varrho(x)\|}\,|d\varrho\wedge dx_1\wedge\ldots\hat{\ }_j\ldots\wedge dx_m| \qquad \text{on } U_j^+\cup U_j^-. \tag{44}$$

First we proof part (ii). Since

$$(|\varrho|+|x-y|^2)^2|x-y|^{m-3-d}\ge|\varrho|^{1-\alpha-\beta}|x-y|^{m-1+2\alpha+2\beta-d},$$

it is sufficient to find a constant $C_1<\infty$ such that, for all $y\in\bar{B}_R$ and $r>0$,

$$\int\limits_{\substack{x\in B_R\\ \varrho(x)<r}} \frac{\|d\varrho(x)\|dx_1\wedge\ldots\wedge dx_m}{|\varrho(x)|^{1-\beta}|x-y|^{m-1+2\alpha+2\beta-d}} \;\le\; C_1r^{\beta}. \tag{45}$$

Since the functions $\varrho, x_1, \ldots, \hat{\ }_j, \ldots, x_m$ are coordinates on U_j^+ and on U_j^-, and (44) holds true, the left hand side of (45) is bounded by

$$2m\int\limits_{\substack{s\in\mathbb{R}^{m-1}\\ |s|<2R}}\frac{ds_1\wedge\ldots\wedge ds_{m-1}}{|s|^{m-1+2\alpha+2\beta-d}}\int_{-r}^{r}|t|^{\beta-1}dt.$$

Since $m-1+2\alpha+2\beta-d<m-1$ and

$$\int_{-r}^{r}|t|^{\beta-1}dt \le 2r^{\beta}, \tag{46}$$

this implies (45).

Now we prove part (iii). We distinguish the integrals over $E(y)$ and over $B_R\setminus E(y)$. Since $\|d\varrho(x)\| \le 2|y|+2|x-y| \le (2/\delta+2)|x-y|$ for $x\notin E(y)$ (cf. point 6 of this appendix), one obtains that

$$\int_{\substack{x\in B_R\setminus E(y)\\ |x|<r}} \frac{\|d\varrho(x)\|^2 |\varrho(x)|^{-\alpha} dx_1\wedge\ldots\wedge dx_m}{(|\varrho(x)|+|h(y,x)|+|x-y|^2)^2 |x-y|^{m-3-d}}$$

$$\leq \left(\frac{2}{\delta}+2\right) \int_{\substack{x\in B_R\\ |\varrho(x)|<r}} \frac{\|d\varrho(x)\| \, |\varrho(x)|^{-\alpha} dx_1\wedge\ldots\wedge dx_m}{(|\varrho(x)|+|x-y|^2)^2 |x-y|^{m-4-d}}$$

for all $y\in\bar{B}_R$ and $r>0$. Since

$$\frac{|\varrho(x)|^{-\alpha}}{(|\varrho(x)|+|x-y|^2)^2 |x-y|^{m-4-d}} \leq \frac{1}{|\varrho(x)|^{1-\beta} |x-y|^{m-2+2\alpha+2\beta-d}},$$

this implies that, for all $y\in\bar{B}_R$ and $r>0$,

$$\int_{\substack{x\in B_R\setminus E(y)\\ |\varrho(x)|<r}} \frac{\|d\varrho(x)\|^2 |\varrho(x)|^{-\alpha} dx_1\wedge\ldots\wedge dx_m}{(|\varrho(x)|+|h(y,x)|+|x-y|^2)^2 |x-y|^{m-3-d}}$$

$$\leq \left(\frac{2}{\delta}+2\right) \int_{\substack{x\in B_R\\ |\varrho(x)|<r}} \frac{\|d\varrho(x)\| dx_1\wedge\ldots\wedge dx_m}{|\varrho(x)|^{1-\beta} |x-y|^{m-2+2\alpha+2\beta-d}}. \tag{47}$$

Since the functions $\varrho, x_1, \ldots, \hat{\ }_j, \ldots, x_m$ are coordinates on U_j^+ and on U_j^-, and (44) holds true, the right hand side of (47) is bounded by

$$\left(\frac{2}{\delta}+2\right)2m \int_{\substack{s\in \mathbb{R}^{m-1}\\ |s|<2R}} \frac{ds_1\wedge\ldots\wedge ds_{m-2}}{|s|^{m-2+2\alpha+2\beta-d}} \int_{-r}^{r} |t|^{\beta-1} dt.$$

Since $m-2+2\alpha+2\beta-d<m-1$ and by (46), this implies that, for some constant $C_2<\infty$,

$$\int_{\substack{x\in B_R\setminus E(y)\\ |\varrho(x)|<r}} \frac{\|d\varrho(x)\|^2 |\varrho(x)|^{-\alpha} dx_1\wedge\ldots\wedge dx_m}{(|\varrho(x)|+|h(y,x)|+|x-y|^2)^2 |x-y|^{m-3-d}} \le C_2 r^{\beta} \tag{48}$$

for all $y\in\bar{B}_R$ and $r>0$. Since $\|d\varrho(x)\|\le 3|y|$ for $x\in E(y)$, it follows from Lemma 10.1 (ii) that

$$\int_{\substack{x\in B_R\cap E(y)\\ |\varrho(x)|<r}} \frac{\|d\varrho(x)\|^2 |\varrho(x)|^{-\alpha} dx_1\wedge\ldots\wedge dx_m}{(|\varrho(x)|+|h(y,x)|+|x-y|^2)^2 |x-y|^{m-3-d}}$$

$$\le \frac{9}{\delta} \int_{\substack{s\in \mathbb{R}^{m-2}\\ |s|<2R}} ds_1\wedge\ldots\wedge ds_{m-2} \int_{-r}^{r} dt \int_{-\infty}^{\infty} \frac{|t|^{-\alpha} du}{(|t|+|u|+|s|^2)^2 |s|^{m-3-d}}.$$

Together with (14), Appendix A, and (48), this implies (20). □

BIBLIOGRAPHY

Ajrapetjan, R.A., and G.M. Henkin 1984:
Integral representations of differential forms on Cauchy-Riemann manifolds and the theory of CR-functions (Russian). Uspehi Mat. Nauk 39 (1984), no. 3, 39-106.

Ajzenberg, L.A. 1966:
The general form of continuous linear functionals on the space of holomorphic functions on a convex domain in $\mathbb{C}^n$ (Russian). Doklady Akad. Nauk SSSR 166 (1966), no. 5, 1015-1018.

Ajzenberg, L.A., and Sh.A. Dautov 1983:
Differential forms orthogonal to holomorphic functions or forms, and their properties. Am. Math. Soc., Providence, Rhode Island 1983.

Akahori, T. 1987:
A new approach to the local embedding theorem of CR-structures for $n \geq 4$, Memoirs of the AMS No. 366, Providence, R.I. 1987

Andreotti, A., and H. Grauert 1962:
Théorèmes de finitude pour la cohomologie des espaces complexes. Bull. Soc. math. France 90 (1962), 193-259.

Andreotti, A., and C.D. Hill 1972:
E.E. Levi convexity and the Hans Lewy problem. I. Ann. Scuola Norm. Super. Pisa 26 (1972), no. 2, 325-363; II. ibid 26 (1972), no. 4, 747-806.

Andreotti, A., and F. Norguet 1966:
Problème de Levi et convexité holomorphe pour les classes de cohomologie. Ann. Scuola norm. sup. Pisa 20 (1966), 197-241.

Andreotti, A., and E. Vesenteni 1965:
Carleman estimates for the Laplace-Beltrami equation on complex manifolds. Publications Mathématiques de l'Institut des Haute Études Scientifiques, 25, 1965.

Barth, W. 1970:
Der Abstand von einer algebraischen Mannigfaltigkeit im komplex--projektiven Raum. Math. Ann. 187 (1970), 150-162.

Bruna, J., and J.M. Burgues 1986:
Holomorphic approximation and estimates for the $\bar{\partial}$-equation on strictly pseudoconvex non smooth domains. Universitat Autonoma de Barcelona, Seccio de Matematiques, Preprint 1986.

Cartan, H. 1953/54:
Séminaire E.N.S., 1953-54, École Normale Supérieure, Paris.

Cartan H., and J.-P. Serre 1953:
Un théorème de finitude concernant les variétés analytiques compactes. C.R. Acad. Sci. Paris 237 (1953), 128-130.

Chern, S.S. 1956:
Complex manifolds. The University of Chikago, Autumn 1955 - Winter 1956.

Čirka, E.M., and G.M. Henkin 1975:
Boundary properties of holomorphic functions of several complex variables (Russian). Current problems in mathematics 4, Moscow 1975, 12-142.

Coen, S. 1969:
Sullomobogia degli a perti q-completi di uno spazio di Stein. Ann. Scuola Norm. Super. Pisa 23 (1969), no. 2, 289-303.

Coleff, N.R., and M. Herrera 1978:
Les courants résidues associés à une forme méromorphe . Springer Lect. Notes Math. 633 (1978).

Dautov, Sh.A. 1972:
On forms which are orthogonal to holomorphic functions with respect to integrating over the boundary of a strictly pseudoconvex domain (Russian). Deklady Akad. Nauk SSSR 203 (1972), no. 1, 16-18.

de Rham, G. 1955:
Variétés différentiables, Hermann, Paris 1955.

Diederich, K., and J.E. Fornaess 1985:
Smoothing q-convex functions and vanishing theorems. Invent. math. 82 (1985), 291-305.

——— 1986:
Smoothing q-convex functions in the singular case. Math Ann. 273 (1986), 665-671.

Dolbeault, P. 1953:
Sur la cohomologie des variétés analytiques complexes. C.R. Acad. Sci. Paris 236 (1953), 175-177.

——— 1956:
Formes différentielles et cohomologie sur une variété analytique complexe. I. Ann. Math. 64 (1956), 83-130; II. Ann. Math. 65 (1957), 282-330.

Fischer W., and I. Lieb 1974:
Lokale Kerne und beschränkte Lösungen für den $\bar{\partial}$-Operator auf q-konvexen Gebieten. Math. Ann. 208 (1974), 249-265.

Folland, G., and J.J. Kohn 1972:
The Neumann problem for the Cauchy-Riemann complex. Princeton University Press 1972.

Folland, G., and E.M. Stein 1974:
Estimates for the $\bar{\partial}_b$-complex and analysis on the Heisenberg group. Comm. Pure Appl. Math. 27 (1974), 429-522.

Fornaess, J.E. 1976:
Embedding strictly pseudoconvex domains in convex domains. Amer. J. Math. 98 (1976), 529-569.

Grauert, H. 1958:
On Levi's problem and the imbedding of real-analytic manifolds. Ann. Math. 68 (1958), 460-472.

——— 1963:
Bemerkenswerte pseudokonvexe Mannigfaltigkeiten. Math. Z. 81, 377-391 (1963).

——— 1981:
Kantenkohomologie. Compositio Math. 44 (1981), 79-101.

Grauert, H., and I. Lieb 1970:
Das Ramirezsche Integral und die Lösung der Gleichung $\bar{\partial}f=\alpha$ im Bereich der beschränkten Formen. Rice Univ. Studies 56 (1970), no. 2, 29-50.

Grauert, H., and R. Remmert 1971:
Analytische Stellenalgebren. Grundl. 176, Springer-Verlag 1971.

——— 1977:
Theorie der Steinschen Räume. Grundl. 227, Springer-Verlag 1977. Theory of Stein Spaces. Transl. by A. Huckleberry, Grundl. 236, Springer-Verlag 1979.

——— 1984:
Coherent analytic sheaves. Grundl. 265, Springer-Verlag 1984.

Greiner, P.C., and E.M. Stein 1977:
Estimates for the $\bar{\partial}$-Neumann problem. Princeton University Press, Princeton, N.Y., 1977.

Griffiths, Ph., and J. Harris 1978:
Principles of algebraic geometry. Wiley, New York 1978.

Gunning, R., and H. Rossi 1965:
Analytic functions of several complex variables. Prentice-Hall, Englewood Cliffs, N.J., 1965.

Harvey, R., and B. Lawson 1975:
On boundaries of complex analytic varieties. I. Ann. Math. 102 (1975), 233-290.

Harvey, R. 1977:
Holomorphic chains and their boundaries. Proceedings of Symposia in Pure Mathematics, 30:1, A.M.S., Providence, R.I. (1977), 309-382.

Henkin, G.M. 1983:
Tangent Cauchy-Riemann equations and the Yang-Mills, Higgs and Dirac fields. Proceedings of the International Congress of Mathematicians, August 16-24, 1983, Warszawa.

——— 1970:
Integral representation of functions in strongly pseudoconvex domains and applications to the $\bar{\partial}$-problem (Russian). Mat. Sb. 82 (1970), 300-308.

——— 1977:
H. Lewy's equation and analysis on pseudoconvex manifolds (Russian). I. Uspehi Mat. Nauk. 32 (1977), no. 3, 57-118; II. Mat. Sb. 102 (1977), no. 1, 71-108.

Henkin, G.M., and J. Leiterer 1984:
Theory of function on complex manifolds. Akademie-Verlag Berlin 1984 and Birkhäuser-Verlag Boston 1984.

——— 1986:
Proof of Grauert's Oka principle without induction over the basis dimension. Preprint P-MATH-05/86, Institut für Mathematik der AdW der DDR, 1986.

Henkin, G.M., and P.L. Poljakov 1986:
Homotopy formulas for the $\bar{\partial}$-operator on CP^n and the Radon-Penrose transform. Isvestija Akad. Nauk SSSR 50 (1986), no. 3, 566-597.

Henkin, G.M., and A.V. Romanov 1971:
Exact Hölder estimates of solutions of the $\bar{\partial}$-equation (Russian). Izv. Akad. Nauk SSSR, Ser. Mat. 35 (1971), 1171-1183.

Herrera, M., and D. Lieberman 1971:
Residues and principal values on complex spaces. Math. Ann. 194 (1971), no. 4, 259-294.

Hirsch, M. 1976:
Differential topology. Springer-Verlag, Graduate Texts in Mathematics 33, New York - Heidelberg - Berlin 1976

Hörmander, L. 1965:
L^2 estimates and existence theorems for the $\bar{\partial}$-operator. Acta Math. 113 (1965), 89-152.

—— 1966:
An introduction to complex analysis in several variables. D. Van Nostrand, Princeton-Toronto- London 1966.

Kodaira, K. 1953:
On cohomology groups of compact analytic varieties with coefficients in some analytic faiscaux. Proc. Nat. Acad. Sci. U.S.A. 39 (1953), 865-868.

Kohn, J.J. 1964:
Harmonic integrals on strongly pseudoconvex manifolds. I. Ann. Math. 78 (1963), 112-148; II. Ann. Math. 79 (1964), 450-472.

—— 1965:
Boundaries of complex manifolds. In: Proc. Conference on Complex Manifolds, Minneapolis 1964, edited by A. Aeppli, E. Calabi, H. Röhrl, Springer-Verlag 1965.

Kohn, J.J., and H. Rossi 1965:
On the extension of holomorphic functions from the boundary of a complex manifold. Ann. of Math. 81 (1965), 451-472.

Kuranishi, M. 1982:
Strongly pseudoconvex CR structures over small balls. I. Ann. Math. (2) 115 (1982), 451-500; II. ibid 116 (1982), 1-64.

Leiterer, J. 1986:
On holomorphic vector bundles over linearly concave manifolds. Math. Ann. 274 (1986), 391-417.

Lieb, I. 1970:
Die Cauchy-Riemannschen Differentialgleichungen auf streng pseudokonvexen Gebieten. I. Math. Ann. 190 (1970/71), 6-44; II. ibid 199 (1972), 241-256.

—— 1979:
Beschränkte Lösungen der Cauchy-Riemannschen Differentialgleichungen auf q-konkaven Gebieten. manuscripta math. 26 (1979), 387-409.

Lieb, I., and R.M. Range 1986:
Estimates for a class of integral operators and applications to the $\bar{\partial}$-Neumann problem. Invent. math. 85 (1986), 415-438.

Martineau, A. 1962:
Indicatrices de fonctionnelles analytiques et inversion de la transformation de Fourier-Borel par la transformation de Laplace. C.R. Acad. Sci. 255 (1962), 1845-1847 et 2888-2890.

——— 1966:
Sur la topologie des espaces de fonctions holomorphes. Math. Ann. 163 (1966), 62-88.

Milnor, J. 1963:
Morse Theory. Princeton University Press, Princeton, New Jersey, 1963.

Milnor, J. 1965a:
Lectures on the h-cobordism theorem. Princeton University Press, Princeton, New Jersey, 1965.

———, 1965b:
Topology from the differential viewpoint. The University Press of Virginia, Charlottesville 1965.

Morrey, C.B. 1966:
Multiple integrals in the calculus of variations. Springer-Verlag, New York, 1966.

Nacinovich, M. 1984:
Poincaré lemma for tangential Cauchy-Riemann complexes. Math. Ann. 268 (1984), no. 4, 449-471.

Narasimhan, R. 1968:
Analysis on real and complex manifolds. North Holland, Amsterdam 1968.

Oka, K. 1984:
Mathematical papers. Transl. by R. Narasimhan, with comments by H. Cartan, ed. R. Remmert, Springer-Verlag 1984.

Øvrelid, N. 1976:
Pseudo-differential operators and the $\bar{\partial}$-equation. In: Springer Lecture Notes 512, 185-192.

Peternell, M. 1985:
Continuous q-convex exhausting functions. Preprint 1985.

Range, R.M. 1986:
Holomorphic functions and integral representations in several complex variables. Graduate Texts in Math. 108, Springer 1986.

Rossi, H. 1965:
Attaching analytic spaces to an analytic space along a pseudoconcave boundary. In: Proceedings of the Conference on Complex Analysis, Minneapolis 1964,edited by A. Aeppli, E. Calabi, H. Röhrl, Springer--Verlag 1965.

Rothschild, L.P., and E.M. Stein 1976:
Hypoelliptic differential operators and nilpotent groups. Acta Mathematica (1976).

Rothstein, W. 1955:
Zur Theorie der analytischen Mannigfaltigkeiten im Raume von n komplexen Veranderlichen. Math. Ann. 129 (1955), no. 1, 96-138.

Scheja, G. 1961:
Riemannsche Hebbarkeitssatze fur Cohomologieklassen. Math. Ann. 144, 345-360 (1961).

Schwartz, L. 1955:
Lectures on complex analytic manifolds. Tata Institute of Fundamental Research, Bombay 1955.

Serre, J.-P. 1955:
Une théorème de dualite. Commentarii Mathematici Helvetici 29 (1955), 9-26.

Siu, Y.-T. 1974:
The $\bar{\partial}$-problem with uniform bounds on derivatives. Math. Ann. 207, 163-176.

Toledo, D., and Y.L. Tong 1976:
A parametrix for $\bar{\partial}$ and Riemann-Roch in Čech-theory. Topology 15 (1976), 273-301.

Wallace, A. 1968:
Differential topology. First steps. W. A. Benjamin, New York - Amsterdam 1968.

Wells, Jr., R.O. 1973:
Differential analysis on complex manifolds. Prentice Hall, Englewood Cliffs, N.J., 1973.

Yau, S.-T. 1981:
Kohn-Rossi cohomology and its applicaion to the complex Plateau problem. Ann. Math. 113 (1981), 67-110.

LIST OF SYMBOLS

$\mathbb{R}$, $\mathbb{C}$	real, complex numbers
$\mathbb{R}^n$, $\mathbb{C}^n$	real, complex Euclidean space of dimension n
$\bar{X}$	topological closure of the set X
$\bar{x}$	$:= (\bar{x}_1,\dots,\bar{x}_n)$ for $x = (x_1,\dots,x_n) \in \mathbb{C}^n$
$\langle x,y \rangle$	$:= x_1y_1+\dots+x_ny_n$ for $x,y \in \mathbb{C}^n$
$\lvert x \rvert$	$:= (\lvert x_1\rvert^2+\dots+\lvert x_1\rvert^2)^{1/2}$ for $x \in \mathbb{C}^n$
$Y \subset\subset X$	means that Y is contained in a compact subset of X
deg f	degree of the differential form f
$\lvert f \rvert$	absolute value of a differential form f of maximal degree; 11
$\lVert f(x) \rVert$	norm of a differential form f at the point x; 12
supp f	support of f,
det	of matrices of differential forms; 13
$\omega(x)$, $\omega'(x)$	15
$\lVert\cdot\rVert_{0,M}$	sup-norm; 16
$\lVert\cdot\rVert_{\alpha,M}$	α-Hölder norm; 16
d	exterior differential operator
∂	holomorphic component of d; 10
$\bar{\partial}$	antiholomorphic component of d; 10
$\langle f \rangle$	the current defined by the form f; 19,20
$\nabla^{\mathbb{C}}$, $\nabla^{\mathbb{R}}$	complex, real gradient; 38
B_D, $B_{\partial D}$	Martinelli-Bochner integral; 22
B'	Martinelli-Bochner integral for currents with compact support; 25
L^v	Cauchy-Fantappie integral; 35, 52
R^v	Cauchy-Fantappie integral; 35
T^v	sum of Cauchy-Fantappie integrals; 36, 52
L^v_K	Cauchy-Fantappie integral; 49
R^v_K	Cauchy-Fantappie integral; 51

S_K	piece of a boundary; 47
$\triangle_K$	simplex; 47
$\lvert K \rvert$	:= l if $K=(k_1,\dots,k_l)$; 46
$K(s\hat{})$	:= $(k_1,\dots,\underset{s}{\hat{}},\dots,k_l)$ if $K=(k_1,\dots,k_l)$; 46
$T_{\mathbb{R},x}(M)$	tangent space of M at x; 59
$T_{\mathbb{C},x}(M)$	complexified tangent space of M at x; 59
$T'_x(X)$	space of holomorphic tangent vectors in $T_{C,x}(X)$; 59
$T'_x(Y)$	:= $T'_x(X)$ $T'_{C,x}(Y)$ if Y is a submanifold of X; 59
$L_\varrho(x)$	Levi form of o at x; 60
$L_{\varrho\mid Y}(x)$	restriction of $L_o(x)$ to $T'_x(Y)$; 60
$p^+_\varrho(x)$, $p^-_\varrho(x)$	number of positive, negative eigenvalues of $L_o(x)$; 60
$p^+_{\varrho\mid Y}(x)$, $p^-_{\varrho\mid Y}(x)$	number of positive, negative eigenvalues of $L_{o\mid Y}(x)$; 60
$p^{\pm}_M(x)$, $p_M(x)$	:= $p^{\pm}_{\varrho\mid M}(M)$, where ϱ is a defining function for M; 62
X^*	dual domain of X; 146

Notations which contain the symbols L^∞, C^α, Z^α, $E^{\beta\to\alpha}$, E^α, c^α, $(0\le\alpha,\beta\le\infty)$, C^{cur}, Z^{cur}, E^{cur}, and $\mathcal{O}$, as for instance $C^\infty_*(M,E)$, $C^\infty_m(M,E)$, $C^\infty_{s,r}(M,E)$, $E^\infty_{s,r}(M,E)$ and $[C^\infty_{s,r}(M,E)]_0$ are introduced in Sections 0.11-0.13.

$\mathcal{O}^E$	sheaf of germs of holomorphic sections of E; 40
$H^q(\mathcal{F})$	Čech cohomology group with coefficients in $\mathcal{F}$
$H^{s,r}_\alpha(X)$	:= $Z^\alpha_{s,r}(X)/E^\alpha_{s,r}(X)$; 30
$H^{s,r}_{cur}(X)$	:= $Z^{cur}_{s,r}(X,E)/E^{cur}_{s,r}(X,E)$; 30
$H^{s,r}_{cur}(X,E)$	:= $Z^{cur}_{s,r}(X,E)/E^{cur}_{s,r}(X,E)$; 30
$[H^{s,r}(X,E)]_0$	:= $[Z^\infty_{s,r}(X,E)]_0/[E^\infty_{s,r}(X,E)]_0$; 201

SUBJECT INDEX

Band 1

Inverse and Improperly Posed Problems in Differential Equations

Proceedings of the Conference on Mathematical and Numerical Methods, held in Halle/Saale (GDR) from May 29 to June 2, 1979

Edited by Gottfried Anger
296 Seiten, 1979 (vergriffen)

Band 2

Fundamentals of Computation Theory FCT '79

Proceedings of the Conference on Algebraic, Arithmetic and Categorial Methods in Computation Theory, held in Berlin/Wendisch-Rietz (GDR) September 17–21, 1979

Edited by Lothar Budach
576 Seiten, 1979

Band 3

Decidability and Generalized Quantifiers

By Andreas Baudisch, Detlef Seese, Hans-Peter Tuschik, and Martin Weese
XII, 235 Seiten, 1980

Band 4

Zur Elastizitäts- und Thermoelastizitätstheorie

Potentialtheoretische Methoden in Anwendung auf verschiedene Modelle

Von Lothar Jentsch und Johannes Maul
157 Seiten, 1980

Band 5

Systems Analysis and Simulation 1980

Proceedings of the International Symposium held at Berlin (GDR) September 1–5, 1980

Edited by Achim Sydow
496 Seiten, 1980 (vergriffen)

Band 6

Nichtlineare diskrete Optimierung

von Eberhard Girlich und Michail M. Kowaljow
218 Seiten, 1981

Band 7

Completeness Problems in the Structural Theory of Automata

By Jürgen Dassow
148 Seiten, 1981

Band 8

Darstellung von Lösungen linearer elliptischer Differentialgleichungen

Von Günther Wildenhain
92 Seiten, 1981

Band 9

Introduction to the Spectral Theory of Operators in Spaces with an Indefinite Metric

By I. S. Iohvidov, M. G. Krein, and H. Langer
120 Seiten, 1982 (vergriffen)

Band 10

Nonlinear Time-discrete Systems

A General Approach by Nonlinear Superposition

By Michael Gössel
115 Seiten, 1982 (vergriffen)

Band 11

Preprimal Algebras

By Klaus Denecke
162 Seiten, 1982

Band 12

Ergodic Theory and Related Topics

Proceedings of the Conference held in Vitte/Hiddensee (GDR) October 19–23, 1981

Edited by Horst Michel
232 Seiten, 1982

Band 13

Singularly Perturbed Differential Equations

By H. Goering, A. Felgenhauer, G. Lube, H.-G. Roos, and L. Tobiska
176 Seiten, 1983

Band 14

Leben wir in einer Volterra-Welt?

Ein ökologischer Zugang zur angewandten Systemanalyse

Von Manfred Peschel und Werner Mende
233 Seiten, 1983

Band 15

Markov Decision Problems with Countable State Spaces

Optimality Criteria – Algorithms – Clustering

By Hans Michael Dietz and Volker Nollau
174 Seiten, 1983

Band 16

Stochastische Verfahren zur Suche nach dem Minimum einer Funktion

Von Ryszard Zieliński und Peter Neumann
133 Seiten, 1983

Band 17

Invertibility and Asymptotics of Toeplitz Matrices

By Albrecht Böttcher and Bernd Silbermann
200 Seiten, 1983

Band 18

Structural Induction on Partial Algebras

Introduction to Theory and Application of Partial Algebras – Part II

By Horst Reichel
205 pages, 1984

Band 19

Algebraic Methods for Toeplitz-like Matrices and Operators

By Georg Heinig and Karla Rost
212 pages, 1984

Band 20

Frege Conference 1984

Proceedings of the International Conference held at Schwerin (GDR), September 10–14, 1984

Edited by Gerd Wechsung

408 pages, 1984

Band 21

Numerical Analysis of Selected Semilinear Differential Equations

Edited by Christian Großmann, Thomas Riedrich, and Hartmut Schönheinz

203 pages, 1984

Band 22

Singular Ordinary Differential Operators and Pseudodifferential Equations

By Johannes Elschner

200 pages, 1985

Band 23

Lotka-Volterra-Approach to Cooperation and Competition in Dynamic Systems

Proceedings of the 5th Meeting of UNESCO's Working Group on Systems Theory held on the Wartburg, Eisenach (GDR) at March 5–9, 1984

Edited by Werner Ebeling and Manfred Peschel

271 pages, 1985

Band 24

Convergence Structures 1984

Proceedings of the Conference on Convergence held in Bechyně (Czechoslovakia), September 24–28, 1984

Edited by Josef Novák, Werner Gähler, Horst Herrlich, and Jan Mikusiński

254 pages, 1985

Band 25

Parcella '84

Proceedings of the II. International Workshop on Parallel Processing by Cellular Automata and Arrays held in Berlin (GDR), September 25–27, 1984

Edited by Wolfgang Händler, Tamás Legendi, and Gottfried Wolf

239 pages, 1985 (vergriffen)

Band 26

Multiobjective and Stochastic Optimization Based on Parametric Optimization

By Jürgen Guddat, Francisco Guerra Vasquez, Klaus Tammer, and Klaus Wendler

175 pages, 1985

Band 27/28

Systems Analysis and Simulation 1985
I: Theory and Foundations
II: Applications

Proceedings of the International Symposium held in Berlin (GDR), August 26–31, 1985

Edited by Achim Sydow, Manfred Thoma, and Robert Vichnevetsky

per volume 495 pages, 1985

Band 29

Parcella '86

Proceedings of the III. International Workshop on Parallel Processing by Cellular Automata and Arrays held in Berlin (GDR), September 9–11, 1986

Edited by Tamás Legendi, Dennis Parkinson, Roland Vollmar, and Gottfried Wolf

279 pages, 1986

Band 30

Fuzzy Sets Applications, Methodological Approaches, and Results

Proceedings of the International Workshop on Fuzzy Sets Applications cosponsored by IIASA and convened in cooperation with IAEA held on the Wartburg, Eisenach (GDR), at March 3–8, 1985

Edited by Steffen Bocklisch, Sergei Orlovski, Manfred Peschel, and Yasushi Nishiwaki

233 pages, 1986

Band 31

Mathematical Methods of Specification and Synthesis of Software Systems '85

Proceedings of the International Spring School, held in Wendisch-Rietz (GDR), April 22–26, 1985

Edited by Wolfgang Bibel and Klaus P. Jantke

245 pages, 1986 (vergriffen)

Band 32

A Model Theoretic Oriented Approach to Partial Algebras

Introduction to Theory and Application of Partial Algebras – Part I

By Peter Burmeister

319 pages, 1986

Band 33

Finite Difference Method on Irregular Networks

A Generalized Approach to Elliptic Problems of Second Order

By Bernd Heinrich

206 pages, 1986

Band 34

Modern Algorithms for Large Sparse Eigenvalue Problems

By Arnd Meyer

125 pages, 1987

Band 35

Parametric Optimization and Related Topics

Edited by Jürgen Guddat, Hubertus Th. Jongen, Bernd Kummer, and František Nožička

407 pages, 1987

Band 36

Discretization in Differential Equations and Enclosures

Edited by Ernst Adams, Rainer Ansorge, Christian Großmann, and Hans-Görg Roos

250 pages, 1987

Bei Fragen zur Produktsicherheit wenden Sie sich bitte an:
If you have any questions regarding product safety,
please contact:

Walter de Gruyter GmbH
Genthiner Straße 13
10785 Berlin
productsafety@degruyterbrill.com